FRONTIERS

FRONTIERS

TWENTIETH-CENTURY PHYSICS

Steve Adams

London and New York

First published 2000
by Taylor & Francis Ltd
11 New Fetter Lane, London EC4P 4EE

Simultaneously published in the USA and Canada
by Routledge Inc
29 West 35th Street, New York, NY 10001

Taylor & Francis Limited is an imprint of the Taylor & Francis Group

© 2000 Steve Adams

Printed and bound in Great Britain by TJ International Ltd, Padstow, Cornwall

British Library Cataloguing in Publication Data
A catalogue record for this book is available from the British Library

Library of Congress Cataloging in Publication Data
A catalogue record for this book is available from the Library of Congress

ISBN 0-748-40840-1

Dedication

This is for my sons,
Joseph, Matthew and Sebastian

My books (which do not know that I exist) are as much a part of me as is this face, the temples gone grey and the eyes grey, the face I vainly look for in the mirror.

Jorge Luis Borges

I am now convinced that theoretical physics is actual philosophy.

Max Born

CONTENTS

Preface..ix

Part 1: Quantum Revolutions.. 1
1.   Old Quantum Theory...3
2.   A New Quantum Theory 1925-30...32
3.   Quantum Mysteries..65
4.   QED...109

Part 2: Explaining Matter...**131**
5.   Atoms and Nuclei...133
6.   The Standard Model...168
7.   Particle Detectors...190
8.   Particle Accelerators..207
9.   Toward a Theory of Everything..225

Part 3: Space and Time..**247**
10.   The Speed of Light...249
11.   Special Relativity...263
12.   General Relativity..292

Part 4: Astrophysics and Cosmology.............................**315**
13.   Observational Astronomy...317
14.   Stars and Distances..335
15.   Cosmology..363

Part 5: Thermodynamics and the Arrow of Time..........**389**
16.   Time, Temperature and Chance...391
17.   Toward Absolute Zero..422
18.   CPT...442

19.   ***Appendices***...**462**
    Appendix 1 The Black-body Radiation Spectrum.....................462
    Appendix 2 The Schrödinger Equation..................................465
    Appendix 3 The Hydrogen Atom...468
    Appendix 4 The Lorentz Transformation Equations 470
    Appendix 5 The Speed of Electromagnetic Waves.................472
    Appendix 6 The Nobel Prize for Physics..............................473
    Appendix 7 Glossary of Important Ideas...............................481
    Appendix 8 Timeline of Major Ideas....................................493
    Appendix 9 Further Reading...496

Index..499

Preface

The twentieth century has been a time of intellectual, artistic and political turmoil. It has also been a time of incredible scientific discovery in which new ideas have changed the way we think about ourselves and the universe and new technologies have revolutionised the way we work, interact and relax. Future historians will probably come to regard this as the golden age of physics, in which special relativity changed our view of space, time and matter and quantum theory undermined the very notions of reality and objectivity. On the largest scale general relativity showed that gravity is but geometry, the distortion of space-time caused by the presence of matter. It equipped us to deal with a universe in which our own Milky way is just one of billions of galaxies rushing apart from some unbelievably violent beginning – the Big Bang.

We have built machines big enough to straddle two countries whose purpose is to smash tiny invisible particles together to create new matter. They help us explore the beautiful connections between fundamental particles and simulate conditions that last existed a fraction of a second after the Big Bang. We can put a satellite in space and use it to look ten billion years back into the past and we can write our names in individual atoms. We contemplate building a computer that carries out its calculations in a stack of parallel worlds, and reputable physicists publish papers outlining serious designs for teleportation devices and time machines. Temperatures lower than any that have ever existed in nature are routinely created in the laboratory and used to explore strange new states of matter. Close to absolute zero weird things happen – superfluid liquids escape from their containers and currents flow without resistance in superconducting wires, magnets levitate and groups of atoms clump together to form exotic quantum blobs called Bose-Einstein condensates. No one could have imagined the peculiar and apparently paradoxical discoveries of the last hundred years.

The twentieth century has provided some of the greatest physicists that have ever lived. Albert Einstein towers above them all, both revolutionary and yet conservative, as enigmatic as the quantum theory he helped to create and yet never came to terms with. He is better known for his two theories of relativity and his lifetime quest to discover a unified field theory - a quest taken up once again at the end of the century by string theorists such as Ed Witten. Then there is Einstein's quantum adversary, Niels Bohr, the architect of the Copenhagen Interpretation. It was Bohr who drew everything together in the 1920s when the most important of all physical theories was emerging from the work of Heisenberg, Schrödinger, Born and others. While theoreticians struggled to build a mathematical model of the atom the great experimentalist, Rutherford, was building on the work of Thomson, Becquerel and the Curies and smashing atoms together to work out what they are made of. Much of high-energy physics is a direct extension of Rutherford's work. Dirac, Feynman, Dyson, Schwinger, Tomonoga, Wheeler and many others extended quantum principles into quantum field theories, paving the way for the unifying models of particle physics that together form the theoretical structure of the Standard Model. Amazing ingenuity has been applied in the development of detectors of ever-greater complexity and subtle precision in order to tease out significant events from the violent collisions induced in giant accelerators. Deeper and deeper layers of structure have been revealed, each one at first baffling but then gradually comprehensible as its internal symmetries and patterns were discovered. First came atoms, then electrons and nuclei, nucleons, and all the baryons, leptons and quarks, and we are still looking for more. New exotic particles may determine whether the universe will eventually collapse or if all matter is ultimately unstable.

Hubble shocked Einstein with his discovery that the universe is expanding. Three quarters of a century later the Hubble Space Telescope sends back remarkable images of the most distant and ancient objects we have ever seen, helping us make measurements that check the rate of expansion and reduce the errors in our estimate of Hubble's constant and the age of the universe. Meanwhile Eddington, Bethe and Fowler worked out the mechanisms of nucleosynthesis in stars and theoretical and observational astronomy added ever more interesting oddities to our catalogues of the heavens – white dwarfs, red giants, neutron stars, pulsar and , most worrying of all, the black hole.

Many years ago I was drawn into physics because it asks and attempts to answer the big questions about the nature of matter, about the origin extent and end of the universe, and about time and space and the nature of reality. In this book I have tried to give an introduction to the main theoretical strands of twentieth century physics and a feel for some of the arguments and ideas that have shaped our present world-view. The work of many Nobel Prize winners has been described and discussed along with many quotations. I have however omitted most of the technology, with the exception of accelerators and detectors in particle physics and various kinds of telescopes since these have extended our ability to pursue these big questions on the largest and smallest scales. I have tried to explain the physics but avoided the temptation to turn this into a textbook. The more involved mathematical sections usually appear in boxes or in the appendices, but the main story can be followed and understood without them. The book is packed with diagrams and pictures and the wonderful pen and ink portraits were drawn by my brother, Nick, to whom I am, once again, extremely grateful. Thanks also to my ex-colleagues in the physics department at Westminster School whose love of physics and willingness to discuss ideas helped iron out many problems. My students too, over many years, have drawn my attention to articles and new developments and asked all the difficult questions that make physics such an exciting subject. I thoroughly enjoyed researching and writing this book and hope you will enjoy reading it.

Steve Adams
July 1999

Part 1
QUANTUM REVOLUTIONS

1 Old Quantum Theory
2 A New Quantum Theory 1925-30
3 Quantum Mysteries
4 QED

1

Classical physics cannot
adequately explain:

specific heats;
black-body radiation;
photoelectric effect;
atomic stability.

Planck, Einstein and Bohr
solve these problems
by ad hoc quantisation.

2

Ad hoc quantisation worked
but no-one knew why – there
was no fundamental theory.

Between 1925 and 1930 three
quantum theories appeared:
Heisenberg's matrix mechanics;
Schrodinger's wave mechanics;
Dirac's operator mechanics.

None of these included relativity
The Dirac equation (1928) did.

3

Quantum theory worked but
no-one knew how to interpret
the equations.

Bohr constructed the Copenhagen
Interpretation, based on Uncertainty
and Complementarity. Einstein
could not accept it. The arguments
continue, but experimental results
support quantum theory.

4

Attempts to explain how light
and matter interact led to
infinities.

Feynman Schwinger and Tomonoga
developed a new quantum field
theory – quantum electrodynamics
or QED in 1948.

1A   In 1962, CERN hosted the eleventh International Conference on High Energy Physics. Among the distinguished visitors were eight Nobel prizewinners. Left to right: Cecil F. Powell, Isidor I. Rabi, Werner Heisenberg, Edwin McMillan, Emile Segre, Tsung Dao Lee, Chen Ning Yang and Robert Hofstadter.

Photo credit: CERN

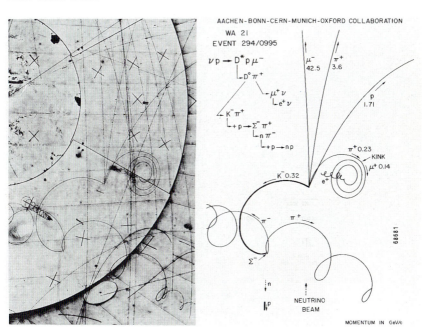

1B   This photograph shows tracks formed in the Big European Bubble Chamber (BEBC) at CERN following the collision of a neutrino (entering bottom centre) with one of the hydrogen nuclei in the chamber. Bubble chambers were very important detectors in the 1970s but have now been superseded by complex electronic detectors.

Photo credit: CERN

2   End view of the ALEPH detector at CERN showing the different layers. Jack Steinberger is second from the left. He shared the 1988 Nobel Prize for his contributions to neutrino physics.

Photo credit: CERN

3   Refurbishing the Hubble space telescope.

Photo credit: CERN

4 These images represent selected portions of the sky as seen in the Hubble Deep Field (HDF) observation. The images were assembled from many separate exposures with the Wide Field and Planetary Camera 2 (WFPC2), for ten consecutive days between December 18 to 28, 1995. The never before seen dimmest galaxies are nearly 30th magnitude. The HDF image covers a speck of sky a tiny fraction the diameter of the full Moon. Though the field is a very small sample of sky, it is considered representative of the typical distribution of galaxies in space because the universe, statistically, looks the same in all directions. The HDF provides important clues to understanding the evolution of the universe. Some of the galaxies may have formed less that one billion years after the Big Bang.

Credit: Robert Williams and the Hubble Deep Field Team (STScI) and NASA

1
OLD QUANTUM THEORY

1.1 A CRISIS IN CLASSICAL PHYSICS

It has been suggested that nineteenth century physicists thought they had more or less reached the end of fundamental physics. After all, they had Newtonian mechanics and gravitation, James Clerk Maxwell's equations, an atomic theory for matter and the laws of thermodynamics. What more could be done but to work out the consequences of these wonderful theories, add a few decimal places to calculated results, probe the details of chemical reactions and apply the ideas to the technology of power? The familiar story then points to an amazing three years in which Wilhelm Röntgen discovered X-rays (1895), Henri Becquerel discovered radioactivity (1896) and J.J. Thomson discovered the electron (1897). These fundamental discoveries are seen in retrospect as heralds of a new age in which relativity and quantum theory revolutionised our views of space, time and matter and particle physics and cosmology gave a wholly new picture of the hierarchy of structures in the universe.

While it is true that physicists did make prophetic pronouncements from time to time (and well past the turn of the century) there were many that were well aware that all was not well with classical physics.

1.1.1 Specific Heats

The first hint of a problem came with the kinetic theory of gases, an attempt to explain macroscopic gas properties in terms of the rapid random microscopic motion of huge numbers of tiny massive particles. Maxwell published his great paper on gas theory in 1859, and showed that many basic properties (like diffusion rates and viscosity) could indeed be accurately modelled using kinetic theory. But he ended the paper by pointing to a problem – predictions for the specific heats of gases do not fit experimental measurements. The problem remained with him, and ten years later he described it as *"the greatest difficulty yet encountered by the molecular theory"*. It is worth looking a little deeper to see *how* kinetic theory fails to explain specific heats.

The specific heat capacity of a substance is the energy required per kilogram to increase its temperature by 1K. The classical approach to specific heats is to consider how the average molecular energy depends on temperature. This in turn depends on the nature of the molecules themselves and the conditions under which they are contained. For gases there are two principal specific heats, c_v and c_p, The specific heats at constant volume and at constant pressure (see Fig. 1.1). Imagine heating a monatomic gas confined in a constant volume container. The gas is prevented from expanding so it can do no work. This means all the heat that flows in increases the internal energy of the gas, in this case by making the molecules, on average, move faster. On the other hand, if the gas was kept at constant pressure, perhaps by placing it in a syringe with a moveable piston, it does work as it expands, so the heat that flows in has to pay for this work as well as for the increase in internal energy. This means that more heat per kilogram of gas must be supplied at constant pressure to raise gas temperature by 1K than must be supplied at constant volume. Another way of saying this is to say that c_p is greater than c_v. Maxwell tried to predict the ratio c_v/c_p (γ) for various gases but failed.

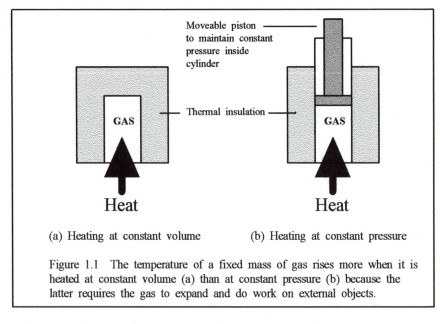

(a) Heating at constant volume   (b) Heating at constant pressure

Figure 1.1   The temperature of a fixed mass of gas rises more when it is heated at constant volume (a) than at constant pressure (b) because the latter requires the gas to expand and do work on external objects.

Maxwell's prediction was based on a theorem of classical thermodynamics called 'equipartition of energy'. This is very easy to understand. It simply says that if the molecules in a gas have many different modes of motion available (for example, flying about, rotating about some axis, vibrating along a bond, etc.) then in thermal equilibrium the internal energy of the gas would be distributed evenly (on average) among them. Each mode was expected to have an average energy ½ kT per molecule, where k is Boltzmann's constant and T is the absolute temperature. For a gas consisting of single atoms (monatomic gas) the only modes of motion are translations in three dimensions, so the average molecular energy should be $3/2$ kT. For more complex molecules (e.g. O_2) rotations and vibrations should be available ($6/2$ kT for O_2), so heat supplied spreads more thinly among the larger number of modes) and the temperature rise is smaller. In other words these gases should have a higher specific heat capacity and a lower ratio of specific heats (because the work done in expanding becomes less significant if there are more internal degrees of freedom).

Experiments showed two significant trends:

- For complex molecules at low temperatures the specific heats were always smaller than expected and γ was always larger than predicted.
- As temperature increases the specific heat increases and γ falls toward the theoretical value.

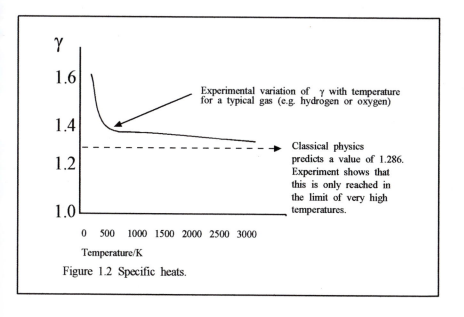

Figure 1.2 Specific heats.

These results suggested to James Jeans (around 1890) that something damped out or 'quenched' some of the degrees of freedom at lower temperatures so that they make no contribution to specific heat. At higher temperatures they suddenly 'switch on' and the specific heat increases. But classical physics could not explain this discontinuous behaviour. This was the first hint that classical physics was fatally flawed, long before Planck's work on black-body radiation or Einstein's work on the photoelectric effect. As the nineteenth century approached the problem sat there as unfinished business, a signpost to things to come.

1.1.2 Black-body Radiation

At the end of a hot day you can feel the heat radiated by a hot stone wall or path. Although our eyes do not detect these long wavelength electromagnetic waves our skin is sensitive to them. All hot bodies (by hot I mean hotter than absolute zero) radiate heat. At relatively low temperatures the vast majority of this radiation is in the infra-red region of the spectrum so it is invisible. It can be made visible using electronic detectors (e.g. photodiodes sensitive to the infra-red) and is used in heat seeking missiles and night sights.

If the temperature of the object is increased it begins to radiate more energy at higher frequencies and the spectrum of thermal radiation may extend into the visible region. At first the only significant radiation is at the red end of the visible spectrum, so the object becomes 'red-hot'. If we continue to heat the object its temperature will continue to increase and the radiation spectrum includes more of the higher frequency radiation, eventually extending across the whole visible spectrum and making the object 'white-hot'. At higher temperatures still significant amounts of ultra-violet radiation are also emitted.

In broad terms this would make sense to a classical physicist. Electromagnetic waves are radiated when charged particles are accelerated. The effect of heating a body is to increase the random thermal motions of the particles it contains. These particles (atoms or molecules) contain electrons, which are charged, so the increased vibration at higher temperature means more thermal energy is radiated. It also means more higher frequency oscillations are taking place, so more of the radiation is at higher frequencies and the gradual transition from red-hot to white-hot makes sense.

Qualitative arguments like this are fine, but the physicist wants to be able to derive the entire thermal spectrum from first principles, and these first principles are the laws of physics. Several physicists attempted to do this in the late nineteenth century. Their starting points were:

- classical mechanics, particularly the mechanics of oscillators;
- classical thermodynamics, to work out how the thermal energy distributes itself among the various modes of oscillation (effectively an application of equipartition of energy again);
- classical electromagnetism, to link the oscillating sources to radiated waves.

On the face of it this seemed a fairly straightforward problem. If there is a certain amount of thermal energy in a body it ought to spread among the available degrees of freedom so that each one gets about $\frac{1}{2}kT$. Simple oscillators have two degrees of freedom, corresponding to the kinetic and potential parts of their total energy, so they should each get about kT. This is a first step toward solving the problem. The second step is to work out how many modes of oscillation lie in each small range of frequencies from f to $f+\delta f$, a result derived in a beautiful piece of work done by Lord Rayleigh in 1900. The final step involves multiplying these two values together to get the energy radiated in each small frequency range f to $f+\delta f$: the spectral energy density for the black-body radiation.

- Mean energy per oscillator: $E = kT$
- Number of oscillators in range f to $f+\delta f$: $n(f) = 8\pi f^2/c^3$
- Spectral energy density: $u(f) = 8\pi f^2 kT/c^3$

There is one major problem with this formula: *it is obviously wrong*. It depends on the square of the frequency, so for high frequencies the energy radiated increases without limit. This outrageous result, firmly based in classical physics, became known as the 'ultra-violet catastrophe'.

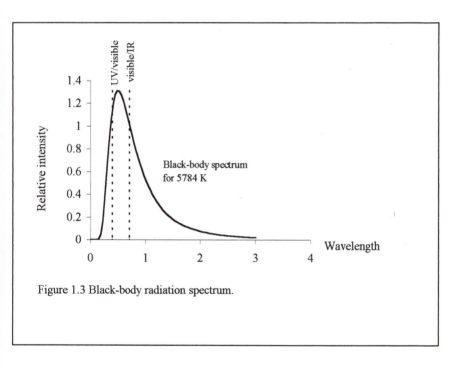

Figure 1.3 Black-body radiation spectrum.

Of course, Rayleigh knew this was not the solution. In the end he adjusted the formula by multiplying it by an exponential function that decays with frequency in order to tame the parabolic growth in the ultra-violet region. But this *ad hoc* approach had no theoretical base and led to poor agreement with experiment in the high frequency region.

The empirical black-body radiation spectrum falls off at high frequencies, so something must reduce the contribution from the high frequency oscillators. They seem to be 'quenched' in a similar way to the 'quenching' of degrees of freedom in gases at low temperatures. But how? There was also the tantalising fact that Rayleigh's formula works very well at the low frequency, long-wavelength end of the spectrum where it correctly predicts that the spectral intensity increases in direct proportion to temperature. This suggested the approach had something going for it and might be successfully modified.

The formula above is usually called the Rayleigh-Jeans Law. Jeans gets a mention because he noticed and corrected a numerical error in Rayleigh's original paper. He did this in an article in Nature published in 1906, six years *after* Planck published his (correct) formula for the black-body radiation spectrum!

Maths Box: Black-body Radiation Equations

Stefan-Boltzmann law: Stefan derived an empirical law linking the intensity I of black-body radiation to its temperature T. Boltzmann derived the same relation from classical thermodynamics:

$$I = \sigma T^4$$
$$\sigma = 5.67 \times 10^{-8} \ \mathrm{Wm^{-2}K^{-4}}$$

Wien's displacement law: Wien used classical thermodynamics and electromagnetism to show that the peak wavelength λ_p in the black-body spectrum is inversely proportional to the temperature:

$$\lambda_p T = \mathrm{constant} = 2.9 \times 10^{-3} \ \mathrm{mT}$$

These results put constraints on the possible form for the black-body spectrum and were both explained by Planck's work.

1.2 QUANTISATION

1.2.1 Planck's Law

Planck too derived the number of modes of oscillation lying between f and $f + \delta f$, and obtained the same result as Rayleigh. However, there was a slight difference in interpretation. Planck worked from the oscillators themselves, whereas Rayleigh saw these as modes of vibration in the radiation field, which is in equilibrium with the oscillators. Nonetheless they both obtained the same very important result. But Planck did not then assume he could apply classical equipartition of energy to the problem, he took a different approach. He asked himself how the mean energy of the oscillators should behave in order to:

- reproduce Wien's displacement law;
- kill off the ultra-violet catastrophe;
- be consistent with the Second Law of Thermodynamics, which links temperature to the rate of change of energy with entropy

His solution was radical. He proposed that the oscillators could not have any value of energy but that their allowed energies were quantised in steps proportional to frequency. The total energy of such an oscillator is then given by: $E = nhf$ where n is an integer, h a constant and f the frequency of oscillation.

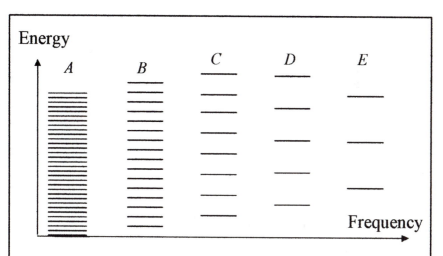

Figure 1.4 This diagram indicates how the spacing of quantised energy levels depends on the frequency of the oscillator. At low frequencies (A) the allowed energy levels are very close together, so the qunatum and classical behaviours are almost identical and the oscillator energy can change almost continuously. At high frequencies (E) this is not the case. It takes a lot of energy to excite the oscillator into its lowest allowed state, so quantum theory quenches the high frequency oscillators, preventing the ultra-violet catastrophe.

At the low frequency end of the spectrum the allowed energy levels in this discrete energy spectrum are packed so closely together (since hf is small) that it is not much different from the continuous energy spectrum assumed by classical physics. Another way of looking at this is to say that if $kT >> hf$ then the thermal energy available to the oscillator can excite it into oscillations that change more or less continuously with temperature. This is the region in which the Rayleigh-Jeans law works. At the other end of the spectrum hf may be comparable to kT or much larger. This means the available thermal energy is unlikely to excite many high frequency oscillators, so the high frequency end contributes much less than in the classical theory and the ultra-violet catastrophe does not happen.

Planck's formula was derived from a combination of statistical mechanics and electromagnetic theory, we shall derive it later using an approach borrowed from Einstein (see appendix), but the formula is:

$$u(f) = \frac{8\pi h f^3}{c^3} \frac{1}{e^{hf/kT} - 1}$$

It has been suggested that Planck actually worked backwards from a result he stumbled upon or else made an informed guess at his final stunning formula. If so he bravely accepted the radically new idea of energy quantisation, even though he had no idea why it made the formula work! At heart Planck was a classical physicist and an unlikely revolutionary. His own discoveries caused him much anguish in the following years:

"My futile attempts to fit the elementary quantum of action somehow into the classical theory continued for a number of years and they cost me a great deal of effort. Many of my colleagues saw in this something bordering on a tragedy. But I feel differently about it. For the thorough enlightenment I thus received was all the more valuable. I now knew for a fact that the elementary quantum of action played a far more significant part in physics than I had originally been inclined to suspect and this recognition made me see clearly the need for the introduction of totally new methods of analysis and reasoning in the treatment of atomic problems."

(Max Planck, *Scientific Autobiography and Other Papers*, Williams and Norgate, London, 1950)

Thirty years later, looking back on his discovery, he described his decision to introduce energy quanta as "*an act of desperation*" but admitted he was "*ready to sacrifice every one of my previous convictions about physical laws*" in order to "*bring about a positive result*".

(Max Planck, letter to R.W. Wood, 1931)

1.2.2 Quanta of Light

Planck's work was revolutionary, but even Planck was not sure what it meant and tried for years to reconcile quantisation with classical ideas. Moreover, Planck had quantised the *oscillators* emitting the radiation, not the radiation itself. He assumed that the radiation could be described by Maxwell's equations. After all, these had been supported by the experiments of Hertz and others and their continuous fields seemed consistent with the continuous wave model of light established by Christiaan Huygens and borne out by experiments involving interference and diffraction (such as Thomas Young's double-slit experiment).

Perhaps Einstein had less to risk by approaching things from a completely new direction. In 1905 he was employed in the patent office in Bern as a 'technical expert third class', but he had a firm grounding in classical physics and was busy producing his own revolution. In this year he published three major papers, each made a crucial contribution to a fundamental area of theoretical physics and any one of them would have been sufficient to establish him as one of the leading physicists of his time. One explained Brownian motion and effectively ended the debate over the 'reality' of atoms. A second contained the special theory of relativity. The third concerned the nature of light. In it Einstein suggested that there might be some phenomena which cannot be explained using the continuous variables of classical electromagnetism, phenomena for which a particle model might work better. It is not clear to what extent he was influenced by Planck's work, but he was certainly aware of it. The beauty of Einstein's approach is that he managed to look directly at physical processes such as emission and absorption of light and built up an explanation from scratch. It is hard to imagine Planck doing this. Here are some lines from the introduction to Einstein's great paper:

"....it should be kept in mind that optical observations refer to values averaged over time and not to instantaneous values. Despite the complete experimental verification of the theory of diffraction, reflection, refraction, dispersion and so on, it is conceivable that a theory of light operating with continuous three-dimensional functions will lead to conflicts with experience if it is applied to the phenomena of light generation and conversion."

(Albert Einstein, Annalen der Physik, 17, 1905)

Planck had considered the oscillators that emit black-body radiation. Einstein went for the radiation itself. He imagined an enclosure containing electromagnetic radiation in thermal equilibrium with the walls. He had a firm grounding in the statistical thermodynamics of Boltzmann and applied this to the radiation. He examined the way the energy and entropy of the radiation changed as the volume and temperature of the enclosure changed, and found that they behaved in a very similar way to the energy and entropy of a gas. It was as if the radiation itself consisted not of continuous waves but of particles bouncing about, and the energy transferred by one of these particles was given by:

$$E = hf$$

an equation reminiscent of Planck's results. Much later these 'particles' would be called 'photons' but we shall adopt the term from now on. (You may be surprised to learn that the term 'photon' was not used until 1926. This was partly because of the unease many physicists felt at resurrecting the discredited particle model for light. The crucial experiments that showed that photons could scatter off electrons like classical particles, and would conserve energy and momentum when they did so, were carried out by Compton and his co-workers in 1923-5).

Einstein suspected that this particle model would be important when radiation is emitted and absorbed and showed that it worked well in three particular cases:

- photoluminescence (emission of light by a material which has previously absorbed light);
- photoionisation of gases (ionisation by light);
- photoelectric effect (ejection of electrons from the surface of a metal when illuminated by light).

In all three cases the particle model worked much better than the continuous wave model and Einstein won the 1921 Nobel Prize *"for his services to Theoretical Physics, and especially for his discovery of the law of the photoelectric effect"*. Planck was first to introduce quantisation but Einstein's work really made physicists take quantisation seriously. It also undermined the classical view of reality, a consequence that Einstein himself found difficult to deal with.

1.2.3 The Photoelectric Effect

The importance of the photon model can be seen if we look at the photoelectric effect. Hertz discovered this when he was verifying Maxwell's predictions about electromagnetic waves. He noticed that a spark gap held just below its sparking potential could be induced to spark if ultra-violet light was shone on one of the electrodes. Somehow the light ejected charged particles (later known to be electrons) from the metal surface, these caused more ionization in the gas, and a spark jumped across the gap. The effect was investigated in more detail by Philipp Lenard (who received the Nobel Prize in 1905 for his work on cathode rays). Lenard discovered that increasing the intensity of light falling on a metal surface has absolutely no effect on the kinetic energy of the ejected electrons (although it does increase the rate of ejection). This seemed strange. If the electrons were ejected as a direct result of the electromagnetic energy absorbed by the metal surface then the classical expectation would be for the electron energies to increase with intensity. In Einstein's photon model the result is obvious. Increase the intensity and more photons per second strike each unit area of the metal, but the energy of individual photons does not change (since this depends on the frequency, not the intensity). If individual photons transfer all their energy to individual electrons then the electron energy depends only on frequency.

The photon model explained other aspects of photoelectric emission too. If the frequency of light is too low no electrons are emitted at all. This is simply because no individual photon has enough energy to knock an individual electron off the surface. Low frequency (low energy) photons excite electrons in the metal but these interact and the absorbed energy simply warms the metal.

If the minimum energy needed to remove an electron from the metal surface is W

(the 'work function') then the energy of an ejected electron will be the difference between the photon energy and the work function:

$$KE_{max} = hf - W$$

There will be a threshold frequency f_o below which the photon energy is less than the work function and incapable of ejecting electrons. This is given by:

$$hf_o = W \quad \text{or} \quad f_o = W/h$$

The threshold frequency is lower for more reactive metals since they lose electrons more easily and so have a smaller work function. For example, potassium (work function 2.23 eV) releases electrons in visible light whereas silver (work function 3.8 eV) requires ultra-violet light.

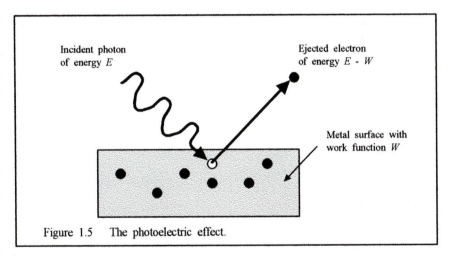

Figure 1.5   The photoelectric effect.

1.3 WAVE-PARTICLE DUALITY

1.3.1 Matter Waves

"After long reflection in solitude and meditation, I suddenly had the idea, during the year 1923, that the discovery made by Einstein in 1905 should be generalised by extending it to all material particles and notably to electrons."

(Louis de Broglie, preface to his re-edited 1924 PhD. thesis, Recherches sur la théorie des quanta, p.4, Masson, Paris, 1963)

The development of high vacuum tubes by Geissler and Crookes allowed a number of physicists to carry out detailed experiments on electrical discharges in low pressure gases. Interest soon focused on the 'cathode rays'. In 1897 J.J. Thomson showed that cathode rays can be explained as a stream of high speed, negatively charged *particles*, later called electrons. He won the 1906 Nobel Prize for this work. His son, G.P. Thomson, also won the Nobel Prize, in 1937. He shared the award with the American, Clinton Davisson *"for their experimental discovery of the diffraction of electrons by crystals"*. Diffraction is a wave property, and cannot be explained if the cathode rays really are a stream of particles. A wave model is needed to explain electron diffraction. So J.J. Thomson won the prize for showing that electrons are particles and his son won it for showing they are waves!

The idea that electrons might exhibit the same 'wave-particle duality' as light was first suggested by Louis de Broglie in 1924. At this time physicists were still struggling to come to terms with the implications of light's strange behaviour, so it might seem that de Broglie's hypothesis would not be particularly welcome. However, it does have the advantage of treating matter and radiation in the same way and, besides, Einstein thought it was a good idea saying, *"I believe it is a first feeble ray of light on this worst of our physics enigmas"*.

The de Broglie equation can be derived quite simply by drawing an analogy between electrons and photons and identifying momentum and wavelength as the defining properties for their particle and wave natures.

Photon energy:

$$E = hf = mc^2$$

Photon momentum:

$$p = mc = E/c = hf/c = h/\lambda$$

De Broglie proposed that the link between momentum and wavelength ($p = h/\lambda$) for a photon is a universal relation applying to electrons and atoms as well as to radiation. G.P. Thomson tested the relation by firing electron beams through thin slices of gold and showing that they

Figure 1.6 **George Paget Thomson** 1892-1975 demonstrated the wave-like nature of electrons when he passed a beam through a thin gold foil and produced diffraction rings. He shared the 1937 Nobel prize for Physics with Clinton Davisson.

Artwork by Nick Adams

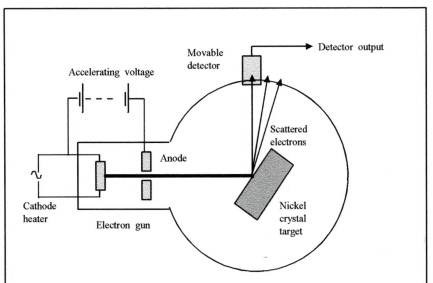

Figure 1.7 Schematic diagrams of the electron diffraction experiemnts of Davisson and Germer (above) and G.P. Thomson (below). Davisson and Germer showed that electrons scattered from the surface of a nickel crystal form diffraction rings just like X-rays. This agreed with de Broglie's predictions.

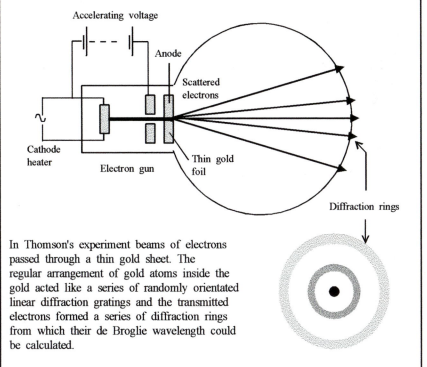

In Thomson's experiment beams of electrons passed through a thin gold sheet. The regular arrangement of gold atoms inside the gold acted like a series of randomly orientated linear diffraction gratings and the transmitted electrons formed a series of diffraction rings from which their de Broglie wavelength could be calculated.

produced clear diffraction rings. Davisson and Germer compared electron diffraction patterns formed when electron beams scattered off the surface of nickel crystals with X-ray diffraction patterns formed in similar ways. Both experiments led to a value for the electron wavelength and in both cases it was in agreement with de Broglie's formula. Electrons *do* behave like waves.

1.3.2 The Compton Effect

In 1923 Arthur Compton carried out a series of experiments which did for photon theory what Young's double slit experiment had done for the wave theory. He directed a beam of monochromatic X-rays onto a graphite target and measured the intensity of scattered X-rays as a function of their wavelength for different scattering angles. The results showed two things:

- The scattered X-rays contained distinct components, one at the original wavelength and the other at a longer wavelength.
- The wavelength of the longer wavelength component increased with scattering angle.

These results could not be understood using a classical wave model. The classical view of scattering would have the incident waves exciting electrons in carbon atoms to oscillate at the original X-ray frequency. These oscillating electrons would then re-radiate at this frequency so that the scattered X-rays should have the same frequency and wavelength as the original waves.

However, if the particle model is used, it all makes sense. If the incident X-ray beam consists of a stream of particle-like photons then a photon scattering off an electron is like a collision between two billiard balls. After the collision the second ball has gained some energy and momentum from the first. This would mean the electron recoils, taking some of the energy and momentum away from the scattered X-ray photon. De Broglie's equation links wavelength to momentum, so the scattered photon, with less momentum, should also have a longer wavelength. Larger angle scattering results in a larger transfer of momentum and so a grater increase in wavelength. This can be analysed in more detail to obtain the following formula, linking the change in wavelength to the scattering angle:

$$\Delta\lambda = \frac{h}{mc}(1 - \cos\theta)$$

The term h/mc is called the '*Compton wavelength*' of the electron. Compton's results were consistent with this equation and gave the most direct evidence that the particle model for light must be taken seriously. It was around this time that the term 'photon' was introduced. It is interesting to note that the theory of the Compton effect was also worked out by Kramers, but he was dissuaded from publishing his results by Niels Bohr who thought that the Compton effect might be an example in which the law of conservation of energy breaks down (an idea Bohr had also considered when trying to explain the beta-ray energy spectrum). Arthur Compton shared the 1927 Nobel Prize for Physics for this discovery.

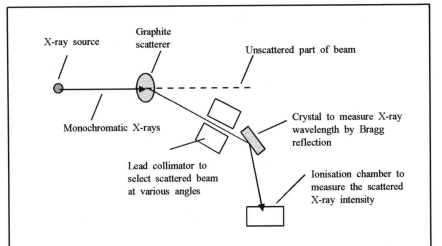

Figure 1.8 Compton's experiment. Compton measured the wavelength and intensity of X-rays scattered from electrons in the graphite target at various angles to the incident beam. The scattered X-rays had a longer wavelength than the unscattered X-rays and Compton was able to derive the shift in wavelength by assuming the photons and electrons collide like particles. The larger the angle of scattering the greater the shift in wavelength.

1.3.3 Interference Effects

The wave model is essential for a clear explanation of diffraction and interference, but the particle model is needed to understand absorption and emission processes. The experimental verification of de Broglie's hypothesis implied that this duality applied equally to matter (e.g. electrons) and radiation (e.g. light). Each model (particle or wave) accurately explains certain properties, but neither model is adequate for a complete explanation of all the properties of matter or radiation. They offer complementary views through two models that are apparently irreconcilable. The difficulties raised by wave-particle duality are clear if we consider a simple interference experiment connected to single photon detectors. In recent years experiments like this have been carried out using light, electrons, neutrons, atoms and even molecules.

The experiment described below is based on a Michelson interferometer and the explanation refers to light and photons. Bear in mind that exactly the same consequences arise if electrons or atoms are passed through an equivalent interferometer. The light source is a laser. Incident light is split by a beam splitter (e.g. half-silvered mirror) so that 50% goes by one route and 50% by the other, eventually reaching a second beam splitter that can direct light into either of two detectors. The paths are set up so that light going via path A to detector X travels exactly one half wavelength further than light going via B. On the other hand, the paths via A and B to detector Y are equal. If we treat the light as continuous waves this means the waves bouncing off the second beam splitter toward X will be 180° out of phase and destructive interference will occur. X will not receive any light. However, waves leaving the second beam splitter and going toward Y will be in phase and will reinforce one another. Y registers a strong signal because of this constructive interference. If the experiment is actually carried out the results are consistent with these predictions.

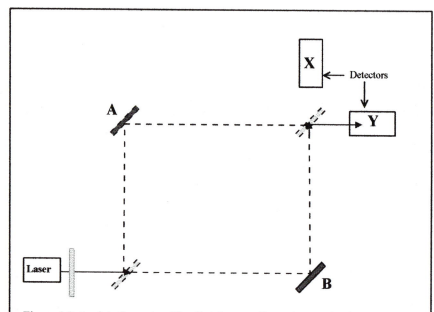

Figure 1.9 An interferometer. The first beam splitter gives each photon a 50% chance of going via A or B. These paths can be adjusted so that the path to detector X differs from the path to detector Y. In this case the paths to X differ by half a wavelength whereas paths to Y are equal. Destructive interference prevents photons from reaching X even when they are sent through the apparatus one at a time.

Now imagine that the two detectors have been replaced by very sensitive light detectors that can respond to single photons (e.g. a photomultiplier based on the photoelectric effect). When the experiment is repeated the same result occurs.

Detector Y clicks away like mad and no photons are recorded at X. This is quite strange. We need the particle model to explain the emission of photons (by stimulated emission) in the laser and also to record their arrival at the photoelectric detector, but it is hard to see how individual indivisible photons can account for the interference pattern. What prevents a photon that has travelled via A, say, from bouncing off the second beam splitter and arriving at X? Perhaps the paths of photons through the apparatus are correlated in some way so that whenever a photon travels via A it meets one that has traveled via B and they interact in some way so that neither can get to X, and both go to Y. This seems rather unlikely, but then so is the experimental result. One way to decide whether photons are 'interfering' with one another in this way is to send them through the apparatus one at a time. This can be done by reducing the intensity of the incident laser light (e.g. by passing the light through an almost opaque filter before it reaches the first beam splitter).

The first person to attempt an experiment like this was G.I. Taylor at the Cavendish Laboratory in Cambridge in 1909. He set up Young's slits apparatus using an extremely low intensity light source and a sensitive photographic plate as a detector. The intensity was so low that the chance of more than one photon being in the apparatus at any one time was small. Of course, with such low intensity light it takes a long time before the plate shows any significant pattern, accumulating points of exposure as photons arrive one by one at various places, and some exposures took several months. Rumour has it he went away on holiday while awaiting his results. When he did develop the plates he found the familiar interference fringes. So single photons do interfere!

Equivalent single quantum experiments have been carried out many times since then, and they confirm that the interference effects are still there. If one photon passes through the interferometer discussed above it will arrive at detector Y, it cannot arrive at X. If many photons are passed through one at a time they all arrive at Y. This is weird. To see just how weird, think about the following modification to the experiment. Imagine that one path, say B, is blocked. In the wave picture this wipes out one of the possible routes and removes the interference pattern, so now any light reaching the second beam splitter will divide equally and both detectors will receive equal amounts of light. This also makes sense with individual photons, if a photon hits the first beam splitter and goes via B it will be absorbed and neither detector will respond. If it goes via A it has an equal chance of going to either X or Y so the detectors both click randomly and (on average) receive equal numbers of photons. Fine, but how does this situation differ from that of a photon which goes via path A when *both* A and B are available to it? If we maintain the idea of individual photons passing through the apparatus we are forced to say that a photon going via path A has different probabilities at the second beam splitter when B is blocked than when B is open. And yet it didn't go via B, so how can the path it did not take affect the possible outcomes of the experiment? This apparent paradox is typical of the logical problems that arise when we try to describe quantum phenomena in classical terms. The results are summarised below:

- both paths available → waves interfere → no light reaches X;
- path B blocked → no interference → light goes equally to X and Y;
- both paths available → photons go by A or B → no photons reach X;
- path B blocked → detected photons go by A → equal numbers of photons reach X and Y.

This shows the counter-intuitive nature of the quantum world. It also emphasizes the limited validity of models, like waves and particles, which are based on analogies drawn from the familiar, macroscopic world of our everyday experience. The wave model gives the overall pattern in which the photons arrive at the detectors, but the arrival itself is discrete. What was needed was a way to interpret wave-particle duality that would allow working physicists to side-step the apparent paradoxes. Max Born provided this interpretation in about 1925.

1.3.4 Born's Statistical Interpretation

Einstein's photon theory treated light like a gas of particles. This approach was motivated by his observation that the fields used in electromagnetism are usually time-averaged and might well conceal particle-like discrete processes when light is absorbed or emitted. The problem with this view is exactly the one that was highlighted above – you can't abandon the wave model in favour of particles because the particle model cannot explain interference.

Einstein's solution for photon theory was to interpret the intensity of electromagnetic waves as a probability density for finding photons, so that high intensity light implies a high rate of arrival of photons. In Young's double slits experiment the maxima and minima can then be interpreted as probabilities resulting in many photons arriving near the maxima and few near the minima. In the interferometer experiment described previously, constructive interference results in a large probability that photons can reach detector Y whilst destructive interference makes the probability of any photon reaching X zero. Since the waves are now treated in a rather abstract way, this approach can also deal with single photon experiments. When one photon passes through an interferometer the probabilities of alternative outcomes (going to detector X or Y) are determined by the wave interference pattern, but the outcomes themselves are always discrete particle-like effects (i.e. a whole photon turns up at a particular detector, rather than parts of a photon at different detectors). This is similar to tossing a coin – the probability of 'heads' is equal to the probability of 'tails' but the outcome of the experiment is always either 'heads' or 'tails'.

In 1926 Born, who was belatedly awarded the Nobel Prize in 1954 *"for his fundamental research in quantum mechanics, especially for his statistical interpretation of the wavefunction"* was trying to make sense of electron waves. He claims that his inspiration came from Einstein:

"I start from a remark by Einstein on the relation between a wave field and light quanta; he said approximately that the waves are only there to show the way to the corpuscular light-quanta, and talked in this sense of a "ghost field" which determines the probability for a light-quantum ... to take a definite path."

(M. Born, Zeitschr. f. Phys., 38, 803, 1926)

Given this background Born's contribution may seem rather obvious. He suggested that the 'intensity' of electron waves is also linked to probability, in this case the probability per unit volume of finding an electron at a particular point in space. To see why this was such a radical and important step we need to put it in its historical context. De Broglie proposed his 'matter-waves' hypothesis in 1923 and Schrödinger published his first paper on wave-mechanics in January 1926. Schrödinger thought he was returning physics to a secure foundation with continuous fields and waves from which correct physical predictions could be made. He constructed a differential equation for electrons whose solutions represented the de Broglie waves. These solutions became known as 'wavefunctions' and are usually denoted by the symbol Ψ. Schrödinger hoped these continuous wavefunctions would give a direct physical representation of the electron.

Born realised that this could not be the case. For a start, Ψ is a complex quantity and all physical observables must be represented by real quantities. He suggested that the square of the absolute magnitude of Ψ (that is $|\Psi|^2$ or $\Psi\Psi^*$) is the link between wavefunctions and observables. If Ψ is the wavefunction for an electron then the value of $|\Psi|^2\delta V$ is a measure of the probability of finding the electron in the small volume δV (we are taking positions as an example here, the same approach can be used to calculate the probability that the electron momentum lies within some small range of momenta, etc.)

It is easy to miss the great shift in interpretation brought about by Born's work. Previously probability entered physics as a way of averaging over alternative states that we cannot directly observe. For example, the classical kinetic theory of gases can be used to calculate the average gas pressure on the walls of a container even though detailed knowledge of the particle configuration is lacking. Born's view is very different. Quantum mechanical probabilities do not arise from our ignorance of the microscopic configuration, they are *fundamental*. If this is true then the idea that individual electrons have well-defined trajectories through space and time or even 'possess' well-defined physical properties such as position and momentum may have to be abandoned. Schrödinger never really accepted all the implications of Born's statistical interpretation, nor could Einstein.

"The motion of particles follows probability laws but the probability itself propagates according to the law of causality."

(M. Born quoted in *Inward Bound*, A. Pais, p.258, OUP, 1986)

1.4 THE BOHR ATOM

1.4.1 Quantising the Atom

The work of Planck, Einstein, de Broglie, Schrödinger and Born have brought us to the threshold of the great synthesis which took place during the second half of the 1920s and which placed quantum theory on a (more or less) firm foundation. However, there were two other major developments taking place in parallel with the arguments over black-body spectra and wave-particle duality. Niels Bohr and Wolfgang Pauli quantised the atom and Satyendra Bose, Albert Einstein, Enrico Fermi and Paul Dirac worked out how to apply statistical principles to quantum particles.

Bohr's work on the atom built on a remarkable nineteenth century observation made by a 60-year-old Swiss schoolteacher, Johann Jakob Balmer. Balmer was looking for a simple mathematical relation between the wavelengths of just four lines in the hydrogen spectrum (others had been measured but Balmer was unaware of them). He came up with:

$$\frac{1}{\lambda_{nm}} = R\left(\frac{1}{n^2} - \frac{1}{m^2}\right)$$

where R is the Rydberg-Ritz constant, $R = 1.097 \times 10^7$ m^{-1} and m and n are both are integers 1, 2, 3 etc. with $m > n$.

Balmer went on to predict that there would be an infinite number of spectral lines corresponding to all integer values of n and m – quite a feat for a 60-year-old publishing his first scientific paper!

Bohr discovered Balmer's work in 1913 and realised almost immediately that it could be combined with quantum theory to construct a new model of the atom. His reasoning ran something like this:

- Isolated atoms emit radiation in a characteristic line spectrum – each atom can emit certain discrete frequencies but not others.
- Discrete frequencies correspond to discrete photon energies.
- Atoms themselves can only lose energy in certain discrete steps.
- Electrons in atoms can only occupy certain orbits, each with a definite energy.
- A photon is emitted when an electron makes a quantum jump from one allowed orbit to one of lower allowed energy.

To us this seems a reasonable way to proceed. However, in 1913 it was quite outrageous. According to Maxwell's equations accelerated charges should radiate electromagnetic waves. Electrons orbiting a nucleus are certainly accelerating (they have a centripetal acceleration toward the nucleus as a result of their circular motion). If they radiate energy they should fall into the nucleus, like a satellite whose orbit decays because of atmospheric friction. This would make the atom unstable. Bohr's idea was built on the assumption that the lowest energy state allowed in his quantum atom *would* be stable because the quantum condition allows no lower energy states. Excited states on the other hand are unstable, since they can decay into lower allowed energy states. However, when they do decay it is in a single quantum jump resulting in the emission of a single photon whose energy E is equal to the energy difference between initial and final quantum states.

1.4.2 The Hydrogen Atom

Bohr added a single quantum condition to the simple idea of an electron orbiting a nucleus as a result of electrostatic attraction. He postulated that the electron orbits can only have angular momentum in integer multiples of $h/2\pi$ (often written as h-bar, $\hbar$). Without the quantum condition the electron could orbit at any radius and have any energy. This leads to the problem of atomic stability – if there are a continuous range of allowed orbits the electron can spiral into the nucleus and radiate energy. However, if the quantum condition is also satisfied this reduces to a discrete (but still infinite) set of allowed energy levels.

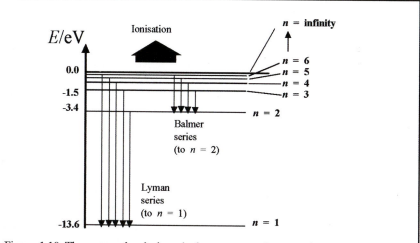

Figure 1.10 The energy levels in a hydrogen atom. Quantum jumps to $n = 1$ generate the Lyman series, those to $n = 2$ generate the Balmer series.

Maths Box: The Bohr Model of the Hydrogen Atom

Electrostatic attraction:
$$F = \frac{e^2}{4\pi\varepsilon_0 r^2} \qquad (1)$$

Bohr's quantum condition:
$$mvr = \frac{nh}{2\pi} \qquad n = 1, 2, 3 \text{ etc.} \qquad (2)$$

Electron energy:
$$E = KE + PE = \frac{1}{2}mv^2 - \frac{e^2}{4\pi\varepsilon_0 r} \qquad (3)$$

These three equations can be rearranged to find an expression for energy that is independent of radius r and velocity v. However, it does depend on n, giving a discrete set of circular orbits characterised by this *principal quantum number*:

$$E_n = -\frac{me^4}{8\varepsilon_0^2 n^2 h^2} \qquad \text{for } n = 1, 2, 3 \text{ etc.} \qquad (4)$$

Figure 1.10 shows these energy levels for hydrogen and shows how quantum jumps between different levels can result in various spectral series. In particular we can derive the Balmer formula by considering a jump from the mth to the nth level:

$$\Delta E_{mn} = E_m - E_n = \frac{me^4}{8\varepsilon_0^2 h^2}\left(\frac{1}{n^2} - \frac{1}{m^2}\right)$$

$$\Delta E_{mn} = hf_{mn} = \frac{hc}{\lambda_{mn}}$$

$$\frac{1}{\lambda_{mn}} = \frac{me^4}{8\varepsilon_0^2 h^3 c}\left(\frac{1}{n^2} - \frac{1}{m^2}\right) = R\left(\frac{1}{n^2} - \frac{1}{m^2}\right)$$

This is, of course, the Balmer formula. (The Balmer series is actually a special case of this formula for $n = 2$).

Bohr's work was still based on unjustified assumptions and contained at its heart an *ad hoc* quantisation. It was clear that a deeper theory was needed from which the quantum conditions would emerge, rather than having to be put in this rather arbitrary way. The success of Bohr's approach led to the gradual incorporation of greater detail into the theoretical model. There were a number of respects in which the original work was only approximate and there was new experimental data to accommodate too.

- Bohr's model treated the nucleus as fixed in space with the electrons orbiting it rather than electrons and nucleus orbiting their mutual centre of mass. Once this was corrected the constant in Bohr's energy formula could be calculated more accurately. In 1932 Harold Urey noticed faint additional lines displaced from the usual hydrogen spectrum. He correctly concluded that these were due to the presence of an isotope of hydrogen with a more massive nucleus, deuterium.

- Although the theory was developed for hydrogen it was also applied successfully to hydrogen-like ions such as He^+ and Li^{++} which consist of a charged nucleus and a single orbiting electron.

- Both electrostatics and gravitation obey an inverse-square law, so an arbitrary electron orbit would be expected to be elliptical (like planetary orbits) rather than circular. Arnold Sommerfield developed Bohr's theory to include elliptical orbits and showed that different eccentricities resulted in slightly different energy levels so that each of the spectral lines predicted by Bohr split into a narrow set of lines. This also introduced a new quantum number l related to the eccentricity of the orbit. This orbital quantum number l can take integer values from 0 to $n-1$, with $n-1$ corresponding to the circular orbit.

- In 1896 Pieter Zeeman at Leiden noticed that spectral lines split into doublets or triplets when the atoms emitting them are in a strong magnetic field. He and Hendrick Antoon Lorentz shared the 1902 Nobel Prize for their (incorrect) explanation of this 'Zeeman effect' based on classical electron theory. In 1916 Sommerfield and Debye showed that the Zeeman effect could be explained using Bohr's theory. To do this they introduced a third quantum number, m (magnetic quantum number) which can take integer values from $-l$ to $+l$. This third quantum number is related to the spatial orientation of the orbit (which in turn affects how much the orbit is affected by the applied magnetic field and hence how far the spectral lines are displaced).

There were problems too. Whilst the Bohr theory predicted the wavelengths of spectral lines it gave no indication of their relative intensity. Another way of saying this is that it could not predict the transition probabilities that determine the rate at which excited states decay. Nor was it able to predict the spectrum of helium or the electronic configurations of atoms more complicated than hydrogen. Yet another problem involved the 'anomalous Zeeman effect'. The Bohr theory predicted that spectral lines should split into doublets or triplets when the source atoms are placed in a magnetic field, but experimental observations had already outstripped this – the sodium D_1 line, for example splits into four components and D_2 into six. The only way this could be explained would be to introduce another quantum number, and Sommerfield did imply the existence of a 'hidden' quantum number. The problem was that it did not seem to correspond to any obvious physical property. The three quantum numbers, n, l, and m at relate to orbital size, shape and orientation, the fourth quantum number seemed completely abstract. Despite these problems, Niels Bohr shared the 1922 Nobel Prize for Physics (with Albert Einstein) *"for his services to the investigation of the structure of atoms and of the radiation emanating from them"*.

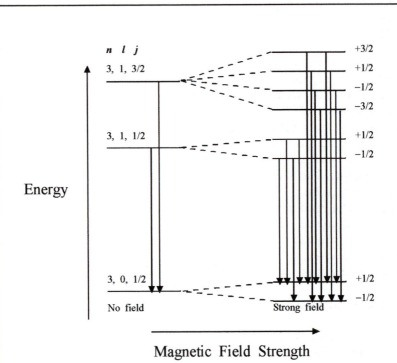

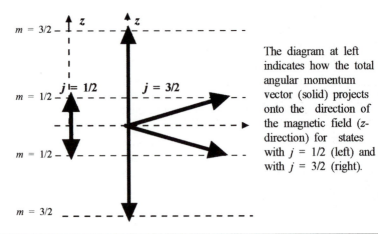

Figure 1.11 The Zeeman effect in sodium. In a zero field sodium D-lines form a doublet. In a strong field these split to form a quadruplet and a sextet, a group of ten lines. This is explained by the interaction of the total angular momentum j with the applied magnetic field. The splitting of the energy levels comes about because the component of total angular momentum is quantised when it is projected onto the direction of the magnetic field. The numbers to the right of the diagram above show how the $j = 1/2$ states split into doublets and the $j = 3/2$ states split into quadruplets.

The diagram at left indicates how the total angular momentum vector (solid) projects onto the direction of the magnetic field (z-direction) for states with $j = 1/2$ (left) and with $j = 3/2$ (right).

1.5 THE EXCLUSION PRINCIPLE

1.5.1 Spin

Each of the first three quantum numbers for electrons in atoms relate to a degree of freedom, size of orbit, shape of orbit, orientation of orbit, etc. The possible existence of a fourth quantum number suggested to Samuel Goudsmit that there was a fourth degree of freedom and that it must be associated with the electron itself. But what? He suggested that electrons have an intrinsic spin of magnitude $h/4\pi$ that can align parallel or antiparallel to any applied magnetic field. This results in additional splitting of spectral lines since the two alignments have different energies. The fourth or spin quantum number s can have a value $\pm 1/2$. (this half integer value still results in integer quantum jumps from $+1/2$ to $-1/2$).

Wolfgang Pauli came to more or less the same conclusion about the need for a fourth quantum number when he tried to explain the subtleties of atomic spectra in 1924. He associated four quantum numbers with each electron and showed that the hierarchical structure of the Periodic Table and the pattern of emissions from atoms made sense if an additional assumption is made:

• No two electrons in any atom can have the same set of quantum numbers.

This is known as the Pauli Exclusion principle and is now known to apply to all particles with half-integer spin (fermions). To see how it generates the hierarchy of atomic structure we simply need to list the four quantum numbers in order and imagine they represent atoms built up from scratch by filling the lowest energy level first and continuing to top them up one electron at a time. In the table below the atoms in the left hand column contain all the electron orbits (sets of quantum numbers) up to and including the one in their own row. The final column gives a symbolic representation of the electronic configuration.

Atom	n	l	m	s	Electronic configuration
$_1$H	1	0	0	$-1/2$	$1s^1$
$_2$He	1	0	0	$+1/2$	$1s^2$
$_3$Li	2	0	0	$-1/2$	$1s^2 2s^1$
$_4$Be	2	0	0	$+1/2$	$1s^2 2s^2$
$_5$B	2	1	-1	$-1/2$	$1s^2 2s^2 2p^1$
$_6$C	2	1	-1	$+1/2$	$1s^2 2s^2 2p^2$
$_7$N	2	1	0	$-1/2$	$1s^2 2s^2 2p^3$
$_8$O	2	1	0	$+1/2$	$1s^2 2s^2 2p^4$
$_9$F	2	1	1	$-1/2$	$1s^2 2s^2 2p^5$
$_{10}$Ne	2	1	1	$+1/2$	$1s^2 2s^2 2p^6$
$_{11}$Na	3	0	0	$-1/2$	$1s^2 2s^2 2p^6 3s^1$

$_{12}Mg$	3	0	0	+1/2	$1s^22s^22p^63s^2$
$_{13}Al$	3	1	−1	−1/2	$1s^22s^22p^63s^23p^1$
$_{14}Si$	3	1	−1	+1/2	$1s^22s^22p^63s^23p^2$
$_{15}P$	3	1	0	−1/2	$1s^22s^22p^63s^23p^3$
$_{16}S$	3	1	0	+1/2	$1s^22s^22p^63s^23p^4$
$_{17}Cl$	3	1	1	−1/2	$1s^22s^22p^63s^23p^5$
$_{18}Ar$	3	1	1	+1/2	$1s^22s^22p^63s^23p^6$

Notice how the structure develops.

- There are 2 orbits with $n = 1$. These form the 1s shell, and account for hydrogen and helium. The stability of helium is linked to its full electron shell. 1s shells have $l = 0$ and are spherically symmetric.
- When we come to $n = 2$ there are 8 possible orbits, 2 corresponding to the $l = 0$ state forming the spherically symmetric 2s shell, and a further 6 in the 2p shell. 2p does not have spherical symmetry ($l = 1$) and there are three possible orientations for this shell (corresponding to $m = −1, 0, +1$). The eight $n = 2$ orbits account for the second period of the Periodic Table, starting with sodium ($Z = 5$), a reactive metal that owes its reactivity to the spare electron in its outer shell ($1s^22s^22p^1$), and ending with neon ($Z = 10$), an inert gas, which has a full outer shell ($1s^22s^22p^6$).
- $n = 3$ has a total of 18 orbits, but only the first 8 correspond to the third period of the Periodic Table. This is because 3s ($n = 3, l = 0$) and 3p ($n = 3, l = 1$) fill but then, rather than the 3d ($n = 3, l = 2$) the lower energy 4s shell begins to fill.
- The fourth period then consists of 4s (2 electrons) followed by 3d (10) and 4p (6) giving 18 elements in the fourth period, from potassium ($Z = 19$) to krypton ($Z = 36$).

The order in which electron sub-shells fill is:

1st period	(2 atoms)	1s
2nd period	(8 atoms)	2s, 2p
3rd period	(8 atoms)	3s, 3p
4th period	(18 atoms)	4s, 3d, 4p
5th period	(18 atoms)	5s, 4d, 5p
6th period	(32 atoms)	6s, 4f, 5d, 4f, 6p

The reason the 4s sub-shell fills before 3d is because of tighter binding for the spherically symmetric states compared to the d and f orbits of larger atoms. The similarity in chemical properties from one period to another (the property that first drew attention to the Periodic Table) arises from similar arrangements of outer electrons (e.g. sodium and potassium both having one 'spare' electron that is easily lost, or neon and argon having full outer shells). The fourth period begins with typical group 1 and group 2 elements (potassium and calcium) but then the 3d sub-shell fills, determining the electronic structure of the ten elements from scandium ($Z = 21$) to zinc ($Z = 30$). The ferromagnetic elements, iron, cobalt and nickel, are in this period. They owe their special magnetic properties to an unusual arrangement of unpaired electrons with parallel spins in the unfilled 3d sub-shell. Once the 3d sub-shell is full, 4p fills with the last six elements of the fourth period. Bromine, for example, fits neatly beneath chlorine (both lack one electron in their respective p sub-shells) and krypton with its full p sub-shell is similar to argon and neon. Elements with incomplete 3d sub-shells have fairly similar properties and those with incomplete 4f and 5f sub-shells are almost indistinguishable. The distinct chemical properties produced by partially filled smaller sub-shells has been lost and the chemistry of the transition elements is fairly uniform.

The limit to the number of elements is not set by electronic structure, but by nuclear structure. Eventually the cumulative effects of long-range electrostatic repulsion between protons cannot be contained by the short-range attraction between nucleons due to the strong force. Uranium is the heaviest element that occurs in any significant quantity on Earth but other heavier elements can be produced artificially even if they do decay with a very short lifetime.

Wolfgang Pauli won the 1945 Nobel Prize for Physics *"for the discovery of the Exclusion Principle, also called the Pauli Principle"*.

1.5.2 Quantum Statistics

Ludwig Boltzmann explained the macroscopic properties of gases (such as temperature, pressure, etc.) using the microscopic hypothesis of tiny particles flying about with rapid random motions. He also provided a microscopic description of the Second Law of Thermodynamics. This law is linked to the 'arrow of time' and states that in all chemical and physical changes the entropy of the universe never decreases. This idea of 'entropy' was very mysterious – entropy changes could be calculated from macroscopic rules but nobody had given a clear explanation of why these rules worked.

Imagine a thermally isolated container of gas divided in two by a barrier that is a good thermal conductor. The gas on one side is at a higher temperature than that on the other. What happens? Gradually the hot gas cools and the cool gas warms up. Heat always flows from hot to cold. But why should this be? Why doesn't the cold gas get colder by passing heat to the hot gas and making it hotter still? After all, this would still conserve energy. Boltzmann's answer was simple. If you imagine all the distinct ways in which the total energy can be distributed among all the molecules you will find that there are far more microscopic distributions (micro-states) corresponding to macro-states close to thermal equilibrium than to macro-states far from thermal equilibrium. If we assume that the evolution of the system is driven purely by random chance then the initial state will evolve toward states which are closer to thermal equilibrium, simply because there are more of them. This is very similar to what happens if you throw 100 coins up into the air together and watch how they fall. The particular combination of heads and tails you get is just 1 from 2^{100} alternatives, any of which could have occurred. However, the ratio of head to tails is likely to be close to 50:50, and almost certainly nothing like 10:90. Why is this? Because the number of ways of arranging heads and tails in a ratio close to 50:50 vastly outnumbers the ways of arranging them 10:90. The individual outcomes (HTHHTTTHHH ...) are micro-states, but we ascribe significance to the macro-states (49H and 51T, etc.) and there are vastly more macro-states corresponding to a 50:50 split than to anything else.

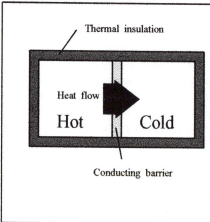

Thermal insulation

Heat flow

Hot   Cold

Conducting barrier

Figure 1.12 Heat flows from hot to cold because the increase in number of ways of distributing energy in the cooler body is greater than the decrease in number of ways of distributing it in the hotter one. The underlying interactions are random so the system moves toward the macroscopic state corresponding to the maximum number of possible microscopic configurations: it maximises its entropy. This is true in both classical and quantum physics, but there are important differences in how the microstates are counted.

Boltzmann's conclusion was that entropy must be linked to the number of ways a system can be configured. The arrow of time emerges as a tendency for random changes to result in more likely macrostates, i.e. for systems to evolve toward the macrostate which can be represented in the maximum number of microscopic ways. This explains the evolution toward equilibrium and the law of increase of entropy (*given a sufficiently low entropy beginning*).

What has this got to do with quantum theory? The link is in how we count the microstates. Boltzmann assumed that the individual gas molecules could be treated like *distinguishable particles* so that a configuration with molecule A in state X and molecule B in state Y is counted as an alternative to B in state X and A in state Y (with all the other molecules in the same states in both alternatives). If we were discussing coin tosses this is like saying that there are 100 ways of getting just 1 head (it could be any of the 100 coins that comes up heads) rather than just 1 way (if different coins are indistinguishable). The black-body radiation spectrum can be derived by treating the radiation as a gas of photons filling the cavity. In order to calculate the energy spectrum we need to work out how many photons are present in each small frequency range, in other words how the available energy is distributed among photons of various frequencies. This is very similar to Boltzmann's approach with a hot gas, but can we count photons in the same way that Boltzmann counted molecules?

The answer to this question is a resounding no, and an Indian physicist, Satyendra Nath Bose, provided it in 1924. He gave an alternative derivation of Planck's black-body radiation spectrum by treating the photons as indistinguishable particles with a tendency to cluster in the same energy states. His work was drawn to the attention of Einstein who generalised it to a method for counting how energy is distributed among a collection of particles (such as photons) all of which have integer spin. Later, integer spin particles were named 'bosons' and the method of counting is known as *Bose-Einstein statistics*. When a large collection of bosons are involved (e.g. superconductivity, superfluidity, laser light, etc.) the probability of finding more than one boson in the same energy state is enhanced by the presence of the others. The reason bosons behave in this way is related to symmetry of the wavefunction that describes them, as we shall see in a later section. As far as photons are concerned, it means that the probability that a photon is emitted into a state already containing n photons is $(n + 1)$ times greater than if there are none already present. This is used in the appendix to derive an expression for the black-body radiation spectrum.

Another consequence of Bose-Einstein statistics is the behaviour of collections of bosons at very low temperatures. With little energy available the bosons 'try' to cluster in the lowest available energy state and undergo a Bose-condensation. Superfluidity in helium-4 (and some other fluids) is an example of this. The motion of helium-4 atoms (which are bosons) proceeds with zero viscosity and results in some amazing physical effects. The origin of superfluidity is emphasised by the fact that liquid helium-3 is not a superfluid (helium-3 atoms have half-integer spin so are not bosons) except at extremely low temperatures, when the fermions form pairs and behave like bosons!

During the same period Enrico Fermi and Paul Dirac showed that particles with half-integer spin (fermions) like electrons in atoms obey yet another form of quantum statistics – *Fermi-Dirac statistics*. Electrons, like photons, are indistinguishable, but Pauli had already shown that the probability of finding two electrons in the same quantum state in the same atom was zero. If a large collection of fermions is involved this prevents them collapsing into the lowest available energy states and is therefore responsible for the structure of atoms, the behaviour of electrons in metals and the 'degeneracy pressure' that supports a neutron star against gravitational collapse (neutrons are also fermions).

Bose-Einstein statistics: probability of adding a boson to n bosons $\propto$ $(n+1)$
Fermi-Dirac statistics: probability of adding a fermion to an occupied state $= 0$

Quantum statistics are discussed in more detail later in the context of low energy physics.

2

A NEW QUANTUM THEORY 1925-30

2.1 QUANTUM THEORIES

Old quantum theory was baffling because it worked so well despite having no firm foundation. The mystery of wave-particle duality and the *ad hoc* quantisations certainly hinted at the existence of an underlying and overarching theory, but up until 1925 no one had the faintest idea what this theory would be. Then a remarkable thing happened. The loose ends of all the ideas that had been so hotly debated were pulled together in *three* different versions of quantum theory. The first of these, matrix mechanics, was proposed in 1925 by a brilliant young German physicist, Werner Heisenberg. This was followed almost immediately by the more 'user-friendly' wave-mechanics of Erwin Schödinger, and then, a little later, by the abstract but powerful operator theory of Paul Dirac.

2.1.1 Heisenberg, Matrix Mechanics and Indeterminacy

"Our scientific work in physics consists in asking questions about nature in the language that we possess and trying to get an answer from experiment by the means that are at our disposal. In this way quantum theory reminds us, as Bohr has put it, of the old wisdom that when searching for harmony in life one must never forget that in the drama of existence we are ourselves both players and spectators. It is understandable that in our scientific relation to nature our own activity becomes very important when we have to deal with parts of nature into which we can penetrate only by using the most elaborate tools."

(W. Heisenberg, in *The Copenhagen Interpretation of Quantum Theory*, 1958, Harper and Row)

Figure 2.1 **Werner Karl Heisenberg** 1901-1976 worked closely with Niels Bohr at Copenhagen in the mid-1920s and developed the first version of quantum theory in 1925: matrix mechanics in which all attempts to visualise atomic processes were abandoned and replaced by a mathematical formalism that allowed physicists to predict the observable outcomes of actual experiments. In 1927 he introduced the Uncertainty Principle and later turned his attention to the problems of nuclear and particle physics. He was one of few top German theoreticians who chose to remain in Germany during the war. His role in directing German nuclear research remains controversial. It is not clear whether he was doing all he could to develop an atomic bomb for Hitler, or remaining in place in order to prevent this happening. He won the 1932 Nobel prize for Physics. Artwork by Nick Adams

Heisenberg did not think that the classical model of orbiting electrons in nuclear atoms was a sensible starting point for theoretical physics. After all, the trajectories of the electrons are unobservable and line spectra implied discontinuous quantum jumps between allowed states. So he rejected this direct representation of electron orbits in atoms and constructed a theory based entirely on observables – in this case the line spectra of excited atoms. This gave him a set of transitions which he arranged in the form of a matrix, each element of the matrix related to the probability of a particular transition. For example, if there are just four possible states then the initial states can be listed vertically and the final states horizontally giving a 4×4 matrix as shown below:

$$
\text{initial states} \quad
\begin{bmatrix}
a_{11} & a_{12} & a_{13} & a_{14} \\
a_{21} & a_{22} & a_{23} & a_{24} \\
a_{31} & a_{32} & a_{33} & a_{34} \\
a_{41} & a_{42} & a_{43} & a_{44}
\end{bmatrix}
$$

final states

The a_{ij}'s are transition amplitudes, the square of their absolute values gives the transition probability.

".... the new theory may be described as a calculus of observable quantities."

(W. Heisenberg, *The Physical Principles of the Quantum Theory*, Dover, New York, 1949)

Heisenberg, Born and Jordan worked out the rules obeyed by these matrices and showed that they could be used to unite many diverse aspects of old quantum theory. Not only this, the mathematical structure of 'matrix-mechanics' bore a certain resemblance to a powerful mathematical form of Newtonian mechanics developed by William Hamilton. This helped explain how classical results emerge in the limit of large energies and also gave hints as to how to formulate classical problems in a quantum mechanical way. Heisenberg discovered another fundamental idea from these matrices, the Uncertainty Principle (or Principle of Indeterminacy).

2.1.2 The Heisenberg Uncertainty Principle

"If one assumes that the interpretation of quantum theory is already correct in its essential points, it may be permissible to outline briefly its consequences of principle ... As the statistical character of quantum theory is so closely linked to the inexactness of all perceptions, one might be led to the presumption that behind the perceived statistical world there still hides a 'real' world in which causality holds. But such speculations seem to us, to say it explicitly, fruitless and

senseless. Physics ought to describe only the correlation of observations. One can express the true state of affairs better in this way: Because all experiments are subject to the laws of quantum mechanics, and therefore to equation (1) [the uncertainty principle] *it follows that quantum mechanics establishes the final failure of causality."*

(W. Heisenberg, *The Physical Content of Quantum Kinematics and Mechanics,* 1927, reprinted in *Quantum Theory and Measurement,* ed. J.A. Wheeler and W.H. Zurek, Princeton, 1983)

Imagine calculating the area of a rectangle by measuring first its width w and then its length l and then multiplying the two values together. You get exactly the same result as if you measure the length followed by the width and multiply these values together. Multiplication of ordinary numbers is commutative, $wl = lw$. However, not all mathematical operations are commutative. Here is an example. Stand in the centre of a room and face in a particular direction. Take two steps forward (call this operation S) and rotate $90°$ to your right (call this operation R). Note where you end up. Now return to your starting position and orientation. This time turn $90°$ to the right and *then* take two steps forward. You do not end up in the same place. The order of the operations matters. This can be written symbolically as:

$$SR \neq RS \text{ or } SR - RS \neq 0$$

The operations S and R are not commutative. Heisenberg found that he could represent operations like measurement of position or momentum by matrix multiplications. However, some of the matrices do not commute, so a position measurement (Q) followed by a momentum measurement (P) might yield different results from a momentum measurement followed by a position measurement. They might also leave the measured system in a different state. If the state on which these measurements are carried out is written as Ψ then $Q\Psi$ represents a position measurement and $P\Psi$ represents a momentum measurement. If position and momentum are measured along the same axis (e.g. in the x-direction) then:

$$Q_x (P_x \Psi) \neq P_x (Q_x \Psi)$$

What is the meaning of this? Heisenberg offered an explanation in terms of a simple (and slightly misleading) thought experiment (see Figure 2.1 and the Maths Box below). Imagine you wish to measure the position of an atom using a microscope. How do you do it? To observe the atom you have to bounce something off it, but the wavelength of visible light is much longer than an atomic diameter so the waves would not give any information about the atom's position (neither will these low frequency photons cause much disturbance to its momentum). You need waves of much shorter wavelength, but the shorter the wavelength the larger the photon momentum (from the de Broglie relation) and therefore the greater the

Figure 2.2 Heisenberg's gamma-ray microscope. To measure the position of an electron very precisely we must use short wavelength radiation. This limits the uncertainty due to diffraction at the aperture. However, shorter wavelengths carry more momentum, so the electron recoils more violently when struck. This recoil is itself uncertain because we do not know the exact path taken by the photon, only that it passes through some part of the aperture.

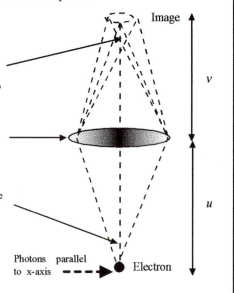

Photons are diffracted as they pass through the lens. This gives them an angular uncertainty which makes their position within the image uncertain too

Convex lens of diameter D used to focus photons

Photons scattered from the electron transfer an unknown momentum to it. However, the photons used to form the image of the electron must enter the lens, so they have a maximum uncertainty in angle given by $1.22D/u$

Photons parallel to x-axis

Image

v

u

Electron

recoil of the atom after it scatters the photon. Increasing the precision of the position measurement (by using light of ever-shorter wavelengths) results in a larger random scattering of the atom so that its momentum after the position measurement is uncertain. This implies that our ability to measure position is intimately linked up with our ability to measure momentum. The uncertainty in our knowledge of one of these two variables is increased if we measure the other one more precisely. This uncertainty principle can be stated more clearly:

$$\Delta x \Delta p_x \geq h$$

"The indeterminacy principle refers to the degree of indeterminateness in the possible present knowledge of the simultaneous values of various quantities with which the quantum theory deals ... Thus suppose that the velocity of a free electron is precisely known, while its position is completely unknown. Then the principle states that every subsequent observation of the position will alter the momentum by an unknown and undeterminable amount such that after carrying out the experiment our knowledge of the electronic motion is restricted by the uncertainty relation. This may be expressed in concise and general terms by

saying that every experiment destroys some of the knowledge of the system which was obtained by previous experiments."

(W. Heisenberg, *The Physical Principles of the Quantum Theory*, Dover, New York, 1949)

Maths Box: Heisenberg's Microscope and the Uncertainty Principle

Photon momentum: $p = \dfrac{h}{\lambda}$

The photon scatters from the electron and passes through a point in the lens. The uncertainty in angle of scatter is subtended by half the diameter of the lens:

$$\delta\theta \approx \pm\frac{D}{2u}$$ where u is the object distance from the lens

This means the photon will impart a momentum parallel to the x-axis that is uncertain to an extent:

$$\delta p_x = \pm p \sin\delta\theta \approx \pm\frac{hD}{2u\lambda}$$

When the photon emerges from the lens diffraction makes its direction uncertain within an angle

$$\delta\phi \approx \pm\frac{1.22\lambda}{D}$$

This means the image could be located within a distance $\pm v\delta\phi = \pm\dfrac{1.22v\lambda}{D}$

However, the actual position of the electron is worked out from this image position and is magnified by a factor $m = u/v$ so that the location of the electron on the x-axis is uncertain to the extent:

$$\delta x \approx \pm\frac{u\lambda}{D}$$

The product of the uncertainty in position and momentum of the electron is:

$$\delta x \delta p_x \approx \frac{1.22u\lambda}{D} \times \frac{hD}{2u\lambda} \approx h$$

This interesting argument is actually misleading. It implies that the electron *does* have a definite position and momentum and that the uncertainty derived above is in our *knowledge* of these values. The uncertainty principle is deeper than this. The position and momentum of the electron are *intrinsically uncertain*. It was this point to which Einstein objected.

I said that this thought experiment was slightly misleading, so is the term 'uncertainty'. Both imply that the atom has a definite position and momentum before we attempt to measure it and that this changes in a definite way through the process of measurement. This would suggest that the uncertainty is in our experimental measurement, and therefore in *our knowledge* of the atom rather than intrinsic to the atom itself. However, the probabilities that enter quantum theory are fundamental, they are not just about our ability to know what the system is doing or how it is configured. This is why 'indeterminacy' is preferable to 'uncertainty', Heisenberg's theory suggests that the properties of position and momentum are themselves indeterminate. They do not possess definite values prior to our measurements.

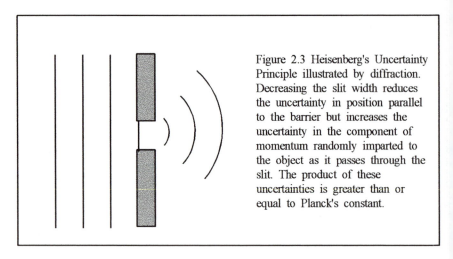

Figure 2.3 Heisenberg's Uncertainty Principle illustrated by diffraction. Decreasing the slit width reduces the uncertainty in position parallel to the barrier but increases the uncertainty in the component of momentum randomly imparted to the object as it passes through the slit. The product of these uncertainties is greater than or equal to Planck's constant.

A simple application of wave-particle duality underlines this far deeper interpretation. Imagine a single electron approaching along a normal to a barrier with a narrow vertical slit. Beyond the barrier is an array of detectors, each of which can detect the electrons. The de Broglie waves will diffract at the slit and spread out sideways. This gives a chance that the electron might be detected by detectors that are not directly opposite the slit. However, the probability is symmetric either side of the centre and there is no way to predict to which side the electron will deflect. As the electron passes through the vertical slit it must get a random horizontal impulse that sends it somewhere in the diffraction pattern determined by the intensity of the diffracted de Broglie waves. How does this link with the Indeterminacy Principle? Make the slit narrower and you reduce the indeterminacy in its horizontal position at the barrier (i.e. it must lie within the width of the slit in order to pass through it). However, a narrower slit results in a broader diffraction pattern, so the probability of getting a larger horizontal impulse increases – the indeterminacy in its horizontal momentum increases. Heisenberg's work emphasised observables – the actual path of the electron is not an observable and it is not possible to reconstruct an unambiguous trajectory for the electron from the observations that are actually made. Perhaps we should accept that the underlying reality is indeterminate.

"Some physicists would prefer to come back to the idea of an objectively real world whose smallest parts exist objectively in the same sense as stones or trees exist independently of whether we observe them. This, however, is impossible."

(Werner Heisenberg, quoted in *Quantum Reality*, Nick Herbert, Rider & Co. 1985)

Maths Box: Diffraction and Uncertainty

When waves of wavelength λ pass through a slit of width d they diffract. This gives them an angular uncertainty

$$\delta\theta \approx \pm\frac{\lambda}{d}$$

and implies that they have an uncertainty in momentum parallel to the barrier (x-direction) of:

$$\delta p_x = p\sin\delta\theta \approx \pm\frac{h}{d}$$

The width of the slit itself is a measure of their uncertainty in position, so:

$$\delta p_x\delta x \approx h$$

Heisenberg's theory is one of the most remarkable discoveries in the story of quantum theory and came after a period of intense intellectual struggle culminating in his great breakthrough during a brief retreat to the island of Helgoland:

"I was deeply alarmed. I had the feeling that, through the surface of atomic phenomena, I was looking at a strangely beautiful interior, and I felt almost giddy at the thought that I now had to probe this wealth of mathematical structures nature had so generously spread out before me. I was far too excited to sleep, and so, as a new day dawned, I made for the southern tip of the island, where I had been longing to climb a rock jutting out into the sea. I now did so without too much trouble, and waited for the sun to rise."

(Quoted in *Second Creation*, R.P. Crease and C.C. Mann, Macmillan, 1986)

In 1925 the idea of representing physical observables by matrices was radically abstract and rather disturbing. Few physicists were familiar with the mathematical formalism and they found it hard to relate to the post-classical world of old quantum theory. Einstein described it as *"a witch's calculus"* and was convinced it could not be *"the true Jacob"*. This 'shock of the new' explains why a more conservative theory proposed by the 38-year-old Erwin Schrödinger was accepted much more readily – even though its mathematical content turned out to be the same as in Heisenberg's theory.

2.1.3 Schrödinger's Wave-mechanics

"My theory was stimulated by de Broglie and brief but infinitely far-seeing remarks by Einstein. I am not aware of a generic connection with Heisenberg. I, of course, knew of his theory, but was scared away, if not repulsed, by its transcendental algebraic methods which seemed very difficult to me ...
The appearance of integers [quantum numbers] *comes about* [in wave mechanics] *in the same natural way as for example the integer quality of the number of modes of a* [classical] *vibrating string. "*

(E. Schrödinger, quoted in *Niels Bohr's Times*, p.281, Abraham Pais, OUP, 1991)

Erwin Schrödinger was never happy with the idea of discontinuous quantum jumps. His physics was firmly rooted in the differential equations of the nineteenth century and his attempt to make sense of de Broglie's matter waves was an attempt to return physics to its classical foundations. He assumed that the de Broglie waves must be solutions to a wave equation just as electromagnetic waves are solutions to the Maxwell equations, so he set about trying to find this wave equation.

The motivation behind this is obvious. If the wave equation (Schrödinger equation) can be found then the de Broglie waves are solutions to this equation and physical problems can be solved in the following way:

- Write down the Schrödinger equation.
- Put in conditions specific to the problem you want to solve (for example, an electron in a hydrogen atom is moving in the potential well of the nucleus, so this must be substituted into the equation).
- Solve the Schrödinger equation subject to the boundary conditions set by the problem to get the appropriate electron wave functions).
- Interpret the electron wavefunctions (i.e. the solutions).

This would provide physicists with a standard technique for solving problems. But how do you set about constructing a wave equation?

Schrödinger had several clues. He knew the equation had to be linear because linear equations obey a superposition principle. In physical terms this corresponds to the interference and diffraction effects characteristic of waves (and demonstrated for electrons by Thomson, Germer and Davisson). He also knew it had to satisfy the de Broglie relations and general conditions for the conservation of energy and momentum. The equation he came up with was a second-order differential equation, the 'time-dependent Schrödinger equation' quoted below:

Figure 2.4 **Erwin Schrödinger** 1887-1961. In 1925-6 Schrödinger developed the ideas of Louis de Broglie into a version of quantum mechanics that became known as wave mechanics. This was welcomed by many physicists, including Einstein, for whom Heisenberg's alternative matrix mechanics seemed too abstract to be a true representation of physical reality. Unfortunately for Schrödinger, the interpretation of his wavefunctions showed that they could not avoid the same counter-intuitive consequences as Heisenberg's theory. In fact the two approaches were equivalent. Schrödinger shared the 1933 Nobel prize for Physics with Paul Dirac. Artwork by Nick Adams

$$\left[-\frac{h^2}{8\pi^2 m}\nabla^2 + V\right]\Psi = \frac{ih}{2\pi}\frac{\partial\Psi}{\partial t}$$

where Ψ is the wavefunction (representing, in some sense, the electron), V is the potential energy (a function of position) and the bracketed term is called the 'Hamiltonian operator' or H. This is linked to the total energy of the electron and is explained in more detail in Appendix 2. Here we will simply mention some of the important characteristics of the equation and its solution. First of all, for situations in which V does not vary with time (e.g. the central electrostatic potential of an atomic nucleus) the equation separates into a time-dependent and a time-independent part and can be written in a shortened form as:

$$H\Psi = E\Psi$$

The general solutions to this equation are travelling waves (if $K = E{-}V$, the kinetic energy, is positive) and decaying exponentials (if $K = E - V$ is negative). In the former case the wavelength associated with the waves is;

$$\lambda = \frac{h}{\sqrt{2m(E-V)}} = \frac{h}{\sqrt{2mK}}$$

as expected for electrons of mass m and kinetic energy K.

In the latter case, which does not arise classically because it corresponds to negative kinetic energy, the wave amplitude decays exponentially with distance. This is responsible for some uniquely quantum mechanical effects such as tunnelling.

Of course the Schrödinger equation must be solved subject to certain constraints on the wavefunctions themselves. These are that Ψ and its gradient must be continuous and the integral of $\Psi\Psi^*$ throughout space must be finite (this normalising condition can be interpreted as saying that the probability of events represented by Ψ must add to 1).

One final point – the wavefunctions themselves are complex. This means they cannot be a direct representation of any physical observable. This was a blow to Schrödinger whose initial hope was that Ψ might be directly related to charge density. Although the Schrödinger equation looked like classical physics it was just as abstract as Heisenberg's matrices. The detailed mathematical connection between matrix mechanics and wave mechanics was demonstrated by several physicists, including Schrödinger in 1926. The key idea was that Heisenberg's matrix elements correspond to transitions between states – e.g. the nth state to the mth state – and these correspond to probabilities of transitions between the different 'wave modes' (eigenstates) in Schrödinger's theory. The diagonal elements of Heisenber'g matrices (e.g. the nn- or mm-matrix elements) correspond to the probability of the nth or mth state respectively.

Schrödinger, like Einstein and de Broglie, had difficulty accepting the 'Copenhagen Interpretation' of quantum theory developed by Bohr and Heisenberg. He was troubled by the discontinuity of the measurement process and the apparent loss of an objective physical reality.

"The rejection of realism has logical consequences. In general, a variable has no definite value before I measure it; then measuring it does not mean ascertaining the value that it has. But then what does it mean? There must be some criterion as to whether a measurement is true or false, a method is good or bad, accurate, or inaccurate – whether it deserves the name of measurement process at all."

(E. Schrödinger, *The Present Situation in Quantum Mechanics*, 1935 quoted in *Quantum Theory and Measurement*, ed. J.A. Wheeler and W.H. Zurek, Princeton, New York, 1983)

2.1.4 Born's Statistical Interpretation

Schrödinger thought he had reduced quantum theory to a classical field theory in which the intensity of the wave-function represents charge density. He hoped in this way to remove discontinuous quantum jumps from physics. Max Born (also in 1926) was first to realise that the wavefunction itself does not represent any real physical observable and that the value of $|\Psi|^2$ must be related to the probability of locating an electron at a particular point in space. This retained the logical problems of wave-particle duality that Schrödinger had hoped to resolve. At the time Born's work was taken pretty much for granted and he had to wait until 1954 before he was awarded the Nobel Prize *"for his fundamental research in quantum mechanics, especially for his statistical interpretation of the wavefunction"*. He was also a co-author, with Heisenberg and Jordan, of the detailed paper on matrix mechanics. He summed up the statistical interpretation as:

"The motion of particles follows probability laws but the probability itself propagates according to the law of causality."

(M. Born, quoted by A. Pais in *Niels Bohr's Times*, OUP, 1991)

2.1.5 Correspondence and Complementarity

"He utters his opinions like one perpetually groping and never like one who believes he is in possession of definite truth."

(Einstein on Bohr, quoted by A. Pais in *Niels Bohr's Times*, OUP, 1991)

Bohr was the great philosopher of quantum physics, always seeking a deeper understanding, a clearer interpretation of the strange new theory. His early work on the hydrogen atom introduced the quantum jumps that so offended Schrödinger and Einstein but he thought the problem lay not in physical reality itself, but in the language we use to describe it. Even before matrix mechanics and wave mechanics he tried to reconcile classical and quantum phenomena. He did this by looking at the conditions in which discontinuous quantum states must begin to blend smoothly with the continuous states allowed in classical theory. For example, as the principal quantum number n is increased in Bohr's atom the separation in energy between adjacent allowed states becomes smaller and smaller until they effectively form a continuum. To Bohr this meant there had to be some kind of analogy or correspondence between classical and quantum theory so that the two are indistinguishable when there is so much energy in the system that the individual quantum jumps are insignificant. This rather vague idea became known as the Correspondence Principle and was used with great skill by Bohr to tease out new results even before the general methods of Heisenberg and Schrödinger became available.

The Correspondence Principle also set Bohr on a path toward his enigmatic philosophy of Complementarity. As early as 1923 he said:

"Every description of natural processes must be based on ideas which have been introduced and defined by the classical theory."

(Quoted by A. Pais in *Niels Bohr's Times*, OUP, 1991)

The idea that we operate in the classical domain and therefore have to describe all experimental results in classical terms implies that we are one step removed from the quantum phenomena we are investigating. In effect our experimental apparatus acts as an interface between the quantum world and us. It therefore makes little sense to speak about the state of the electron or atom unless we also give the exact experimental arrangement used to carry out the observation. This shifts the emphasis and purpose of physics:

"It is wrong to think that the task of physics is to find out how Nature is. Physics concerns what we can say about Nature."

(Quoted in *The Cosmic Code*, Heinz Pagels, Simon and Schuster, 1982)

This approach gives a neat, but controversial, solution to the problems of wave-particle duality. Experiments designed to produce interference patterns are set up in such a way that they reveal the underlying wave-like properties of phenomena. On the other hand, experiments to detect the photo-electrons or gamma-ray clicks in a geiger counter are set up to reveal particle-like properties. These two types of experiment are themselves mutually exclusive, so we must choose which aspect we wish to observe. In classical physics it would be impossible for something to be both a wave and a particle, but the problem need not arise if we adopt Bohr's philosophy of complementarity since the results of mutually exclusive experimental set-ups are themselves mutually exclusive. This idea was seen as an unacceptable 'soothing philosophy' by Einstein, but it became central to the so-called Copenhagen Interpretation of quantum theory, which soon gained wide popularity among quantum physicists. It ties in neatly with Heisenberg's uncertainty principle – position and momentum are complementary properties for an electron, so are the energy and time of a quantum jump. The idea itself was honed in discussions with Einstein, and Bohr came to regard it as his most important contribution. He also applied the principle outside physics and incorporated it into his coat of arms. Bohr and Heisenberg removed the idea of a hidden mechanistic reality and placed indeterminism and complementarity at the heart of the new theory.

2.2 THE METHOD

2.2.1 A Recipe for Discovery

The beauty of Schrödinger's wave-mechanical approach was that it looked like the familiar physics that most practising physicists had learnt in their university days. They were confident with the mathematics and the solutions were analogous to classical systems such as waves on strings or harmonic oscillators. The interpretation was controversial, but at least the theory worked! The wavefunctions themselves were well behaved solutions to a second order differential equation and problems in quantum theory could be solved by following a standard 'recipe':

- Get the classical Hamiltonian (effectively the total energy of the system).
- Convert it to mathematical operators using standard rules.
- Put it into the Schrödinger equation.
- Solve the Schrödinger equation.

Einstein was delighted. He welcomed Schrödinger's method whilst shying away from Heisenberg's, and so did many other physicists who found Heisenberg's methods abstract and difficult to apply. By contrast, wave mechanics seemed to offer a practical solution to the quantum problem and they immediately applied the new method to a wide range of physical phenomena. Paul Dirac even remarked that the method was so powerful and universal that a lot of first rate work was done by second rate physicists during this time, and that may be true, but it is also true that physics had entered a new era. Suddenly there was a way forward.

The method of wave mechanics can be illustrated using simple examples (readers who wish to avoid the mathematics should skip to section 2.3):

- electrons in free space;
- electrons trapped in an infinite one-dimensional (1D) potential well;
- the quantum harmonic oscillator (electron trapped in a quadratic potential well);

- the hydrogen atom;
- alpha decay (an example of barrier penetration, or the 'tunnel effect').

Before applying the method to these examples we ought to remind ourselves of how to interpret the wavefunctions we obtain. Schrödinger and Einstein at first thought they were direct wave-like representations of the electron. It was Born, as we have seen, who provided a sensible statistical interpretation. The square of the absolute magnitude of the spatial wavefunctions is related to the probability per unit volume of finding the electron in a particular region of space.

2.2.2 Electrons in Free Space and in an Infinite Potential Well

The Schrödinger equation was constructed so that de Broglie waves would be solutions, so this is a trivial problem. Electrons in free space can be represented by 1D wavefunctions of the form:

$$\Psi = \Psi_0 \exp 2\pi i \left(\frac{Et}{h} - \frac{px}{h} \right)$$

This is an infinite travelling wave of wavelength h/p and frequency E/h. The fact that it must be infinite in the x-direction implies that we have no information about the x-position of the electron. This is consistent with the uncertainty principle – if we fix the wavelength we have effectively fixed x-momentum so the uncertainty in position grows without limit. In general wavefunction for any free electron can be made up from a superposition of wavefunctions like this having a spectrum of energies and momenta. In this way there is a range of possible values for momentum and position limited by the uncertainty principle.

If the electron is trapped in some region of space then the probabiltiy that it is found outside this region is vanishingly small, so the amplitude of its de Broglie waves must fall to zero at the boundary of the region. The wavefunctions behave like standing waves on a stringed instrument and we go from the continuous spectrum above to a discrete spectrum of allowed energies reminiscent of the Bohr atomic model. This time, however, the quantisation arises naturally from the theory and does not have to be put in as an *ad hoc* assumption.

Electrons in atoms are trapped in a potential well created by their electrostatic interaction with the nucleus, so we should expect electron wavefunctions to form a discrete set by analogy with the harmonics on a stringed instrument. The 1D infinite potential well gives us a simple illustration of a similar problem. The diagram shows how the potential energy of an electron varies with position along the x-axis. In the range $a \geq x \geq -a$ this potential energy is zero. Outside this range it increases suddenly to infinity. In practice this means the electron can only exist between limits $x = \pm a$ since it is hardly likely to acquire an infinite amount of energy. A classical analogue would be a golf ball trapped at the bottom of an

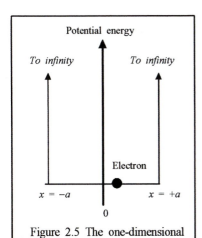

Figure 2.5 The one-dimensional infinite potential well.

infinitely deep hole. No matter how much kinetic energy it is given, it cannot escape.

The potential energy is independent of time, so we can solve the time-independent Schrödinger equation to find spatial wavefunctions.

$$H\psi_s = E\psi_s$$

$$-\frac{h^2}{8\pi^2 m}\frac{\partial^2\psi_s}{\partial x^2} + V\psi_s = E\psi_s$$

The only way to solve this equation in the region where V blows up to infinity is for ψ_s to be zero here. This means $\psi_s\psi_s^*$ is also zero so the probability of finding the electron outside the well is zero (just like the classical case). It also means that ψ_s must fall to zero at both edges of the well ($x = \pm a$), so this fixes the boundary condition for solutions inside the well. In this region $V = 0$ so the Schrödinger equation reduces to:

$$-\frac{h^2}{8\pi^2 m}\frac{\partial^2\psi_s}{\partial x^2} = E\psi_s$$

This has the same mathematical form as the familiar equation for displacement in simple harmonic motion. Compare the SHM equation:

$$\frac{d^2 x}{dt^2} = -\omega^2 x$$

with

$$\frac{\partial^2\psi_s}{\partial x^2} = -\frac{8\pi^2 Em}{h^2}\psi_s$$

This leads to a general solution for ψ_s of the form:

$$\psi_s = Ae^{ikx} + Be^{-ikx} \quad \text{where} \quad k = \frac{2\pi\sqrt{2mE}}{h} \quad -a < x < +a$$

The real parts of the two complex exponentials represent sinusoidal spatial variations and are mathematically equivalent to:

$$\psi_s = A'\sin kx + B'\cos kx$$

The condition that these solutions fall to zero at $x = \pm a$ is equivalent to:

$$ka = \pi, 2\pi, \dots n\pi \qquad \text{(for the sines)}$$

$$ka = \frac{\pi}{2}, \frac{3\pi}{2}, \dots \frac{(2n+1)\pi}{2} \qquad \text{(for the cosines)}$$

If these two conditions are combined we obtain:

$$ka = \frac{n\pi}{2}$$

$$\frac{2\pi a\sqrt{2mE}}{h} = n\pi$$

$$E = \frac{n^2 h^2}{8ma^2} \qquad n = 1, 2, 3, \dots$$

Notice that we have ended up with a discrete spectrum of allowed energy levels. Notice also that $n = 0$ is not allowed. That is because sin 0 does not represent an electron in the well, so it cannot be a valid solution to the problem. This leads to an interesting result, which is linked to the stability of atoms. The lowest allowed energy level has $n = 1$ and $E_1 = h^2/8ma^2$. The 'ground state' is *not* a state of zero energy. This is called the *zero point energy* and it is a purely quantum mechanical result.

It is interesting to think about the origin of this zero point energy. If the width of the well is increased it will approach the classical result of zero, so it must be something to do with the fact that the electron has been confined in a potential well of finite size. In fact it is linked to Heisenberg's Indeterminacy Principle. If the well has width $2a$ then the uncertainty in the electron's position is of the order $2a$. A rough and ready argument shows that this implies an uncertainty in electron energy about the same size as the zero point energy:

$$\Delta x \approx 2a$$

$$\Delta p \approx \frac{h}{\Delta x} \approx \frac{h}{2a}$$

$$\Delta K \approx \frac{\Delta p^2}{2m} \approx \frac{h^2}{8ma^2}$$

(Don't take this too seriously!)

The underlying idea is this. By confining the electron in a small region of space the uncertainty in its position is small. This introduces a large uncertainty in momentum and hence kinetic energy, so the zero point energy is associated with a certain zero point motion. This implies that the classical idea that all molecular motion ceases at absolute zero is incorrect. It also has practical consequences – the zero point energy of helium atoms prevents solidification (except under intense pressure) even at extremely low temperatures.

In the derivation above we completely neglected the time variation of the wavefunction. This is given by:

$$\psi_t = e^{\frac{2\pi i E t}{h}}$$

The complete solutions are then $\Psi = \psi_s\,\psi_t$ and represent de Broglie waves bouncing backwards and forwards inside the well. The boundary conditions at $x = \pm a$ make these positions nodes for standing waves in the well, so the hierarchy of discrete energy levels is analogous to the harmonic series of waves on a stretched string. This gives a very clear link from quantum physics to familiar classical models and was one of the main reasons Schrödinger's method caught on so quickly.

2.2.3 The Quantum Harmonic Oscillator

This is a similar problem to that of the infinite square well, but now the potential energy of the electron varies in proportion with the square of its distance from the centre of the well. This is a quantum mechanical version of the simple harmonic oscillator problem in classical mechanics. The link between the classical and quantum oscillators is quite clear.

In classical physics simple harmonic motion results whenever a mass m is acted upon by a force F that is proportional to displacement from some fixed point and directed back toward that point.

$$F = -kx$$

This results in a potential energy that varies with x^2:

$$V = \frac{1}{2}kx^2$$

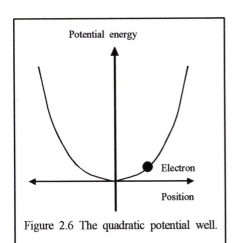

Figure 2.6 The quadratic potential well.

The potential energy is time-independent so once again it is a matter of solving the time-independent Schrödinger equation to find the spatial wavefunction.

Physically the solution to this problem differs from the infinite potential well in two respects:

- The wavefunction falls gradually to zero as the potential energy rises and x increases.
- The solutions again represent standing waves in the well, but now the wavelength of the waves (and hence the momentum of the electrons) changes with position. (This is not surprising – it is exactly what happens to the momentum of a classical simple harmonic oscillator).

The solution of the Schrödinger equation involves some tricky mathematics, which we will not go into here, but the result is interesting. Once again (unsurprisingly) the solutions form a discrete spectrum of allowed energy levels. The energies are given by:

$$E_n = \left(n + \frac{1}{2}\right)hf \quad \text{where} \quad f = \frac{1}{2\pi}\sqrt{\frac{k}{m}}$$

Notice the steps between allowed energy levels are $\Delta E = hf$, exactly the quanta Planck introduced for his harmonic oscillators when explaining the black-body radiation spectrum. Notice also that the quantum oscillator is never still, it too has a zero-point energy, in this case $hf/2$.

2.2.4 The Hydrogen Atom

The first major success of wave-mechanics was a solution to the problem of the hydrogen atom. This is similar to the examples above, in which the electron is confined in a potential well, but now the well is due to the electrostatic interaction between the electron and nucleus, with a potential energy given by:

$$V = \frac{-e^2}{4\pi\varepsilon_0 r}$$

This is a three-dimensional (3D) spherically symmetric potential well, so the first step is to recast the Schrödinger equation in polar co-ordinates, r, θ, ϕ. This exploits the symmetry, making the equation easier to solve and its solutions easier to interpret. The potential is time-independent so the equation to solve is:

$$\nabla^2 \psi(r, \theta, \phi) = -\frac{8\pi^2 m}{h^2}(E - V)\psi(r, \theta, \phi)$$

The great advantage of using polar co-ordinates is that the equation separates into three independent parts, representing the spatial variation with each of the three polar co-ordinates. The solution itself is therefore a product of three functions:

$$\psi_s(r, \theta, \phi) = R(r)\Theta(\theta)\Phi(\phi)$$

Furthermore, just as the 1D wavefunctions for electrons in potential wells led to a discrete energy spectrum determined by a quantum number, so each of the three components of ψ_s generate distinct states determined by their own quantum numbers.

- The total or principal quantum number, n, determines the variation of ψ_s with distance from the nucleus and comes from the radial part of the equation. It is linked to the total energy and the size or radius of the orbit.
- The orbital quantum number l determines the variation of ψ_s with angle θ. It relates to the allowed angular momentum states of the atom (remember how it was introduced to explain eccentricity of orbits in the Bohr model).
- The magnetic quantum number m determines the variation of ψ_s with angle ϕ. It is related to the orientation of the orbit's axis (as it is in Bohr's theory).

Larger values of n result in a larger number of allowed orbitals. The number of orbitals for a particular value of n is called the *degeneracy*. The values of l and m are linked to n in the same way as in Bohr's theory, but this time these links arise directly from the solution:

$$l = 0, 1, 2 \dots (n-1)$$
$$m = -l, -(l-1) \dots 0 \dots +(l-1), l$$

For $n = 1$ this gives a single state with $n = 1$, $l = 1$, $m = 0$. This is the 1s shell. But hold on a moment, *two* electrons can inhabit the 1s shell. Even Schrödinger's theory requires us to put in an extra internal degree of freedom 'by hand'. Spin does not come out of the equation. The two 1s electrons differ in some way that is not predicted by the 3D spatial wavefunction. This tells us two things – quantum mechanical spin does not have a simple classical analogue, and Schrödinger's equation cannot be the final solution to the problem of the hydrogen atom. We have already discussed how the hierarchy of atomic structures can be built up by adding electrons one at a time to these allowed energy states and using the Exclusion Principle to prevent them all falling into the lowest level.

The solutions themselves give a total energy dependent only on the principal quantum number, n:

$$E = -\frac{me^4}{8\varepsilon_0^2 h^2} \frac{1}{n^2}$$

This agrees with Bohr's formula and explains the spectral series in a similar way. The great advantage of Schrödinger's method over Bohr's is that is gives us the wavefunctions of the elctrons inside the atom. These wavefunctions can be interpreted using Born's interpretation and used to work out the electron probability density in the atom.

The table below lists some of the spatial wavefunctions. Z represents the atomic number of the nucleus because the theory works equally well for all *one electron* atoms.

n	l	m	Spatial wavefunction
1	0	0	$\psi_{100} = \dfrac{1}{\sqrt{n}} \left(\dfrac{Z}{a_0}\right)^{\frac{3}{2}} e^{-Zr/a_0}$
2	0	0	$\psi_{200} = \dfrac{1}{4\sqrt{2\pi}} \left(\dfrac{Z}{a_0}\right)^{\frac{3}{2}} \left(2 - \dfrac{Zr}{a_0}\right) e^{-Zr/2a_0}$
2	1	0	$\psi_{210} = \dfrac{1}{4\sqrt{2\pi}} \left(\dfrac{Z}{a_0}\right)^{\frac{3}{2}} \dfrac{Zr}{a_0} e^{-Zr/2a_0} \cos\theta$
2	1	± 1	$\psi_{21\pm1} = \dfrac{1}{8\sqrt{\pi}} \left(\dfrac{Z}{a_0}\right)^{\frac{3}{2}} \dfrac{Zr}{a_0} e^{-Zr/2a_0} \sin\theta\, e^{\pm i\phi}$
3	0	0	$\psi_{300} = \dfrac{1}{81\sqrt{3\pi}} \left(\dfrac{Z}{a_0}\right)^{\frac{3}{2}} \left(27 - 18\dfrac{Zr}{a_0} + 2\dfrac{Z^2 r^2}{a_0^2}\right) e^{-Zr/3a_0}$
3	1	0	$\psi_{310} = \dfrac{\sqrt{2}}{81\sqrt{\pi}} \left(\dfrac{Z}{a_0}\right)^{\frac{3}{2}} \left(6 - \dfrac{Zr}{a_0}\right)\dfrac{Zr}{a_0} e^{-Zr/3a_0} \cos\theta$
3	1	± 1	$\psi_{31\pm1} = \dfrac{1}{81\sqrt{\pi}} \left(\dfrac{Z}{a_0}\right)^{\frac{3}{2}} \left(6 - \dfrac{Zr}{a_0}\right)\dfrac{Zr}{a_0} e^{-Zr/3a_0} \sin\theta\, e^{\pm i\phi}$
3	2	0	$\psi_{320} = \dfrac{1}{81\sqrt{6\pi}} \left(\dfrac{Z}{a_0}\right)^{\frac{3}{2}} \dfrac{Z^2 r^2}{a_0^2} e^{-Zr/3a_0} \left(3\cos^2\theta - 1\right)$
3	2	± 1	$\psi_{32\pm1} = \dfrac{1}{81\sqrt{\pi}} \left(\dfrac{Z}{a_0}\right)^{\frac{3}{2}} \dfrac{Z^2 r^2}{a_0^2} e^{-Zr/3a_0} \sin\theta\cos\theta\, e^{\pm i\phi}$
3	2	± 2	$\psi_{32\pm2} = \dfrac{1}{162\sqrt{\pi}} \left(\dfrac{Z}{a_0}\right)^{\frac{3}{2}} \dfrac{Z^2 r^2}{a_0^2} e^{-Zr/3a_0} \sin^2\theta\, e^{\pm 2i\phi}$

The wavefunctions above are all written interms of a_0, the Bohr radius:

$$a_0 = \frac{\varepsilon_0 h^2}{\pi m Z e^2} = 5.29 \times 10^{-11} \, \text{m}$$

This turns out to be the radius at which the probability density for the electron in the $n = 1$ orbital has its maximum value.

The probability density is linked to the radial wavefunction but it is not the same thing. Look at the wavefunction for $n = 1$. This is clearly a decaying exponential. It has its *maximum* value at $r = 0$, but this does not mean that the probability of finding the electron at $r = 0$ is a maximum at $r = 0$. The reason for this is that the intensity of the radial wavefunction RR^* is linked to the probability of finding the electron *per unit volume* between r and $r + \delta r$, but the volume between r and $r + \delta r$ increases with radius. In fact it is represented by a spherical shell of thickness δr and radius r. This has a volume $4\pi r^2$. So the probability of finding the electron between r a nd $r + \delta r$ is given by:

$$P = p(r)\delta r = RR^* 4\pi r^2 \delta r$$

where P is the probability and $p(r)$ the probability density. The function $p(r)$ falls to zero at the origin and at large values of r and has its maximum value at the Bohr radius. This represents a spherically symmetric probability distribution centred on the nucleus with the greatest probability of finding the electron close to the Bohr radius, but giving no hint of any well-defined orbital trajectory.

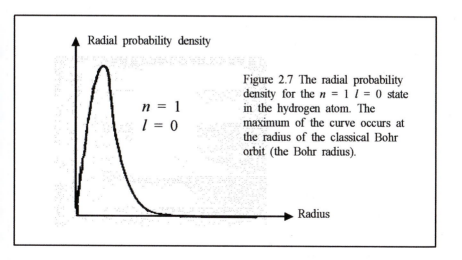

Radial probability density

$n = 1$
$l = 0$

Figure 2.7 The radial probability density for the $n = 1$ $l = 0$ state in the hydrogen atom. The maximum of the curve occurs at the radius of the classical Bohr orbit (the Bohr radius).

Radius

2.2.5 Barrier Penetration and the Tunnel Effect

The solutions of the Schrödinger equation for electrons in free space are travelling waves. If these waves encounter an infinite potential barrier they reflect off it. This is how standing waves are set up in the infinite potential well. However, if the barrier is of finite height and width the amplitude of the wavefunction decays into the barrier and there is a chance that electrons can pass through it. This is a new, quantum mechanical effect that explains the mechanism of alpha decay and is utilised in many electronic components.

Think of a situation where electrons of total energy E encounter a potential barrier of 'height' $V > E$ and width a. The time-independent Schrödinger equation inside the barrier is:

$$\frac{h^2}{8\pi^2 m}\frac{\partial^2 \psi_s}{\partial x^2} = (E - V)\psi_s$$

$$\frac{\partial^2 \psi_s}{\partial x^2} = \frac{8\pi^2 m}{h^2}(E - V)\psi_s = \frac{8\pi^2 mK}{h^2}\psi_s$$

where K is the electron kinetic energy and can be replaced by $K = \dfrac{p^2}{2m} = \dfrac{h^2}{2m\lambda^2}$

where p is the electron momentum and λ is the de Broglie wavelength. Hence:

$$\frac{\partial^2 \psi_s}{\partial x^2} = \frac{4\pi^2}{\lambda^2}\psi_s$$

This equation has simple decaying exponential solutions:

$$\psi_s = \psi_0 e^{-\frac{2\pi x}{\lambda}}$$

The value of ψ_0 must equal the amplitude of the wavefunction at the edge of the barrier to ensure continuity with the incoming and reflected waves. The overall picture then is of an incoming wave of amplitude ψ_0 decaying exponentially into the barrier. The amplitude will fall to half at a distance $x_{1/2}$ beyond the edge of the barrier:

$$e^{-\frac{2\pi x_{1/2}}{\lambda}} = 0.5$$

$$x_{1/2} = \frac{\lambda \ln 2}{2\pi} \approx 0.1\lambda$$

If the barrier width is not too great (e.g. less than the de Broglie wavelength) there will be a small but significant amplitude remaining at the far side of the barrier. This means that $\psi_s\psi_s^* \neq 0$ beyond the barrier. There is a probability that the electron is found on the far side of the barrier. This has no classical analogue. It is like finding yourself trapped in a garden surrounded by high walls you could never possibly jump over. You walk backwards and forwards in the garden approaching and leaving the walls then, suddenly, you find yourself on the far side of the wall!

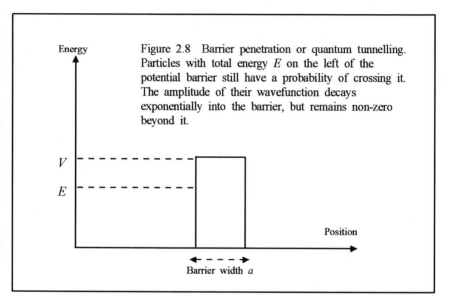

Figure 2.8 Barrier penetration or quantum tunnelling. Particles with total energy E on the left of the potential barrier still have a probability of crossing it. The amplitude of their wavefunction decays exponentially into the barrier, but remains non-zero beyond it.

The effect described is 'barrier penetration', but the alternative name, the *quantum tunnel effect* is more evocative. It is as if the electron, which did not have enough energy to pass *over* the potential barrier actually tunnelled *through* it. If we treat the electron as a particle with definite energy at each point then while it passed through the barrier its potential energy was greater than its total energy so it kinetic energy must have been *negative* – a situation that can never arise in classical physics.

There are many important applications of the tunnel effect.

- In some heavy nuclei (e.g. uranium-238) the energy released when alpha particles form within the nucleus (i.e. when two protons and neutrons bind together temporarily) is sufficient to boost that alpha particle into a positive total energy state. However, it should not have enough energy to escape because it is confined between potential barriers formed by the combination of long range coulomb repulsion and short range attraction by the strong nuclear force. Classically it is trapped, but quantum mechanics gives it a small but finite probability of penetrating the barrier, so nuclei like uranium-238 are unstable and decay by emitting alpha particles. This was first explained by

Gamow, Condon and Gurney in 1928. The theory also suggests that alpha particles with higher energy will have a higher probability of emission, so short half-life alpha-emitters emit high energy alpha particles.

- When Hans Bethe and others worked out the details of nuclear fusion reactions converting hydrogen to helium in the Sun, they realised that the core temperature was too low to initiate proton-proton fusion reactions simply making the particles collide violently. The problem with fusion is getting the protons close enough for the short-range strong nuclear force to kick in and bind them together. Electrostatic repulsion pushes them apart, so extremely high temperatures are needed. However, if the temperature is not quite high enough the protons will be separated by a coulomb barrier at their point of closest approach. The tunnel effect does the rest, allowing fusion reactions to take place at a lower (but still very high) temperature.

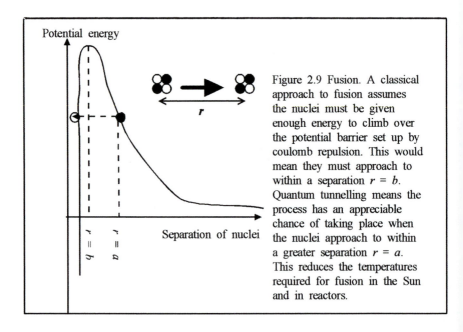

Figure 2.9 Fusion. A classical approach to fusion assumes the nuclei must be given enough energy to climb over the potential barrier set up by coulomb repulsion. This would mean they must approach to within a separation $r = b$. Quantum tunnelling means the process has an appreciable chance of taking place when the nuclei approach to within a greater separation $r = a$. This reduces the temperatures required for fusion in the Sun and in reactors.

- Several electronic components (e.g. the tunnel diode) employ the tunnel effect in order to facilitate rapid electronic switching. The basic idea is that a junction region between two semiconductors acts as a potential barrier. The width of the barrier can be controlled by changing small external bias voltages, this controls the rate of barrier penetration and the current across the junction.

2.3 DIRAC'S 'MORE ELABORATE THEORY'

2.3.1 Atoms and Light

By the late 1920s Paul Dirac was searching for a 'more elaborate theory' of the interaction of atoms and light. His main guiding principle in this quest was to look for mathematical beauty and simplicity in his equations – an approach that has continued to yield results of great physical significance and power.

" ... *it is more important to have beauty in one's equations than to have them fit experiment ... It seems that if one is working from the point of view of getting beauty in one's equations, and if one has really a sound insight, one is on a sure line of progress.*"

" ... *Just by studying mathematics we can hope to make a guess at the kind of mathematics that will come into the physics of the future ... If someone can hit on the right lines along which to make this development, it may lead to a future advance in which people will first discover the equations and then, after examining them, gradually learn how to apply the ... My own belief is that this is a more likely line of progress than trying to guess at physical pictures.*"

(*The Evolution of the Physicist's Picture of the World*, Scientific American, 1963)

Dirac realised that an explanation of emission spectra based on quantisation of the atom was only half the story. If a full understanding of the interaction of radiation and matter was wanted then the radiation fields had to be quantised too. Dirac identified three essential components to a deeper theory:

- the state of the atom;
- the interaction of the atom and the electromagnetic field;
- the state of the electromagnetic field.

The general approach to a problem of this kind, as we have seen, is to construct the total energy operator or Hamiltonian and then solve the resulting Schrödinger equation. This Hamiltonian should have three parts, each one corresponding to one of the components identified above.

$$H = H(\text{atom}) + H(\text{interaction}) + H(\text{field})$$

$H(\text{atom})$ is just the Hamiltonian for the isolated atom, the one Schrödinger had used to solve the hydrogen atom. $H(\text{interaction})$ was constructed using equations from classical electromagnetism describing the energy of an electron in an electromagnetic field. $H(\text{field})$ was new. To get this Dirac had to work out the energy of the field in terms of a model that could be easily converted to a quantum mechanical form. To do this he treated the electromagnetic field like a collection of oscillators which might or might not be oscillating – rather like the surface of a smooth sea on which waves of all frequencies can be impressed, or the springs of a mattress which can resonate when the bed is shaken. He then quantised the oscillators. (This is similar to Einstein's approach to black-body radiation).

Rather than tackle the whole problem all at once, Dirac tackled it a bit at a time. First he considered what happens to the field in the absence of an atom. This involves only the H(field) part of the Hamiltonian and led Dirac to introduce two new operators, a^+ and a^-, the *creation and annihilation operators*. The action of a^+ is to create a new oscillator in the field, a^- removes an oscillator from the field. These oscillators are an alternative way to think of photons of particular frequency and polarisation. So a^+ creates a photon, a^- annihilates one. In this view the electromagnetic field is seething with repeated creations and annihilations as photons come into existence and disappear. (This has become a central idea in all quantum field theories.)

The problem of isolated atoms had been solved (up to a point) by Schrödinger, so Dirac turned his attention to the interaction between atom and field. A full explanation for this requires the complete Hamiltonian. He did not know how to set this up exactly, so he used a tried and tested approximation – a *perturbation*. This starts from the assumption that the atom and field are pretty well independent of each other and that the effect of the interaction is to cause a slight disturbance (a perturbation) in the states of each. This method is still the only practical way to approach problems in quantum electrodynamics, and it is capable of generating incredibly accurate predictions. It usually produces a series of terms providing finer and finer corrections that converge to the result required. The accuracy of the calculation improves as more and more terms are included. It is rather like analysing a car crash and starting from the assumption that the cars act like Newtonian point masses. The next step might be to add in the effects of the mass distributions in the vehicles, then the effects of moving parts in the engine, wheels, etc. eventually tiny details (like air resistance on wing mirrors) will have a negligible effect on the outcome, so they are neglected.

Dirac's initial calculations used a first order approximation that effectively meant he was only dealing with the interaction between the atom and a single photon in the field. Nevertheless his results were promising, and he managed to derive formulae for spontaneous and stimulated emission from first principles. When he extended this approach to higher order terms in the series he found that transitions between well-defined initial and final states passed through undefined intermediate states, some of which seemed to violate energy conservation. He realised that this could be interpreted as an example of Heisenberg's Uncertainty principle, in this case linking energy and time.

$$\Delta E \Delta t \approx h$$

A loose interpretation of this formula is that the intermediate state can violate energy conservation by an amount of order ΔE as long as this energy is 'paid back' within a time $\Delta t < h/\Delta E$. These ephemeral states are called 'virtual states' and they give us a remarkable picture of the electromagnetic field in a vacuum. Far from being empty, space is filled with virtual particles bursting into existence and disappearing a very short time later. The more energetic the photon the shorter its virtual lifetime. For example, a virtual photon of visible light might exist for about 2×10^{-15} s.

Although the idea of a vacuum state seething with virtual particles may seem rather strange, it has real physical consequences. One of these is the '*Casimir effect*' a small but measurable force of attraction between two parallel metal plates in a vacuum. It is very simple to explain in terms of virtual photons. Any virtual photons created between the plates will bounce back and forward like waves trapped on a string. The only waves that do not cancel by destructive interference are those which form standing wave patterns. This reduces the probability of all other virtual photons between the plates to zero. The overall effect is that there is less intense bombardment of the plates from virtual photons between them than from virtual photons outside the plates. This results in a small force whose size can be predicted and which, when measured agrees with the theoretical prediction.

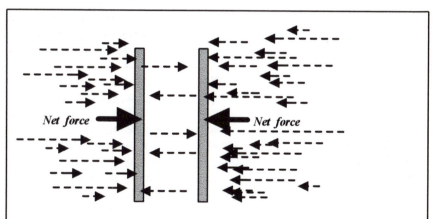

Figure 2.10 The Casimir effect. The field strength must fall to zero on the surface of a conducting plate. This constraint limits the wavefunctions that can exist between the plates so that the flux of virtual photons is less on the inside than outside, resulting in a small but measurable force of attraction.

The idea of a quantum field is not restricted to electromagnetic fields and photons, it also applies to the lepton field from which electrons and positrons emerge. The strong electric field around an electron surrounds it with a 'cloud' of virtual electrons and virtual positrons. The virtual electrons are pushed away from the real electron and the virtual positrons are attracted toward it. These virtual positrons will shield the negative charge of the real electron from the outside world, so any experiment to measure the charge on an electron is bound to give the 'wrong' answer. The charge we measure is the shielded charge, not the 'bare' charge, which is inaccessible to our experiments. The existence of a cloud of virtual charges around any real charge is called *'vacuum polarisation'*.

Dirac's picture of vacuum polarisation around the bare charge of an electron gives a glimpse of one of the problems which beset all attempts to create a genuine quantum field theory. There were reasons for thinking that the bare charge should be infinite, a result that must throw doubt on the theory. However, if the vacuum polarisation this produced was also infinite it might be possible for the two infinities to cancel almost exactly, leaving just a small finite difference that we take to be the effective charge of an electron! This idea, that infinities arising in the theory might be cancelled by negative infinities is called *'renormalisation'*, something Pauli disparaged as 'subtraction physics'.

Dirac's work set the scene for the third quantum revolution – the theory of quantum electrodynamics constructed by Feynman, Schwinger and Tomonoga. But Dirac wasn't satisfied with the Schrödinger equation either, and set about showing how it could be brought into line with special relativity.

2.3.2 The Dirac Equation

Einstein's special theory of relativity showed that Newtonian mechanics breaks down for particles travelling at speeds comparable to the speed of light. Schrödinger's equation was based on a classical view of space and time and on Newtonian mechanics (as far as energy and momentum were concerned). This was obviously a problem, especially as the motion of electrons in atoms was known to involve very high-speed motion. Dirac realised that the Schrödinger equation could not be a final description of electrons in atoms. Besides, the Schrödinger equation was mathematically 'ugly' and this upset Dirac. So he set about constructing another equation which would give a relativistic theory of electrons in atoms. The result, in 1928, was the Dirac equation, an equation so powerful that many physicists regarded the problem of the electron as completely solved.

You will recall that the Schrödinger equation is closely related to the classical non-relativistic equation for kinetic energy,

$$KE = \frac{p^2}{2m}$$

From here Schrödinger converted the energy and momentum terms to differential operators and ended up with his famous equation.

$$\langle KE \rangle \Psi = \left\langle \frac{p^2}{2m} \right\rangle \Psi$$

$$\langle E - V \rangle \Psi = \left\langle \frac{p^2}{2m} \right\rangle \Psi$$

Figure 2.11 **Paul Adrien Maurice Dirac** 1902-1984 shared the 1933 Nobel Prize for Physics with Erwin Schrödinger. In 1926 he developed a general mathematical formalism for quantum theory and in 1928 he created a relativistic equation for the electron – the Dirac Equation. This gave the first natural explanation for spin, a quantum mechanical property that had to be added to Schrödinger's equations. It also led to the radical prediction of the existence of anti-matter. According to Dirac's theory, electron and anti-electron (positron) pairs could be created from high energy gamma-rays when they interact with matter. Furthermore, the collision of an electron and a positron should result in annihilation. These predictions were soon confirmed in experiments. Dirac also took the first crucial steps toward quantum electrodynamics and made major contributions in many other areas of theoretical physics.   Artwork by Nick Adams

The key new ingredient in Dirac's approach was to start from the relativistic equation for energy and momentum:

$$E^2 = m_0{}^2 c^4 + p^2 c^2 \tag{1}$$

where E is total energy, p is linear momentum, m_0 is rest mass and c the speed of light. This too can be converted into a differential equation using a similar mathematical approach, changing the energy and momentum terms to operators. This leads to:

$$\langle E \rangle^2 \Psi = \langle p \rangle^2 c^2 \Psi + m_0{}^2 c^4 \Psi$$

However, Dirac was not happy with this. He wanted an equation, like Schrödinger's, in which the energy operators were not squared – a linear equation. He tried the equation below:

$$\langle E \rangle \Psi = \alpha \langle p \rangle c \Psi + \beta m_0 c^2 \Psi \tag{2}$$

Of course, this has to be consistent with equation (1). That is where the new terms α and β come in. To make equations (1) and (2) consistent α, β and Ψ all turn out to be rather exotic mathematical objects. Ψ is 4 component column matrix, β is a 4×4 matrix and α is a three component vector made up of 4×4 matrices!

   Very few physicists would have ventured this far into speculative mathematics, and those who did would probably have rejected the result as obviously unphysical. Dirac, on the other hand, managed to interpret every part of the equation and came up with two unexpected theoretical discoveries. The first of these was that electron spin is a natural consequence of the equation and not something that has to be added as an afterthought to agree with experiment (as was the case with the Schrödinger equation). Of course, spin had already been introduced, so was not really a prediction of the theory. The other discovery, however, was a revolutionary prediction, and is one of the most important breakthroughs in twentieth century physics. Dirac had derived his equation from a kind of square root of the relativistic energy and momentum equation, and everybody knows that numbers have two square roots, a positive one and a negative one. In a similar way, there are two possible ways to construct the Dirac equation:

$$\langle E \rangle \Psi = \alpha \langle p \rangle c \Psi + \beta m_0 c^2 \Psi$$

or

$$\langle E \rangle \Psi = -\alpha \langle p \rangle c \Psi - \beta m_0 c^2 \Psi \tag{3}$$

If solutions to equation (2) represent electrons in positive energy states, then those to equation (3) must represent *negative* energy states. But this raises an immediate problem. We saw that the hydrogen atom is stable in its ground state because there are no lower energy levels allowed or available – this prevents the electron from making a quantum jump to lower energy and radiating. If there are an infinite set of vacant negative energy states then no positive energy electrons can be in stable states, they can all make quantum jumps to negative energy states and disappear from the universe in a flash of electromagnetic radiation. This is a serious problem.

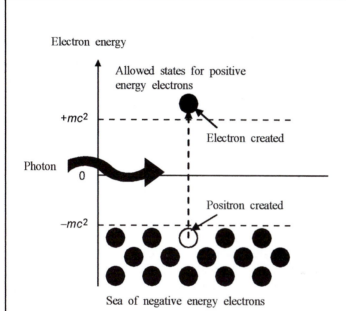

Figure 2.12 Creation of an electron-positron pair. Dirac assumed all negative energy states were occupied by electrons. However, one of these unobserved negative energy electrons can be promoted to a positive energy state by an energetic photon. The hole left behind acts like a positive particle, the anti-electron or positron.

Dirac suggested an ingenious (and rather incredible) solution to this problem. Electrons in higher orbits in large atoms do not fall into lower orbits because those orbits are already occupied and the Pauli Principle forbids two electrons from having the same set of quantum numbers. If all the negative energy states are occupied then positive energy electrons can be stable for the same reason. Of course, this does imply that the universe is filled with an infinite sea of negative energy electrons. These are ubiquitous so they have no direct observable effects, they form a uniform background against which the rest of physics takes place.

The infinite negative energy electron sea solved the problem of stability for atoms and positive energy electrons, but it also led to another intriguing possibility. Can a negative energy electron absorb enough energy to make a quantum jump into a positive energy state and emerge, as if from nowhere in the 'real' world? Dirac thought it could. What is more, if it did, it would leave a 'hole' in the negative energy sea that would look a lot like a positive energy particle of the same mass spin and magnitude of charge as the electron. So Dirac's theory predicted that gamma rays of high enough energy might suddenly 'disappear' as they are absorbed by a negative energy electron. In their place would appear a pair of particles – an electron and an anti-electron (a positron). This process is called 'pair production' and was first observed in 1932 by Carl Anderson in a cloud chamber photograph of cosmic rays.

The positron was the first particle of antimatter to be discovered, but Dirac's theory implies that all particles have corresponding anti-particles, and recently anti-atoms have been created at CERN. These anti-hydrogen atoms consist of positrons orbiting anti-protons. There is no doubt that a whole anti-Periodic Table of elements is possible, and one of the as yet unsolved problems is why our universe seems biased toward matter rather than anti-matter.

3
QUANTUM MYSTERIES

3.1 EINSTEIN AND BOHR

3.1.1 Problems of Interpretation

Einstein played a leading role in the development of the *ad hoc* 'old quantum theory', but he was uneasy about it. Heisenberg's matrix mechanics seemed too abstract to be a true representation of reality, however well it fitted experimental results. On the other hand, Schrödinger's wave mechanics looked promising until Born came along and spoiled it with his 'statistical interpretation'. In many ways Einstein had been happier with physics in the early 1920s when it was clear that major discoveries had been made and equally clear that these had not yet been adequately explained. Once Bohr had drawn the threads of the emerging theories together, and Dirac had shown that they are all mathematically equivalent, Einstein found himself in a reactionary minority, unable to accept the full implications of the theory and yet able to offer no credible alternative.

"You believe in the God who plays dice, and I in complete law and order in a world which objectively exists, and which I, in a wildly speculative way, am trying to capture ... Even the great initial success of the quantum theory does not make me believe in the fundamental dice game, although I am well aware that our younger colleagues interpret this as a consequence of senility."

(Letter from Einstein to Born, 1944)

Throughout this period he proposed a number of thought experiments intended to reveal the consequences of quantum theory to its creators and show that, in some respect, it could not be the 'final theory' of physical reality. Niels Bohr responded to Einstein's ideas with a series of brilliant arguments based on what came to be known as the *'Copenhagen Interpretation'* of quantum theory. His brilliance and the theory's obvious successes established quantum theory, particularly Schrödinger's wave mechanical version and the Copenhagen Interpretation, as the dominant paradigm in physics for fifty years.

Einstein was not the only physicist to object to the implications of quantum theory. Schrödinger initially believed his wave mechanics would return physics to an era of continuous differential equations and waves – something that would appeal to any classical physicist. He was appalled at the discontinuities implied by quantum jumps and the measurement process. De Broglie and others tried to retain some physical reality in both the waves and particles of wave-particle duality and opened the door to a number of 'hidden-variables' theories eventually perfected by David Bohm. Later, even more peculiar interpretations were proposed, and taken seriously. Among them was Feynman's brilliant 'sum-over-histories' and Hugh Everitt III's 'many-worlds' theory. Recently John Cramer's transcational interpretation, involving influences passing back and forth in time, has made an impression. The debate over the interpretation of quantum theory is not over and no account of twentieth century physics would be complete without some reference to it.

3.1.2 The Copenhagen Interpretation

Many physicists contributed to the Copenhagen Interpretation, but it is most closely associated with the name of Niels Bohr because it was he who gave it a philosophical base and he who defended it so devoutly against Einstein's ingenious attacks. It was Bohr who established the Institute for Theoretical Physics in Copenhagen and was its first director, and Bohr who, in the 1920s, acted as the magnet for young creative physicists (like Heisenberg) to visit Copenhagen and try out their latest ideas on the old master.

- Bohr contributed the correspondence principle and complementarity.
- Heisenberg, with Jordan and Born, created matrix mechanics.
- Heisenberg added the uncertainty principle.
- Schrödinger created wave mechanics.
- Born provided the centrepiece, a statistical interpretation of the wavefunction.

In recent years the Copenhagen Interpretation has been challenged by other more exotic interpretations, but it is still the starting point for most discussions about the 'meaning' of the theory and the nature of physical reality. Its main ideas are summarised below (Dirac notation is used for states from here on):

- The state of a system (e.g. electron, photon, etc.) is defined by its wavefunction, $|\Psi\rangle$.
- The wavefunction evolves continuously and deterministically (obeying the Schrödinger equation) until a measurement is made.
- Measurements involve an interaction between the system and the measuring apparatus, so the measurement process plays a significant part in determining the possible outcomes of the measurement.

Figure 3.1 **Niels Hendrik David Bohr** 1885-1962 began his research career with Ernest Rutherford at Manchester. In 1913 he explained the stability of the Rutherford atom by applying a quantum condition to the electronic orbits. He used this model to explain the spectrum of radiation emitted by the hydrogen atom and in so doing introduced the idea of discontinuous quantum jumps. In 1918 he became the first director of the Institute of Theoretical Physics in Copenhagen which soon became the prime focus of attempts to formulate and interpret the new quantum theories. His work with Werner Heisenberg and others throughout the 1920s led to the Copenhagen Interpretation, which he defended successfully against ingenious attacks by Albert Einstein. The central ideas of the Copenhagen model were Heisenberg's Uncertainty Principle and Bohr's principle of complementarity. In the 1930s he introduced the liquid drop model of the nucleus with which he explained the essential physics of nuclear fission. Although he worked on the Los Alamos Project to build the first atomic bomb he was one of the first to oppose its use and after the war he campaigned for the control of nuclear weapons. He won the Nobel prize for Physics in 1922 for his work on atomic theory.   Artwork by Nick Adams

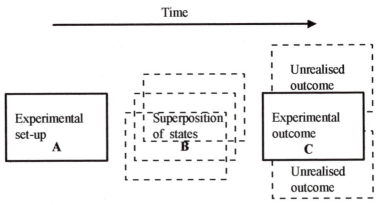

(i) CLASSICAL MECHANICS

Time ⟶

Experimental set-up **A**	Smooth mechanical evolution through unique states **B**	Experimental outcome **C**

A defines the state at time $t = 0$.
B system evolves reversibly according to laws of mechanics..
C unique outcome at time t.

(ii) COPENHAGEN QUANTUM THEORY

Time ⟶

Experimental set-up **A**	Superposition of states **B**	Unrealised outcome / Experimental outcome **C** / Unrealised outcome

A defines the wavefunction at time $t = 0$.
B wavefunction evolves reversibly.
B-to-C collapse of the wavefunction.
C one outcome from many possibilities.

Figure 3.2 Classical mechanics versus the Copenhagen Interpretation of Quantum Mechanics.

- Prior to experiment the wavefunction represents a superpostion of all possible but as yet unrealised outcomes of the experiment.

$$|\Psi\rangle = a_1|\psi_1\rangle + a_2|\psi_2\rangle + \ldots$$

  Think of this as a weighted sum of alternative states labelled 1, 2, etc.
- Measurement results in the instantaneous collapse of the wavefunction into a state representing the actual outcome:

$$|\psi_1\rangle \quad \text{or} \quad |\psi_2\rangle, \quad \text{etc.}$$

- Collapse of the wavefunction takes place in a random non-determinstic and unpredictable way.
- Although the precise results of measurements cannot be predicted with certainty, the probability of each alternative outcome can be calculated from the intensity of its respective wavefunction using the Born interpretation.

$$p_1 = a_1 a_1^* = |a_1|^2 \qquad p_2 = a_2 a_2^* = |a_2|^2 \quad \text{etc...}$$

  where p_1 is the probability of state 1 etc....

The implications for the nature of physical reality become clear as soon as this approach is applied to a simple example. Imagine a single radioactive atom that decays by beta emission. As time goes on the state of the system is represented by a superposition of two wavefunctions. One represents the undecayed atom, the other is a combination of the decayed atom and the emitted beta particle. The longer we wait the smaller the amplitude of the first part ($a_1(t)$) becomes and the larger the amplitude of the second part ($a_2(t)$).

$$|\Psi\rangle = a_1(t)|\psi_1\rangle + a_2(t)|\psi_2\rangle$$

where $|\psi_1\rangle$ is the undecayed state and $|\psi_2\rangle$ is the decayed state.

At any particular time the probability that it has decayed is $a_1 a_1^*$ and the probability it has not decayed is $a_2 a_2^*$ (assuming the state is normalised so that $a_1 a_1^* + a_2 a_2^* = 1$).

If a measurement is carried out then the system will not be found in a partially decayed state, but *either* decayed *or* undecayed. In other words, the effect of this measurement has been to cause a discontinuous change in the wavefunction (a quantum jump) from its continuously evolving superposition into one or other definite state. This is called the *'collapse of the wavefunction'* and has always been the most contentious aspect of the Copenhagen Interpretation:

$$\text{Before}: \quad |\Psi\rangle = a_1(t)|\psi_1\rangle + a_2(t)|\psi_2\rangle$$
$$\text{After}: \quad |\Psi\rangle = |\psi_1\rangle \quad \text{OR} \quad |\Psi\rangle = |\psi_2\rangle$$

Now think about the emitted beta particle after the atom decays. This is emitted

Figure 3.3   Beta decay. An unstable nucleus emits a high energy electron.

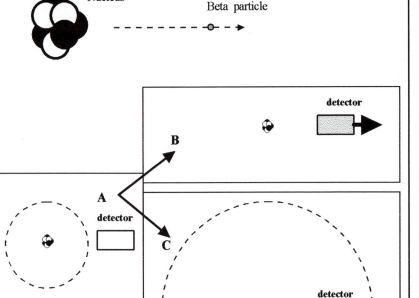

A: A short time after the decay. The beta particle is represented by a spherical wavefront centred on the nucleus. This gives an equal probability that the electron is anywhere on the surface of the wave.

B and C represent alternative outcomes at a later time.

B: If the detector detects the electron the wavefunction collapses everywhere, becoming zero everywhere except inside the detector.

C: if the detector fails to capture the electron then the wavefront continues to expand everywhere except in the region blocked by the detector. Once again it has collapsed, but now the only part that is zero is the part that corresponds to detection.

at random and could travel off in any direction with equal probability. This is represented by spherical de Broglie waves spreading out from the source, their amplitude would be equal over the wave surface and the amplitude would fall with distance from the source. The amplitude is related to the probability per unit area that the electron is at that point. Now imagine that a detector is placed to one side of the source at distance r and that its collecting area is 1% of the total surface area of a sphere of radius r. The probability that an emitted beta particle will be recorded in the detector is 0.01, and the probability that it is not detected is 0.90. Immediately prior to a possible measurement the wavefunction for the electron must satisfy the following equations:

$$\int_{det} \langle \Psi | \Psi \rangle \, dA = 0.01$$
$$\int_{A-det} \langle \Psi | \Psi \rangle \, dA = 0.99$$

where 'det' means that the surface integral is taken over the area of the detector and 'A–det' means it is taken over the rest of the area of the wavefront at radius r. A moment later the detector has either detected the beta particle or not, so the wavefunction now satisfies different conditions:

If it is detected : $\quad \int_{det} \langle \Psi | \Psi \rangle \, dA = 1 \quad \int_{A-det} \langle \Psi | \Psi \rangle \, dA = 0$

If it is not detected : $\int_{det} \langle \Psi | \Psi \rangle \, dA = 0 \quad \int_{A-det} \langle \Psi | \Psi \rangle \, dA = 1$

This discontinuous collapse of the wavefunction (notice that it collapses whether we detect the particle or not) is similar to what happens when the atom itself decays, but this time there is an extra surprise. Consider the case in which the electron is detected. Prior to detection the electron wavefunction is spread over a sphere centred on the source. When the wavefunction collapses it must do so everywhere on the sphere all at once. So everywhere except at the detector itself the probability of detecting the electron changes suddenly from something to nothing. In the detector it changes from a small value to a large value at the same instant. No wonder Einstein was worried – detecting the electron at one place has had an instantaneous effect on the state of the wavefunction at all other places. This instantaneous action-at-a-distance ('spooky action-at-a-distance') is called 'non-locality' and is built into the Copenhagen Interpretation. And it must be instantaneous. If this was not the case then the wavefunction far from the detector would remain non-zero for a short time after the detector had registered an electron. This would leave a small chance of detecting an electron elsewhere even after the electron has been detected in a particular detector. In some experiments this would result in one electron being emitted and two being detected. This would be even more disturbing than non-locality!

The examples above show how quantum theory begins to entangle the observer, or at least the observer's measuring apparatus, in the experiment itself. This makes it difficult to separate an objective physical reality from the experimental set-up and led Bohr to suggest that the two are in fact inseparable.

"The discovery of the quantum of action shows us, not only the natural limitation of classical physics, but, by throwing a new light upon the old philosophical problem of the objective existence of phenomena independently of our observations, confronts us with a situation hitherto unknown in natural science. As we have seen, any observation necessitates an interference with the course of the phenomena, of such a nature that it deprives us of the foundation underlying the causal mode of description. The limit, which nature herself has thus imposed upon us, of the possibility of speaking about phenomena as existing objectively finds its expression as far as we can judge, just in the formulation of quantum mechanics."

More succinctly he said:

"It is wrong to think that the task of physics is to find out how Nature is. Physics concerns what we can say about Nature."

(Niels Bohr, quoted in *Quantum Theory and Measurement*, Wheeler and Zurek, Princeton, 1983)

This is a radical shift in philosophy. The success of Newtonian mechanics, which was based on a particle model of matter, had encouraged classical physicists to think that the microscopic particles themselves exist in the same way as stones or trees in the macroscopic world. This *naive realism*, in which macroscopic effects emerge from a similar but invisible microscopic world was rejected by Bohr and Heisenberg. The uncertainties of quantum theory are not just a measure of our ignorance about particular microscopic configurations, they are intrinsic characteristics of the microscopic world. Electrons, atoms, photons simply do not possess definite properties such as position and momentum. The most complete picture of the microworld is captured by the wavefunction, and this is necessarily probabilistic.

Determinism is also rejected. Newton's laws predict a unique evolution in time for any mechanical system. If the positions and momenta of all particles are known at a particular instant then their future (and past) arrangements are completely determined. Of course, it is impossible for us to measure positions and momenta with infinite precision so we can never make a perfect prediction about the future. However, as Laplace pointed out, an omniscient being with detailed knowledge of the exact state of the universe at one moment could predict all future and past states with absolute certainty. Uncertainties arise in our predictions *because of our lack of knowledge*, not because of any intrinsic indeterminacy. Quantum theory, however, is different. The indeterminacy is fundamental and the future is not determined. The same initial conditions can result in a wide range of final outcomes, so quantum theory leaves the future open and inhibits our ability to know the past. It is hardly surprising that this radical shift in the interpretation of physical reality should shock and surprise so many physicists. It is probably the deepest and most significant paradigm shift of the twentieth century.

3.1.3 Schrödinger's Cat

Schrödinger was disturbed by these ideas, particularly by the superposition of unobserved states and the discontinuous 'collapse of the wavefunction'. Bohr had stressed that experiments involve an amplification of quantum effects by some suitable measuring equipment so they can be registered in our macroscopic world. Schrödinger published a paper in 1935 in which the amplification from quantum events (the decay of an unstable nucleus) involves a conscious creature, a cat, whose fate depends on the detailed way in which the wavefunction collpases. By confining the cat in a box the Copenhagen Interpretation implies that the cat must 'exist' in a superposition of live and dead states until another observation is carried out, i.e. until the experimenter opens the box and observes the state of the cat. The Schrödinger cat thought experiment is the most famous in quantum theory and is described in detail (ok, it's a pig ...) for both the classical and quantum perspectives in the boxes on the next two pages.

"One can even set up quite ridiculous cases. A cat is penned up in a steel chamber, along with the following diabolical device (which must be secured against direct interference by the cat) in a Geiger counter there is a tiny bit of a radioactive substance, so small, that perhaps, in the course of one hour one of the atoms decays, but also, with equal probability, perhaps none; if it happens, the counter tube discharges and through a relay releases a hammer which shatters a small flask of hydrocyanic acid. If one has left this entire system to itself for an hour, one would say that the cat still lives if meanwhile no atom has decayed. The first atomic decay would have poisoned it. The ψ-function of the entire system would express this by having in it the living and the dead cat (pardon the expression) mixed or smeared out in equal parts.

It is typical of these cases that an indeterminacy originally restricted to the atomic domain becomes transformed into macroscopic indeterminacy, which can then be resolved by direct observation. That prevents us from so naively accepting as valid a 'blurred model' for representing reality. In itself it would not embody anything unclear or contradictory. There is a difference between a shaky or out-of-focus photograph and a snapshot of clouds and fog banks."

"Reality resists imitation through a model. So one lets go of naive realism and leans directly on the indubitable proposition that actually ... there is only observation, measurement."

(E. Schrödinger, 1935, *The Present Situation in Quantum Mechanics,* translated in Proceedings of the American Philosophical Society, 124, 1980)

Figure 3.4 Schrödinger's pig!

A pig is confined in a windowless box for one hour. The box contains a
fiendish apparatus that will administer a lethal dose of poison if triggered by
the decay of a single radioactive atom.

If decay is detected
the pig is automatically
injected with poison

Radiation detector ———

Radioactive atom
with 50% chance of
decay in 1 hour

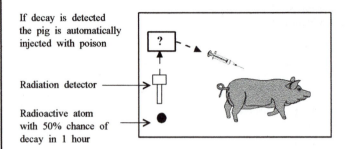

After one hour Schrödinger's assistant opens the door of the box and sees one
of two possible outcomes:

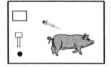   OR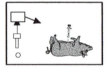

STATE 1. Atom has not   STATE 2. Atom has decayed and
decayed, pig is still alive.   pig is dead.

Classically the system is in one or other of these states at all times during the
hour. If it is observed in STATE 2 at the end of the hour it must have jumped
to that state as a result of a decay at some instant during the hour. These two
classical possibilities are illustrated graphically below.

STATE 1 ━━━━━━━━━━━━━━━━━━━━━━━━━━━━━━━━━ ─ ─

STATE 2 ─
        Start   1 hour

STATE 1 ━━━━━━━━━━━━━━━━─ ─ ─ ─ ─ ─ ─ ─ ─ ─ ─ ─ ─
                        │
                        │ decay
STATE 2 ─ ─ ─ ─ ─ ─ ─ ─ └━━━━━━━━━━━━━━━ ─ ─
        Start   1 hour

continued on p.75

Figure 3.4 Schrödinger's pig, continued.

According to quantum theory the unobserved system is described by a wavefunction which is a superposition of STATEs 1 and 2 at all times up until the moment the box is opened and the pig observed. This implies that the unobserved system exists in an indeterminate state and the outcome is determined at the moment of observation, at which time the wavefunction collapses and one or other alternative is realised.

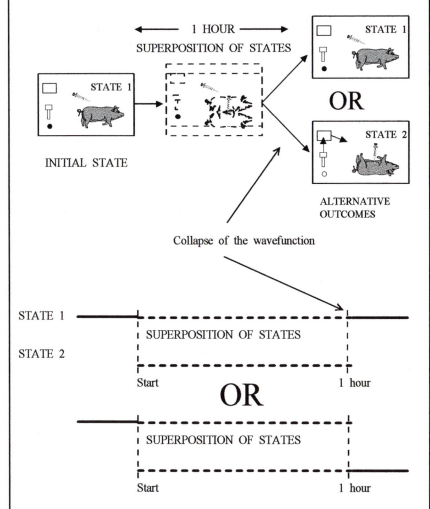

The idea that the atom is in a superposition of decayed/undecayed states and the pig in a superposition of live/dead states prior to observation is disturbing. However, if superposition does not take place it is impossible to explain the appearance of interference effects in related experiments (e.g. Young's slits).

3.1.4 Defining reality – the EPR 'Paradox'

Einstein could not accept that quantum theory meant the end of an objective physical reality. He thought that quantum theory applied only to the average behaviour of ensembles of particles, and was not a complete theory of the behaviour of individual particles. In an attempt to show how strange quantum theory is and to show how it could not be a complete theory of physical reality he constructed an ingenious sequence of thought experiments culminating in the famous Einstein-Podolsky-Rosen (EPR) paper of 1935.

Einstein believed that the physical world of atoms, electrons and photons exists independently of us and that the individual properties of particles are independent elements in an objectively real world. His main aim in attacking the Copenhagen doctrine was to show that quantum theory could not support this kind of reality and so must be incomplete. In order to defend reality Einstein had to define it. In the EPR paper the authors give the following 'definition':

"If, without in any way disturbing a system, we can predict with certainty (i.e. with probability equal to unity) the value of a physical quantity, then there exists an element of physical reality corresponding to this physical quantity."

(EPR paper, Physical Review, 47, p.777-80, 1935)

In order to attack quantum theory as incomplete a description of completeness is required. EPR provided the following:

"Whatever the meaning assigned to the term 'complete', the following requirement for a complete theory seems to be a necessary one: every element of physical reality must have a counterpart in the theory."

(EPR paper, Physical Review, 47, p.777-80, 1935)

This prepares the ground very clearly:

Realism: Properties we are able to predict (because we have measured them) must be real – they must be elements of physical reality.
Completeness: A complete theory must contain elements corresponding to each element of reality.

The Copenhagen view asserts that the wavefunction is the bottom line, giving a complete description of any physical system. However, this description includes the indeterminacy principle, which implies that accurate knowledge of one variable (say x-position) precludes accurate knowledge of another linked (conjugate) variable (such as x-momentum). The EPR thought experiment set out to show that an experiment carried out on a pair of particles that have interacted and then separated *can* provide accurate information about two conjugate variables for either particle. *If* this is possible then both properties are, by the EPR criterion, elements of reality. This would be true despite the fact no independent elements of the wavefunction correspond to these properties and so quantum theory is therefore an incomplete theory of physical reality.

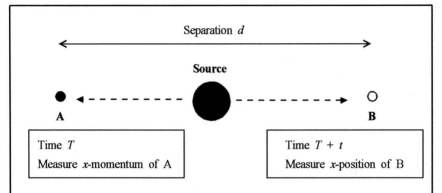

Figure 3.5 EPR-type experiments. The properties of A and B are correlated when they are emitted at the source. If A's *x*-momentum is measured we can use the result to work out B's *x*-momentum. If B's *x*-position is then measured we know both its position and momentum to any accuracy. If the measurement on B is carried out a time $t < d/c$ after the measurement on A then no signal travelling from A to B at or below the speed of light can alert B to the measurement disturbance at A. It looks like a water-tight argument, but it is wrong.

There are many variations on the original EPR thought experiment, any one of which could in principle be used to test quantum theory, but the basic idea is this. Imagine a central source that emits two particles in a single quantum process. The properties of the particles become correlated in order to satisfy conservation laws. For example, the momentum of a particle ejected to the right must be equal in magnitude to the momentum of its partner emitted to the left in order to satisfy the law of conservation of linear momentum. This means that a measurement carried out on one of the pair determines both its own linear momentum *and that of the other particle*, even when a great distance separates the particles. If the distance between the particles is great enough then it seems reasonable to assume they are no longer interacting with one another. The measurement of A's momentum is hardly likely to affect the properties of the distant particle B. But it does tell us B's momentum just as accurately as we know A's. We can now arrange to carry out an accurate measurement of B's position and end up knowing both the position and the momentum of B to any degree of accuracy. In other words we have a practical technique that apparently violates the uncertainty principle. But there is more - if this technique is possible then, regardless of whether or not we choose to measure them, B's position and momentum are both, by the EPR criterion, elements of physical reality. This is despite the fact that the wavefunction contains no elements corresponding to simultaneous values for position and momentum:

"One is thus led to conclude that the description of reality as given by a wavefunction is not complete."

(EPR paper, Physical Review, 47, abstract, 1935)

If the wavefunction, and by association quantum theory, do not give a complete description of physical reality then there must be a deeper theory which does. Within this theory one would expect to find other elements of reality that predispose the particles to interact with measuring apparatus in the way do. These new elements of reality are called *'hidden variables'*. Einstein was really proposing a hidden variables interpretation of quantum theory that would leave its statistical predictions with a status similar to that of statistical thermodynamics, and reduce quantum probabilities to measures of our lack of detailed knowledge of the hidden variables.

The EPR paper stunned Bohr, but his response was masterful. He agreed that a measurement on particle B in no way implied a mechanical disturbance of A. However, it does disturb *the system as a whole*. If instead of treating the wavefunctions for A and B as independent systems, the wavefunction for the system as a whole is considered, *it* is certainly affected by a measurement on B. When the momentum at B is measured the system wavefunction collapses into a new state (corresponding to the state revealed by the measurement) and the possible outcomes of future measurements of A change. The fact that A and B interacted when they were emitted has *entangled* their properties in such a way that it makes no sense to talk about independent elements of physical reality tied uniquely to either particle. However, the entanglement introduces correlations between particle properties that seem to be fundamental. In this sense quantum theory is as complete as it can be because the correlations *do* have corresponding elements in theory and can be derived from the system wavefunction.

Bohr's view differs from Einstein's in one very important respect. For Einstein, the properties of the individual particles are fixed during their brief period of interaction at the source and they carry these objectively real properties with them until they are detected. If Einstein's detector tells him that particle A has momentum p then he concludes it left the source with this momentum. For Bohr, the most complete description of the system is given by the two-particle wavefunction, which contains the probabilities of a range of correlated momenta for the particles. If his detector tells him that particle A has momentum p then he concludes that the measurement process has resulted in the collapse of the wavefunction into this state. Furthermore, the collapse of the wavefunction into a state of definite linear momentum leaves the system in new state in which the position of both particles has a large uncertainty.

In arguing this way Bohr was merely restating the fundamental ideas of the Copenhagen Interpretation of quantum theory. He was prepared to accept the 'spooky action-at-a-distance' this seemed to imply – the idea that a measurement on A can affect the possible result of future measurements on its distant partner, B. He was not prepared to accept the EPR criterion for reality and he defended this position in a paper published in the Physical Review later in 1935:

"My main purpose ... is to emphasise that ... we are not dealing with an incomplete description characterised by the arbitrary picking out of different elements of physical reality at the cost of sacrificing other such elements, but with a rational discrimination between essentially different experimental arrangements and procedures which are suited either for an unambiguous use of the idea of space location, or for a legitimate application of the conservation theorem of momentum. Any remaining appearance of arbitrariness concerns merely our freedom of handling the measuring instruments, characteristic of the very idea of experiment."

(Niels Bohr, Physical Review, 48, 696-702, 1935)

This is an interesting point of view, Bohr called it *complementarity*, an idea he never defined but which he developed almost to the level of a philosophical system. Where EPR had assumed that it made sense to pick properties of individual particles in isolation, even when those particles belonged to a larger entangled system, Bohr believed that the entire physical system including the experimental arrangement must be taken into account when discussing particular properties. Prior to a measurement there is no unambiguous way to define the position or momentum of either particle, so these are not valid elements of reality. However, once a decision has been made to measure the momentum of A then B's momentum is also determined, but the wavefunction has changed in such a way that position is completely uncertain. An experimental arrangement designed to measure momentum excludes arrangements designed to measure position, the two arrangements are complementary, and we have to choose one or the other. For EPR physical reality exists *out there*, independent of our means of observation. For Bohr the measuring instrument itself is inseparable from the phenomena under observation. He regarded this as a *"principal distinction between classical and quantum-mechanical description of physical phenomena"*.

Bohr proposed a simple thought experiment to illustrate the idea of complementarity. Imagine a particle passing through a narrow slit in a rigid barrier. Its de Broglie waves diffract outwards so its momentum perpendicular to the slit becomes uncertain as it emerges on the far side (the diffracted waves spread out so the probability that the particle deflects sideways increases). The narrower the slit the smaller the uncertainty in the position of the particle as it passes through, so the greater the uncertainty in its sideways momentum (this is Heisenberg's Uncertainty Principle). This experimental arrangement is effectively a position-measuring device. If the slit has width Δx then we know the particle's x-co-ordinate to this accuracy, but at the expense of introducing an uncertainty into

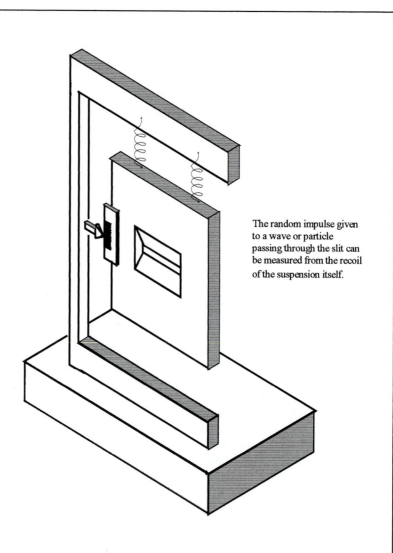

The random impulse given
to a wave or particle
passing through the slit can
be measured from the recoil
of the suspension itself.

Figure 3.6 Bohr's thought experiment to illustrate complementarity. Waves passing through the slit diffract. This gives the associated particles a random sideways impulse, making their momentum parallel to the barrier uncertain. If the slit is suspended and attached to a sensitive measuring device it will be possible to determine this impulse (from the recoil of the suspension) and therefore calculate the transmitted particle's momentum more precisely. However, the price of achieving greater precision in the momentum measurement is that the position of the slit itself becomes uncertain. This increases the uncertainty in the position of the particle. The product of the uncertainties in position and momentum is still greater than or equal to Planck's constant, as predicted by the Uncertainty Principle.

its x-component of momentum. Now consider a modification of the apparatus. If the slit is in a delicately balanced diaphragm then the recoil of the diaphragm after the particle passes through can be used, with momentum conservation, to measure the particle's x-component of momentum. However, this now reduces the effectiveness of the slit as a position-measuring device because its own position is no longer fixed. The experimental arrangement for measuring position is complementary to that for measuring momentum. Although this discussion is entirely qualitative it can be made quantitative and remains valid.

This discussion of the EPR 'paradox' was useful for clarifying exactly what quantum theory implies, but neither Einstein nor Bohr were proposing that it could be used to make a direct test between the quantum and classical views of physical reality. However, in 1964 John Bell showed that the classical and quantum interpretations lead to different predictions in EPR-type experiments. This opened the door to a series of increasingly rigorous tests of quantum theory which could distinguish between Einstein's local hidden variables (naive realism) and conventional quantum theory.

3.1.5 The Bell Inequality

John Bell showed that quantum theory and local hidden variables theories lead to observable differences in the results of certain EPR-type experiments. Bell's theory was based on an idea for an EPR-type experiment proposed by David Bohm in 1951. Bohm simplified the ideas of EPR by applying them to a system involving correlated atomic spins. This had the advantage of giving a discrete set of outcomes (spin axis directions) rather than a correlation between continuous variables like position and momentum. More recently Alain Aspect and others have carried out similar experiments using pairs of photons emitted in opposite directions from an excited atomic source. Conservation laws which apply to the photons at the point of emission force them to have perpendicular polarizations – if one is emitted to the left and is polarized vertically, its partner will be emitted to the right and polarized horizontally. Bell analysed the correlation between polarization measurements carried out on each of the emitted photons.

Hidden variables theories assert that the photon polarisation states are objective real properties of the individual photons determined at the moment of emission. Quantum theory assumes that the system of two photons remains in a *superposition of states* until a measurement is carried out, at which point the wave function describing this superposition collapses and the polarisation states are determined. Thus quantum theory leads to the conclusion that a measurement on one photon determines the instantaneous polarization state of both photons. This generates a correlation between distant parts of the system which is not present in the classical case.

Consider the experiment in the diagram. The source emits pairs of photons. Polarizing 'filters' some distance from the source measure the polarization states of these photons. They do this by scattering incoming photons into one of two 'channels', we shall label these + and −. When a photon emerges in the + state it is polarized parallel to the filter axis, one emerging in the − channel is polarized perpendicular to the filter axis. Each photon must emerge in one or other channel, the filter (which is assumed ideal) does not absorb any photons. The experiment is set up so that filters on either side of the source are parallel to one another. Let us say that they are aligned with their axes vertical.

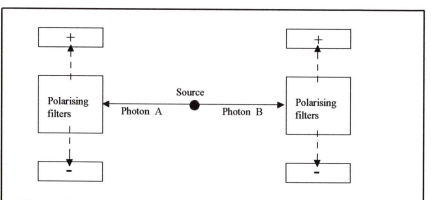

Figure 3.7 An EPR-type experiment with photons. When a photon hits one of the sets of polarising filters it is deflected into one or other channels. If it is polarised parallel to the filter axis it is scattered into the + channel. If it is polarised perpendicular to the filter axis it is scattered into the − channel.

The difference between the classical and quantum cases hinges on how and when the photons assume their measured polarization states. Quantum theory implies that they only adopt these states when the measurement is made, i.e. when a photon interacts with a filter. If photon A is scattered into the + channel of the left hand filter, it must be vertically polarized. This forces photon B to be horizontally polarized and so B must be scattered into the − channel of the right hand filter. If A scatters into the − channel of the left hand filter then B must scatter into the + channel of the right hand filter. Quantum theory means that the outcomes + − or − + are equally likely and for a large number of photons these would appear at random, but with approximately equal frequencies. However, the coincidence of both filters scattering paired photons into the same channel, + + or − − would never occur. The probabilities of each outcome are:

+ +   probability = 0
− −   probability = 0
+ −   probability = 1/2
− +   probability = 1/2

In the classical case the interpretation and outcomes are different. The photon polarisation states are determined at the instant of emission, and, although each member of a pair is polarised at $90°$ to the other one, the line along which they are polarised is oriented at random from one pair to another. Now consider photon A as it approaches the left-hand filter: it will in general be polarised at some angle θ to the vertical and so has a probability equal to $\cos^2\theta$ that it will be scattered into the $+$ channel and $\sin^2\theta$ that it will scatter into the $-$ channel. Whether or not it is actually transmitted has no bearing whatsoever on photon B which is a long way away from A and polarized at an angle $(90 - \theta)$ to the direction of the right hand filter. Thus B has a probability equal to $\sin^2\theta$ of scattering into the $+$ channel and $\cos^2\theta$ of scattering into the $-$ channel. It is quite clear that all combinations of outcomes are possible:

$++$	probability $= \cos^2\theta\sin^2\theta = \frac{1}{4}\sin^2 2\theta$
$--$	probability $= \sin^2\theta\cos^2\theta = \frac{1}{4}\sin^2 2\theta$
$+-$	probability $= \cos^4\theta = \cos^2\theta - \frac{1}{4}\sin^2 2\theta$
$-+$	probability $= \sin^4\theta = \sin^2\theta - \frac{1}{4}\sin^2 2\theta$

If a large number of photons are involved and the polarization of each pair occurs at random then the overall probabilities are:

$++$	probability $= 1/8$
$--$	probability $= 1/8$
$+-$	probability $= 3/8$
$--$	probability $= 3/8$

Comparing these results with those derived from quantum theory shows how the correlation between distant photons restricts the possible outcomes and predicts different statistics to classical theory. The example above emphasises the effect of entanglement in quantum theory, but it could be explained in a classical way if the experimental arrangement somehow acted back on the source to force emitted photons to be polarized parallel or perpendicular to the filter axes. This new interaction would depend on hidden variables but would allow the photons to possess real polarization states (in the EPR sense) prior to measurement.

Bell derived an inequality involving the frequencies of the different experimental results that would be satisfied by *any* classical hidden variables theory, but violated by quantum theory. This gave experimental physicists a simple prediction to test.

3.1.6 The Aspect Experiment

The Bell inequality applies to correlations between distant measurements on pairs of particles emitted from a common point. For classical reality the Bell inequality holds, for quantum theory it is violated. If an experiment can be carried out to measure this correlation its results will distinguish classical from quantum reality.

Bohm's thought experiment involved measurements on a pair of correlated spin-1/2 particles, but subsequent experimental work has used photons with correlated polarizations, so we shall describe a modified optical version of Bohm's experiment. The basic idea is summarized in the diagram: S is a source of a pair of perpendicularly polarized photons, X and Y are linear polarizing filters with their polarization axes set at some angle ϕ to one another. Photons scattered into either channel by the filters are detected by photomultipliers. The pulses from each photomultiplier are then fed back to an electronic counter that records the number of coincident pulses. Every coincidence corresponds to a situation in which both photons have been scattered into the same channel of their polarising filters.

Since any pair of photons have perpendicular polarisation directions the

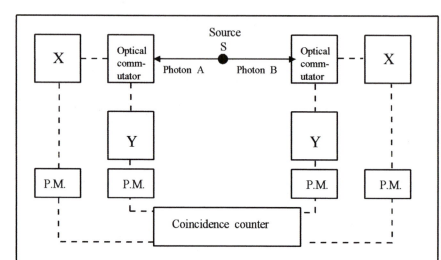

Figure 3.8 The Aspect experiment. The optical commutators are switched rapidly, independently of one another, so that the decision as to which measurement is to be made (X or Y) is determined after the photons leave the source.

coincidence rate will be 0% when the filters are aligned parallel to one another, and 100% when they are perpendicular. Both quantum theory and classical realism agree at these points. If the filters were orientated at some intermediate angle the coincidence rate R would be in the range $0 < R < 1$. The Bell inequality shows that this coincidence rate must be less than some maximum value at any particular angle if the assumptions of local hidden variables theories are true. However, the predictions of quantum theory lead to a violation of the Bell inequality at certain angles. Thus the coincidence rate recorded in this experiment provides a direct test between the two theories.

During the 1970s a number of EPR-type experiments were used to test the Bell inequality. The results were not absolutely conclusive, but gave very strong support to quantum mechanics and it was generally accepted, sometimes rather begrudgingly, that the Bell inequality *is* violated in these experiments.

At first sight it would seem that this settles the question about the nature of physical reality in favour of the quantum mechanical description by ruling out the possibility that the photons exist in objectively real polarisation states prior to an observation. However, there remained a subtle loophole which gave the diehard hidden variables theorists a slender chance of maintaining their model. Certainly the local hidden variables theory was untenable, but what about a *'separable hidden variables theory'*? In order to understand the nature of this distinction, and hence the significance of the Aspect experiment we must first consider the idea of causality.

In the naive realism of classical physics the properties of a body possess an objective existence and are only affected by *locally* acting influences. This does not rule out the possibility of 'action-at-a-distance' as apparently occurs in a gravitational or electromagnetic field, but interprets this action as the result of locally transmitted disturbances through the field itself. What is not allowed is an *instantaneous* action-at-a-distance whereby events at one point are instantly affected by events elsewhere in the universe. Einstein's theory of relativity puts an upper limit on the velocity at which these causal influences can propagate – the velocity of light in a vacuum. If any influence travels faster than this it results in situations in which 'effects' precede 'causes' for some observers. These ideas can be illustrated using a 'space-time' diagram. For the sake of clarity only one of the three spatial directions is shown. The diagonal lines crossing at the origin represent the trajectories of impulses travelling at the velocity of light. An event occurring at the origin will only be able to affect future events by means of interactions propagated at sub-light velocities. These events are contained in the area above the horizontal axis – the *'future light cone'* of the original event. Similarly this event could only have been affected (or caused) by events contained within its *'past light cone'*. Events outside this shaded area cannot be causally connected to the event at the origin of this coordinate system.

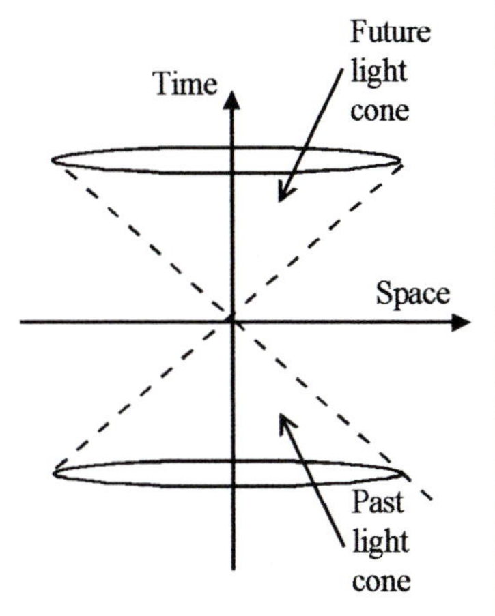

Figure 3.9 The past and future light cones contain all space-time events that could be causally connected with an event at the origin.

The early tests of EPR involve arrangements in which the polarizers and detectors are set up *well in advance* of any measurements being made, so there is the unlikely (but not impossible) chance that some interaction takes place between the distant devices and generates a correlation in their results. Thus it is possible to maintain the hypothesis of a hidden variable theory in which some as yet undiscovered influence is transmitted at or below the velocity of light between different parts of the apparatus and generates the quantum correlations. This is called a *'separable hidden variables theory'*.

In 1976 Alain Aspect proposed an experiment which could distinguish between the predictions of separable hidden variables theories and quantum theory. The basic idea is to randomly and independently select the orientation of each polarizing filter while the photons are 'in flight' from the source. This means that causal influences from the measuring devices travelling at the speed of light cannot influence the initial states of the photons emitted from the source. It also implies that information about the measurement to be carried out at one polarizer cannot be transmitted to the other polarizer in advance of its own measurement. If the Bell inequality holds in the Aspect experiment it implies that separable hidden variables theories are possible, whereas if it is violated it means that any hidden variable theory would have to incorporate instantaneous action at a distance, the famous *'non-locality'* of quantum theory.

The experimental scheme is shown in the diagram. For each photon there is a pair of polarizers set up in different orientations. Which of these is to carry out the measurement on the photon is determined at the last moment by a rapid optical commutator (or switch) that is controlled by its own independent generator. Ideally these optical switches are operated completely randomly and deflect the incident photon to one or other polarizer. The polarizers are used in conjunction with a photomultiplier and coincidence counter as before. The experiment was finally carried out in 1982. *It showed a clear violation of the Bell inequality.*

3.1.7 The Optical Commutators

The loophole left by experiments carried out prior to 1976 was closed in the Aspect experiment by introducing rapid optical switches whose state is determined *after* the photons have been emitted from the source. The technological problem here is the extremely high switching frequency required because of the short time of flight of the photons. For photon paths of the order of a few metres the time of flight will be given by $t = s/c$ where s is the path length and c the velocity of light. Since s is typically a few metres the switching time needs to be of the order of nanoseconds.

The switch used in the Aspect experiment consisted of a small vial of water that could be forced to oscillate by a high frequency generator. These oscillations produced standing waves in the water making it behave like a diffraction grating. Incident photons emerged along one or other of two specific directions sending them to one or other of the polarizing filters, which are preset in different orientations. Completely independent generators drive each of the two optical

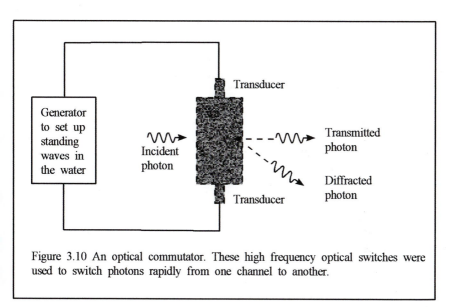

Figure 3.10 An optical commutator. These high frequency optical switches were used to switch photons rapidly from one channel to another.

commutators.

In the experiment itself the separation of the two commutators was 12 m and the commutators were switched every 10 ns. Since it takes light 40 ns to travel 12 m, this switching rate is fast enough to distinguish between the predictions of separable hidden variables theories (incorporating Einstein locality) and quantum theory.

One possible objection to this method is that although the generators are independent of one another, they switch the commutators periodically rather than randomly. However, as Aspect pointed out, in order for this to affect the validity of their conclusions we should have to assume that the polarizers have a memory for previous results and adjust their future behaviour in order to produce the quantum correlations.

3.1.8 Photon generators

Pairs of photons are generated by an 'optical cascade' in excited calcium atoms. The excitation is produced by firing a tuned laser beam at a beam of calcium atoms. As a result an electron in the calcium atom is promoted to a higher energy level from which it returns to the ground state via an intermediate level, releasing two optical photons in the process. The two photons are polarised perpendicularly to one another.

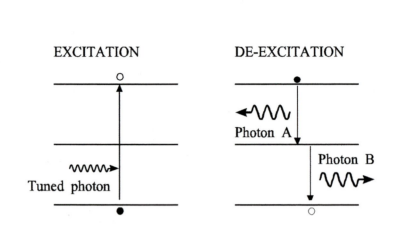

EXCITATION   DE-EXCITATION

Figure 3.11 Generating photon pairs in an optical cascade. The horizontal lines represent energy levels in calcium. A tuned laser excites electrons to the higher level. When de-excitation occurs it does so via a short-lived intermediate level, emitting two correlated photons. The experimental arrangement is shown below.

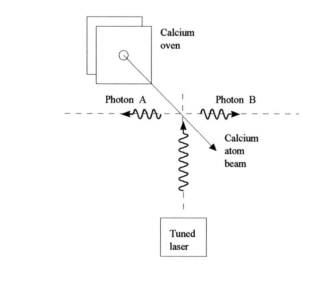

3.1.9 Summary of the EPR experiments

- EPR: If elements of physical reality are assumed to possess independent, objective, locally determined properties then quantum mechanics is incomplete and we must seek a more complete theory of hidden variables.
- BELL'S INEQUALITY: Quantum theory predicts that particles which have interacted at some point in the past become 'entangled' in such a way that the results of measurements on the separate particles in the future produce correlations which are not predicted by local hidden variables theories. Quantum mechanics violates the Bell inequality.
- BOHM'S THOUGHT EXPERIMENT: These correlations can be measured in coincidence experiments on systems in which a pair of particles are emitted in opposite directions with correlated spin (or polarisation) states.
- COINCIDENCE EXPERIMENTS PRIOR TO ASPECT: The Bell inequality is violated. Thus *local* hidden variables theories are eliminated, the correlations are in agreement with the predictions of quantum theory.
- EINSTEIN LOCALITY: Causal interactions propagate at or below the velocity of light. In all experiments prior to that of Aspect the settings of the measuring devices are determined well in advance of the emission of the photons. This means that they could become correlated by locally transmitted influences without violating causality or implying quantum non-locality.
- SEPARABLE HIDDEN VARIABLES THEORIES: These accept the violation of Bell's inequality in the early coincidence experiments but assume that the distant measurements become correlated by causal interactions which do not violate Einstein locality.
- THE ASPECT EXPERIMENT: This experiment can test between the predictions of separable hidden variables theories (for which Bell's inequality holds) and quantum mechanics (for which it is violated). The Bell inequality is violated in the Aspect experiment and the results are in agreement with the predictions of quantum theory.

Aspect described his experiment as an attempt to test the non-separability of quantum mechanics. He carried out the experiment with Dalibard and Roger at the Sorbonne in 1982 and it yielded results which violated the Bell inequality by five standard deviations whilst remaining in good agreement with the predictions of quantum theory. The immediate conclusion must be that separable hidden variables theories are not valid representations of the underlying physical reality.

The original motivation behind the construction of hidden variables theories seems to be an attempt to return the foundation of physics to the concepts of classical reality. The Aspect experiment implies that, if we persist with a description in terms of hidden variables, we must additionally assume that they may be affected by instantaneous action-at-a-distance, a conclusion hardly likely to be welcomed by the 'classical realist'. Alternatively we must accept quantum mechanics, with its property of nonseparability or nonlocality, as a complete description of the nature of physical reality. Some people have suggested that the existence of non-local connectivity means that causality will be violated. In a restricted sense this may be true, but since these non-local interactions only result in *correlations* between random sequences of events at separate detectors, the nature of the correlation (i.e. information) may only be retrieved by comparing sets of results from different points. This comparison is only possible if we physically transmit information from one point to another, a process limited by the speed of light. So there is no violation of causality. The statistical nature of the quantum correlations prevents them from being used to transmit messages instantaneously from one place to another.

The Aspect experiment demonstrates that the nature of physical reality is not subject to the constraints of Einstein locality. It shows that the results of measurements on physical systems correspond to the predictions of quantum theory rather than those of classically based hidden variables theories. It provides convincing evidence that quantum theory may be a complete description of physical reality and puts tight constraints on the nature of any new theory that might replace it. All of this underlines the radical departure of quantum physics from classical physics.

3.2 AGAINST COPENHAGEN

3.2.1 Bohm's Ontological Interpretation

"All that is clear about the quantum theory is that it contains an algorithm for computing the probabilities of experimental results. But it gives no physical account of individual quantum processes, Indeed, without the measuring instruments in which the predicted results appear, the equations of the quantum theory would be just pure mathematics that would have no physical meaning at all ... That is to say, it seems, as indeed Bohr and Heisenberg have implied, that quantum theory is concerned only with our knowledge of reality and especially of how to predict and control the behaviour of this reality ... quantum theory is primarily directed towards epistemology which is the study that focuses on the question of how we obtain our knowledge (and possibly on what we can do with it).

It follows from this that quantum mechanics can say little or nothing about reality itself. In philosophical terminology, it does not give what can be called an ontology for a quantum system."

(David Bohm and Basil Hiley, in *The Undivided Universe*, Routledge, 1993)

Bohm began as a believer in the Copenhagen orthodoxy, which is hardly surprising, since he studied under Bohr in Copenhagen. However, he became worried about the status of quantum theory and its failure to provide realistic models of objects in the physical world. Following discussions with Einstein he set about constructing an alternative realistic *hidden-variables* interpretation of quantum theory based on an idea that had first been pursued (and then abandoned) by Louis de Broglie in the late 1920s:

- Quantum objects (e.g. electrons) are objectively real and possess well-defined properties of position, momentum, etc., at every instant in time.
- All quantum objects are accompanied and linked together by real but unobservable *pilot-waves*. These pilot-waves are affected by the whole experimental set-up and change everywhere instantaneously whenever the experimental arrangement is changed.
- When a measurement is carried out on a quantum object its pilot-waves change and so possible outcomes of future measurements on related properties will also be affected.

Bohm developed this approach through a number of papers published between the 1950s and his death in 1992. He was able to show that a hidden-variables theory *can* reproduce all the results of conventional theory and retain the reality of quantum objects. However, this does not signify a return to the naive realism of classical physics. Bohm had to pay a price. When the pilot wave changes it does so everywhere at the same time, and this needs influences to travel through the field faster than the speed of light. In other words Bohm's model of reality has non-locality built in. This is why physicists have not raced to embrace hidden-variables and is linked once again to Bell's inequality.

The fact that the Bell inequality is violated forces us to accept one of two conditions on any interpretation of the physical world:

- If we insist that quantum objects are objectively real, in the sense that they possess definite properties independent of any attempt to measure or observe them, then non-local influences (i.e. influences propagated faster than the speed of light) *must* be included.
- If we insist that physics is local (i.e. no super-luminal connections are allowed) then we must abandon the idea of objective reality (in the sense described above).

Put more simply, the outcome of the Aspect experiment implies that a local realistic model of the physical world is impossible.

Although Bohm's original papers referred to a 'hidden-variables' theory, he later called it an 'ontological interpretation' to emphasise the fact that it retained a realistic view of quantum objects. Bohr developed the Copenhagen Interpretation into a complete philosophical system built around the concept of complementarity. Bohm (with Basil Hiley) did a similar thing with the ontological interpretation, using it as the foundation for a holistic view of physics. The central idea was one of a unifying 'implicate order' in which the unity of nature is manifest as opposed to the revealed or explicate order around which most of the present laws of physics have evolved.

Bohm's ideas remain an interesting diversion in twentieth century physics, but seem unlikely to be adopted as orthodoxy. Meanwhile other more radical interpretations of quantum theory are beginning to displace the Copenhagen view.

3.2.2 The Many-worlds Interpretation

In 1957 an American physicist, Hugh Everett III submitted a thesis to Princeton University as part of his Ph.D. It was a 'Relative State Formulation of Quantum Mechanics' (Reviews of Modern Physics, 29, 454-62, 1957) and its purpose was to supply a more general formulation of quantum theory from which the conventional version can be derived. The motivation for this was to produce a meta-theory within which the old theory could be analysed more clearly in the hope that it might make it easier to see how to apply quantum theory to general relativity. The ideas put forward in this highly influential paper are now known as the Many-Worlds Interpretation of quantum theory.

Everett was particularly worried about the two fundamentally different ways in which state functions change with time:

- Process 1: the discontinuous collapse of the wavefunction when an observation is made;
- Process 2: the continuous evolution of the wavefunction between observations.

These two processes can be accommodated as long as we are dealing with an observer who stands *outside* the system he observes. But this is impossible if we are trying to apply the laws of quantum theory to the universe as a whole, particularly if that universe is closed. Conventional quantum theory works for systems subjected to external observations, but offers no clear picture if the observer is part of the system. Everett's solution was to take the wavefunction itself as a complete mathematical model for every isolated physical system and to assume that all cases of external observation are themselves embedded within a larger system which includes all observers. This is where the idea of a relative state comes in: no individual subsystem can be in a definite state independent of all other parts of the composite system – for any definite state of one subsystem there is a corresponding set of relative states for all other parts. In conventional theory the observer making a measurement causes the collapse of a wavefunction so that all other subsystems jump into definite states relative to the observer. However, if the total system is represented by a total wavefunction then it must consist of a superposition in which all possibilities for all observers are included. In this super-system all results of all possible observations are equally real and in some sense continue to coexist. Any particular observer experiences a unique subset of these states which themselves appear to make discontinuous jumps, but this is simply a consequence of their reference frame, rather than a new physical process.

The relative state formulation can be interpreted as follows. The total

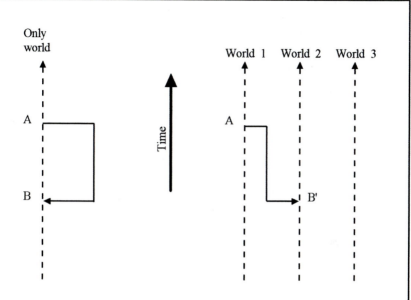

Only world

World 1   World 2   World 3

Time

A: Bob steps into a time machine and heads for the past.

B: Bob murders his own grandfather.

This implies Bob's father and Bob himself cannot ever be born. So who committed the murder?

A: Bob in world 1 steps into a time machine and heads for the past.

B': Bob from world 1 murders a man who would, in world 2 have lived to father a man like Bob's father, but in world 2. Bob's own grandfather (still in world 1) is oblivious to all this.

There are no logical problems in either worlds 1 or 2 and all other worlds in which variants of Bob's family live, are unaffected.

Figure 3.12 Time travel, the grandfather paradox and the many worlds theory of quantum mechanics.

wavefunction for the universe evolves in a continuous way according to the usual rules of quantum theory and consists of a superposition of all possible states. When an observation is made all possible outcomes are included as separate components of the total wavefunction which continues to evolve. There is no collapse of the wavefunction, and no discontinuous transition from the possible to the actual – all alternative realities coexist. However, a conscious observer (e.g. you or I) is represented many times over in the total wavefunction, once for each alternative set of possibilities so the individual experience will consist of a sequence of branching 'paths' through the maze of alternative realities. At each branch the subjective experience will be of a discontinuous change even though it is clear that no such process exists in the physics of the total wavefunction.

This strange idea that each observer follows a unique trajectory through parallel universes is known as the many-worlds hypothesis and is becoming one of the most widely used interpretations of quantum theory, especially among quantum cosmologists! It also offers some surprising solutions to traditional paradoxes. For example, it has been argued that time travel into the past must be impossible because you could use a time machine to travel back to a time prior to your father's conception and kill your grandfather. On the face of it this leads to a logical contradiction because your father cannot then be born and nor can you, so you would not exist and could not have travelled back to do the dirty deed! Many other equivalent stories can be constructed, but this version is known as the grandfather paradox. If the many-worlds theory is correct it offers a neat solution to the grandfather paradox. When you travel back in time you cross to an alternative reality in a parallel universe *in which you will not be born*. This does not lead to a paradox because there are still many parallel universes in which you *do* get born, including the one in which you climbed aboard a time machine and travelled back into the past.

3.3 QUANTUM INFORMATION

3.3.1 Qubits

As we move into the twenty-first century quantum theory retains its ability to confuse and surprise. By the 1990s advances in technology and experimental techniques enabled many of the thought experiments of the 1920s and 1930s to be carried out in the laboratory and opened the door to a variety of intriguing applications using the same counter-intuitive ideas that challenged Einstein. Examples include quantum cryptography, quantum computing, quantum teleportation and non-interactive measurements. These incursions into weird science suggest that the twenty-first century may well be host to some peculiarly quantum phenomena. It is possible that we shall come to rely on quantum theory to secure our financial transactions and to allow rapid computations on super quantum computers, or even to form images of objects we have never observed!

Most of these new possibilities are linked to the strange way in which information can be encoded and transmitted in a quantum system, so before we look at particular examples and experiments it is worth considering the links between quantum mechanics and information theory.

Although all modern computers rely on quantum theory for their electronics they still encode process and transmit information in a classical way. In classical binary logic the starting point is the bit, a fundamental 'particle' of information that can exist in one of two states, denoted '0' and '1' and usually represented by distinct electronic states separated by a significant energy barrier. Of course, there are many alternative ways to encode information and any two-state quantum system can be used. When this is done the two states are written:

$$\text{logic '1':} \qquad |1\rangle$$

$$\text{logic '0':} \qquad |0\rangle$$

These '*qubits*' have been represented by polarisation states of photons (e.g. linear and horizontal), spin states of electrons or nuclei, energy levels in atoms or even the propagation directions of particles. The new feature that enters information theory when we use qubits is the possibility that of a superposition of states. Unlike their classical counterparts, any linear superposition of qubit states is also a state that can be used to encode information. To see how this works think about encoding information onto a pair of quantum particles, each of which can carry one bit of information. The classical possibilities for a two-bit system would be $(0,0)$, $(0,1)$, $(1,0)$ and $(1,1)$, all of which have quantum counterparts:

$$|0\rangle|0\rangle$$

$$|0\rangle|1\rangle$$

$$|1\rangle|0\rangle$$

$$|1\rangle|1\rangle$$

(The two terms correspond to the first and second particle respectively).

This choice of qubit states is equivalent to the classical situation, but it is also possible to build up alternative qubit states from linear superpostions of these 'classical' states. There are many ways in which this can be done, the superposition states listed below have been used in a number of experiments and are known as the 'Bell states'. Notice that there are still just four distinct states so, just as with the classical states, these can carry two bits of information:

$$\left|\Psi^+\right\rangle = \frac{1}{\sqrt{2}}\left(|0\rangle|1\rangle + |1\rangle|0\rangle\right)$$

$$\left|\Psi^-\right\rangle = \frac{1}{\sqrt{2}}\left(|0\rangle|1\rangle - |1\rangle|0\rangle\right)$$

$$\left|\phi^+\right\rangle = \frac{1}{\sqrt{2}}\left(|0\rangle|0\rangle + |1\rangle|1\rangle\right)$$

$$\left|\phi^-\right\rangle = \frac{1}{\sqrt{2}}\left(|0\rangle|0\rangle - |1\rangle|1\rangle\right)$$

Once again the order of terms is important, so that $|\Psi^+\rangle$ for example, is a superposition of two states – in one of which the first qubit is 0 and the second is 1, and in the other the first is 1 and the second is 0. It is not possible to ascribe definite states to either qubit individually, the information is encoded in their combined state, not in the individual particles. This is an example of *entanglement* and opens up a number of intriguing possibilities. Entanglement is closely linked to non-locality, and is at the heart of the counter-intuitive quantum mysteries that upset Einstein and others.

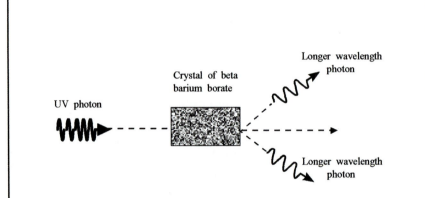

Figure 3.13 Type II parametric down-conversion. One of the emitted photons is polarized vertically and the other is polarized horizontally. If the cones in which these photons are emitted overlap the polarization of photons in the region of overlap is an entangled state. These photons can be used to carry quantum bits of information.

If we wish to generate entangled states in the laboratory it is essential that we cannot know, even in principle, what the states of the individual particles are. There are various ways in which this can be done. Many of the experiments based on entanglement have used pairs of photons. These are generated using non-linear crystals in a process called type-II parametric down-conversion. When an ultra-violet photon hits a crystal of barium borate there is a small probability that it will decay to a pair of longer wavelength photons emitted symmetrically about the original ultra-violet photon direction. Furthermore the two photons have perpendicular polarisation directions, so if one is horizontally polarised the other is vertically polarised. These polarisation states can be identified with qubit states:

$$\text{horizontal polarisation} = |1\rangle$$
$$\text{vertical polarisation} = |0\rangle$$

However, there is no way of knowing which photon is in which polarisation state, so the system of two photons is an entangled state and all we can say about the individual photons is that their polarisations states are different to one another.

3.3.2 Quantum Cryptography

The aim of cryptography is to allow information to pass from one party to another whilst preventing access to that information for any third party. One of the most famous classical methods for doing this is the 'one-time pad' system proposed by Gilbert Vernam at AT&T in 1935. In this system the person sending the message (usually called Alice) adds the message itself bit by bit to a randomly generated key. She then transmits this encrypted message (which contains no information) to someone (usually called Bob). Bob receives a string of random bits which is meaningless without access to the same random key. However, if Bob shares the key (one-time pad) he can subtract the random bits from his scrambled message to decode it. Let us assume that a third party wants to break into this system and

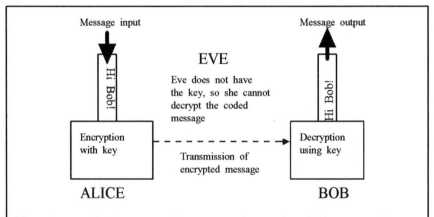

Figure 3.14 A simple cryptographic system. The security of this system relies on the secrecy of the key. The most secure method is to use a one-time code pad shared by Alice and Bob. Public key encryption systems rely on the difficulty of factorising large numbers. If some short-cut to factorisation could be found then these systems would be compromised.

intercept and interpret the message; we call her Eve. What can Eve do? If she intercepts the scrambled message it will consist only of random bits from which she can extract no information. If Alice and Bob use the same key more than once then Eve could gradually build up a profile of the key by comparing different messages and looking for correlations between them. This is why it must be a one-time key only. Assuming Alice and Bob are careful to use the key once and once only and have a secure method of sharing that key then the message is secure. This method is fine for spies but hardly useful for the security of credit card numbers or bank accounts!

In 1976 Whitfield Diffie and Martin Hellman at Stanford proposed a 'public key crypto-system' based on 'one-way functions'. The basic idea is that the key depends on the relation between a function $f(x)$ and the variable x. There are many examples in which it is easy to calculate $f(x)$ from x but very difficult (i.e. involves a large amount of computation) to calculate x from $f(x)$. The relationship between a large number and its prime factors can be used to create a public key. The system works like this. Bob, who is going to receive encrypted messages, chooses a private key (x) which he uses to construct a public key $(f(x))$. He publicises $f(x)$ and waits. Alice prepares her encrypted message using the public key and transmits it to Bob. Bob decodes it using his private key. Eve, who intercepts the message and knows the public key, is faced with the problem of working out Bob's private key from the public key. Since the (classical) computing operations needed to factorise an integer increase exponentially with the number of bits used to describe it, Eve has a formidable challenge. In fact, if the integer Bob has used is large enough it will be impossible (given finite time and resources) for Eve to crack the code. Systems like this are now widely used and are partly responsible for the security of information transmitted across the Internet.

Quantum cryptography allows Bob and Alice to create and share a random secret key without having to meet or use a go-between. It also uses the sensitivity of entangled quantum systems to reveal to Alice or Bob whether Eve has managed to break into their encrypted communications. Quantum cryptography makes use of the fact that any measurement on a quantum system perturbs the system and turns this apparent limitation to great advantage. Charles Bennett of IBM has been behind many of the developments in quantum information technology and, with Gilles Brassard from the University of Montreal proposed the BB84 communication protocol in 1984. It works like this. Bob and Alice are connected by two communications channels. One of these is used to transmit qubits, usually photon polarisation states transmitted along an optical fibre. The other can be any classical connection (including a telephone line, but usually again an optical fibre) and can be a public channel. The difference between the two channels of communication is that the classical one deals with hundreds of quanta per bit of information whereas the quantum channel deals with one quantum per bit. Alice's message consists of a string of 1s and 0s. She encodes these in photon states using two alternative sets of polarizing filters. These could be horizontal and vertical, or diagonal. As the message is sent she encodes each photon by selecting a polarizing filter at random and recording the filter she used to send the photon. At the receiving end Bob also has two sets of polarizing filters and he too selects these at random for each incoming photon, recording the sequence of filters he has used to receive the message. If Alice uses the vertical/horizontal filters to send a particular photon and Bob happens to use his vertical/horizontal filter to detect that photon then he receives the same information that Alice sent; if it was a '1' he receives a '1' if it was a '0' he receives a '0'. If, however, Alice uses vertical/horizontal coding and Bob selects a diagonal filter then he will receive either a 1 or a 0 with equal probability (whatever Alice is transmitting). This is because the incoming photons (be they vertically or horizontally polarized) are in a superposition of *both* polarization states with respect to the diagonal polarizing filters. At the end of the transmission Bob will have received a string of bits, some of which are the same as the bits that Alice sent and some of which are different. He will not know which is which. However, he will also have a record of what polarization directions were selected as he received each bit and Alice will have a record of the polarization filters she used to transmit each bit. Next Bob lets Alice know (via the public channel) the sequence of settings he used to receive the message (but not the results). Alice compares this sequence with her record of transmission settings and sends a message to Bob to tell him when the transmission and reception polarizers were compatible (i.e. the occasions on which Bob received the bit Alice actually sent). Once again she does reveal the actual message. Alice and Bob now have a secure sequence of bits that they both agree on. This can be used to generate a key that can be used like the one-time pad in Vernam's original system.

They can also detect whether their key has been intercepted. An eavesdropper, Eve, would need to receive the same signals as Bob. The public part is openly accessible, so she has this already. She also needs the messages sent by the quantum channel. To get these she could cut the cable between Alice and Bob, read the message, and generate a string of photons to send forward to Bob in order to conceal what she has done. Unfortunately for Eve this cannot be done without affecting the message that is forwarded. In order to read the intercepted signal Eve has to pass it through some kind of polarization analyser. If we assume that she is expecting either vertical/horizontal polarization or diagonal polarization then her polarizer is likely to agree with Alice's for 50% of the bits. The other 50% will be received by Eve with the wrong polarizer, so they will be 1s and 0s at random and Eve will not be able to correct them before passing them on to Bob. This means that some of the bits Bob uses to generate his key will disagree with Alice's (since they have been altered *en route* as a result of Eve's eavesdropping). This will be discovered if Alice and Bob systematically compare a fraction of the bits they exchange to generate the key. This comparison takes place over the public channel, so the bits used for comparison cannot then be used for encryption.

This process is complicated, but it could be automated using a computer to receive, record and compare data and to adjust the polarization filters on each channel. The first prototype was built at IBM in 1989 – polarized photons were used as qubits and a signal was sent 30 cm through the air. In the 1990s several groups succeeded in sending quantum encrypted messages and it is now possible to use this method for secure communication over several kilometres. Optic fibres provide a convenient medium for the channels but the signals cannot be amplified because amplification would have a similar effect to eavesdropping and destroy information. One of the main problems for quantum cryptography is overcoming errors of transmission that appear to Bob like the garbled messages that would be passed on by Eve. If the error rate is too high it will not be possible to distinguish noisy signals from compromised signals, making the system insecure.

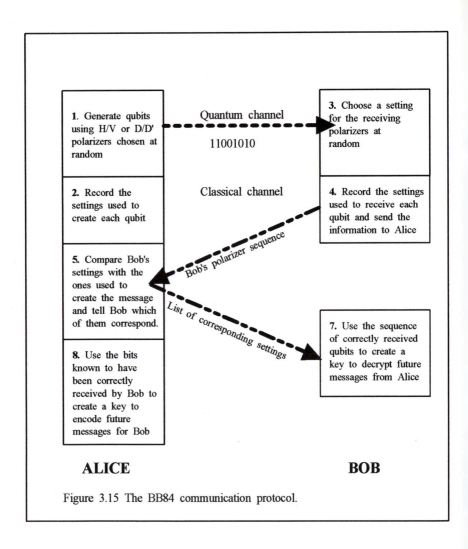

Figure 3.15 The BB84 communication protocol.

3.3.3 Quantum Computing

"Quantum theory describes a much larger reality than the universe we observe around us. It turns out that this reality behaves just like multiple variants of the universe we see, and that these variants co-exist and affect each other through interference phenomena ... an observed particle is just one aspect of a tremendously complex entity that we cannot detect directly. Quantum computation is all about making the invisible aspects of the particle – its counterparts in other universes – work for us."

(David Deutsch and Artur Ekert, Physics World, March, 1998)

We have already seen how qubits can be used to store information either classically (where each qubit is in a state which is either 1 or 0) or quantum mechanically (using a superposition of states so that no particular qubit is in any definite state and the information resides in the correlation between qubits). If we take the many-worlds view seriously then a superposition of states represents two different parallel worlds. This means that a memory made of two qubits can store 2^2 different numbers all at once. It can do this by existing in a superposition of states, such as:

$$\Psi = \frac{1}{2}\left(\,|0\rangle|0\rangle + |0\rangle|1\rangle + |1\rangle|0\rangle + |1\rangle|1\rangle\,\right)$$

This is quite different to a classical memory, which can store only one of the four possible numbers at any particular time. The idea in a quantum computer is to carry out mathematical operations on all the numbers at once, in other words to build a massive parallel computing system. In practice this may turn out to be rather difficult, but in principle quantum theory makes it possible, and several research groups are already preparing the ground. One particular problem that it might 'solve' is how to factorise large numbers quickly (and hence crack existing encryption systems), another is in carrying out rapid sorting procedures. On the other hand, these techniques will not lead to a massive increase in computer memory systems because any attempt to read the data from a quantum memory results in a single output (effectively collapsing the wave function in the Copenhagen view, or branching into a particular universe in the many-worlds view). Quantum computers will be useful solving problems where the results of many parallel computations can interfere to give information in one universe that depends on the outcomes of calculations in all the other universes.

One possible consequence of quantum computing is that we may have to change what we consider to be a valid mathematical proof. In a classical computation proof consists of a sequence of propositions that are either axioms or else follow from earlier propositions in the sequence by standard rules of inference. Proofs carried out by quantum computation will not be like this; it will not be possible to follow each step in the proof and generate a simple sequence of propostions. Deutsch has extended the classical theory of computation to quantum computers and concludes that a proof must now be identified with a process, the computation itself rather than a record of steps taken in carrying out the proof.

Quantum computers may also lead us into a maze of logical curiosities as we begin to acquire information in this universe that arose in others. Richard Jozsa, a mathematician at the University of Plymouth has speculated on the *quantum non-computer*, a device that has the potential to give answers but is never actually switched on. This potential to compute may be sufficient to produce real results! To understand how this may work we need to consider how quantum theory can allow us to make non-interactive measurements.

3.3.4 Interaction Free Measurements

In a classical system any observation requires an interaction. For example, in order to see the page you are reading photons must bounce off it and enter your eye. In quantum theory the possibility that an interaction may take place can be sufficient to provide observational information even if the interaction does not in fact take place. This is called an interaction-free measurement. Information can be gained from counterfactual events, things which could have happened but did not.

In 1993 Abshalom Elitzur and Lev Vaidman proposed an intriguing thought experiment. Imagine a factory that produces 'Q-bombs'. A Q-bomb has a very sensitive light operated detonator, so sensitive that just a single photon is sufficient to set it off. However, setting up such a sensitive trigger is difficult and some of the detonators are duds. If light hits them it simply reflects off and the bomb does not explode. Of course, it would be simple to discover which bombs were duds – shine a light at each one in turn and the duds are the ones that don't explode. The operational bombs could be found in the same way, but at the expense of blowing them up. The key question posed by Elitzur and Vaidmann was this: *is it possible to identify sound Q-bombs without exploding them?*

On the face of it this seems to be an impossible task. To test the bomb a photon has to hit it, and if it is a sound bomb it explodes. However, if we phrase the question in a slightly different way it opens up to a counterfactual quantum process:

- *"If I direct a photon at the detonator will it absorb it and explode?"*

is equivalent to the slightly different question:

- *"If I carry out a Q-bomb test in which the detonator of a particular bomb could have been hit by a photon **but was not** would it have exploded if it had been hit?"*

Phrased like this, we are asking about a *counterfactual*, something that would have happened in a possible world different from the actual world in which we happen to be. If we can set up an experiment in which photons could hit the Q-bomb detonator the mere possibility may be enough to interfere with events in our universe, tell us if the bomb is a sound and yet not explode it.

In the double slit experiment, the behaviour of a photon going through one slit depends on whether another slit *through which it does not pass* is open or closed. In beam splitting experiments the alternative possibilities interfere to determine what outcomes are possible, so the paths a photon does not take do help determine what it may do in the future. Elitzur and Vaidman's ingenious solution to the Q-bomb problem makes use of this behaviour. The proposed solution involves a uniquely quantum mechanical process, the *interaction-free measurement* or IFM.

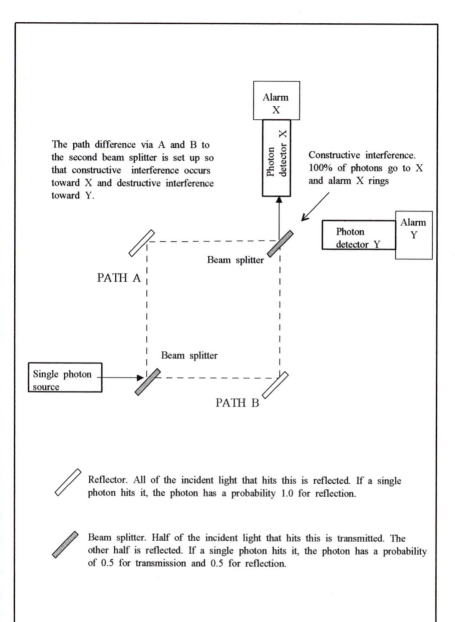

The path difference via A and B to the second beam splitter is set up so that constructive interference occurs toward X and destructive interference toward Y.

Alarm X

Photon detector X

Constructive interference. 100% of photons go to X and alarm X rings

Alarm Y

Photon detector Y

Beam splitter

PATH A

Beam splitter

Single photon source

PATH B

Reflector. All of the incident light that hits this is reflected. If a single photon hits it, the photon has a probability 1.0 for reflection.

Beam splitter. Half of the incident light that hits this is transmitted. The other half is reflected. If a single photon hits it, the photon has a probability of 0.5 for transmission and 0.5 for reflection.

Figure 3.16 A photon interferometer. The optical paths A and B can be adjusted to introduce any desired path difference. Here it has been set up so that constructive interference occurs toward X and destructive toward Y. This is an example of an experiment where there is no way, even in principle, to determine by which of the two routes the photon travels, so in some sense it travels along both routes and intereference occurs.

In a single photon experiment, quantum interference occurs if there is no way, *even in principle*, to determine by which of two alternative paths the photon travelled. On the other hand, if some observation could determine the photon trajectories (even if we do not choose to look at the results of such an experiment) then the interference effects disappear. Elitzur and Vaidman used the Q-bomb detonator as a reflector in one path of a Mach-Zehnder interferometer. Photons entering the interferometer may travel via either of two alternative paths to reach either of two detectors X or Y. The interferometer arms are slightly different in length so that interference results in constructive interference toward detector X and destructive toward detector Y. However, if either path is blocked interference effects are destroyed and each photon can reach either detector with equal probability.

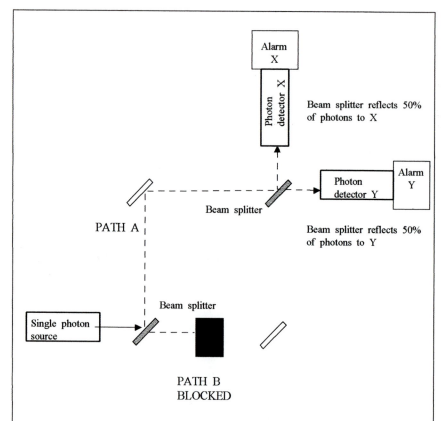

Figure 3.17 If either path is blocked any photons that go that way are absorbed. This absorption removes intereference effects at the second beam splitter. If the interferometer has been set up as it was in the previous diagram then the arrival of photons at Y prove that one path must be blocked.

- If the bomb is dud photons can reflect off it and they have two possible paths, this results in interference. All photons must go to detector X.
- If the bomb is good then it absorbs any photons sent toward it and explodes. This removes the possibility of interference so photons passing by path A, when there is a bomb in path B, can arrive at *either* detector X or detector Y.

Bomb status	Photon paths available	Photon path	Interference?	Observed outcome	Conclusion
Dud	A&B	Both	Yes	X rings Y never rings	Ambiguous
Live	A or bomb	A	No	X rings for 50% photons Y rings for 50% of photons	Ambiguous Live bomb
Live	A or bomb	To bomb	No	Bang!	Bomb ok but ...

The essential summary is:

- If X goes off, the bomb may be a dud or a live bomb, but we cannot tell which.
- If Y goes off, the bomb is certainly live.
- If the bomb explodes it *was* live.

Of course, the process described here is not terribly efficient. On average, 50% of the times a live bomb is present the photon will detonate it. For the remaining 50% of live bombs the photon goes by route A but reflects to detector X. This outcome is indistinguishable from the outcome selected by constructive interference when a dud bomb is present, so alarm X is an ambiguous outcome that cannot be used to directly identify live bombs. However, the remaining photons travelling via A when a live bomb is present in path B, will reflect to detector Y. This outcome can *only* occur when a live Q-bomb is present, so it gives an unambiguous interaction-free measurement that identifies a live Q-bomb. It is true that we only identify 25% of the live bombs we produce, but it is remarkable that *any* can be unambiguously identified at all. (Refinements of these methods, e.g. recycling bombs that result in alarm X going off, can increase the efficiency of detection, but the principles and their implications are the same).

We can define a figure of merit Q for a Q-bomb detector. It is the ratio of number of outcomes detecting the bomb without setting it off (IFMs) to the total number of unambiguous detections (IFMs + bangs). This gives:

Q = probability of an IFM / (probability of an IFM + probability of a bang)
Q = 0.25 / 0.75 = 0.33

So the figure of merit for the tests described above is 0.33. It is possible to raise this almost to 1.0.

What use is this? In 1998 Paul Kwiat at Los Alamos demonstrated that a similar

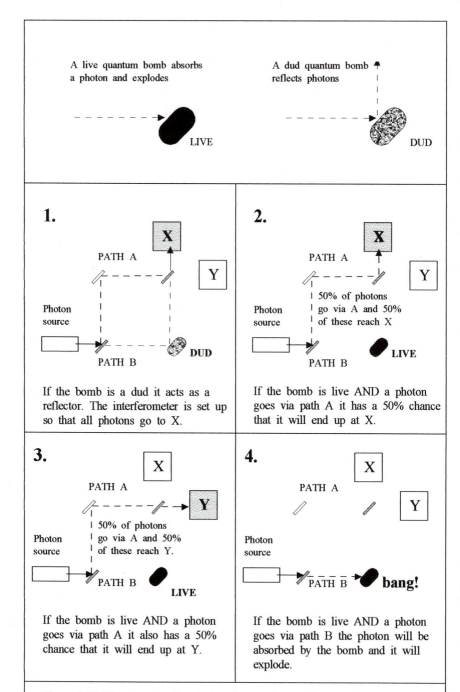

Figure 3.18 The Quantum Bomb thought experiment illustrates the counter-factual nature of quantum theory.

technique can be used to prove the presence of an object without interacting with it, and may be able to generate images without disturbing objects. One of the most outrageous claims for this type of technique has already been mentioned, Josza's quantum non-computer that can yield the result of a computation simply because it could potentially carry that computation out but actually has not been switched on. This idea is based very closely on the Elitzur Vaidmann Q-bomb experiment, but now the parallel computations that could be carried out by the quantum computer are analogous to the alternative paths that might be taken by photons in an interferometer. The beam splitters which allow these alternative photon paths are analogous to switches that can be set into a superposition of 'on' and 'off' states. The output memory of the quantum computer is analogous to the detectors where the photons end up and this memory can therefore end up containing information about computational trajectories that did not actually take place. Deutsch would explain this by saying that the universe we see, and in which we gain information is linked via interference effects to other parallel universes in which those computations were actually carried out. They are the source of our information.

3.3.5 Quantum Teleportation

In 1993 Charles Bennett of IBM showed how non-locality and entanglement could be used to teleport a quantum system from one place to another. This is an idea familiar from science fiction (and especially Star Trek). An object dematerialises at some place and rematerialises elsewhere, with every atom and molecule copied exactly and reassembled in the same relative position and state as the original. On first consideration it would seem that the uncertainty principle must rule out any such process since it is impossible to gather exact information about all the quantum states of all the particles that make up the original object. In addition to this it is not clear why or how the original gets destroyed. In fact quantum teleportation gets around both of these problems.

Imagine Alice has an object she wishes to transmit to Bob. This object can be described completely by an unknown quantum state $|\psi\rangle$. For simplicity assume this is a photon in a particular polarization state (although in principle it could be a complex object). Alice and Bob begin by sharing an entangled pair of photons. Alice then carries out a joint Bell-state measurement on the photon she wishes to teleport with a photon from the ancillary pair. This will result in Alice's photon pair going into any one of the four Bell states. However, since the ancillary photon is entangled with Bob's photon this measurement projects Bob's ancillary photon into a state that is directly related to that of the photon Alice wants to teleport. If Alice now tells Bob what the result of her Bell state measurement was, he can carry out the appropriate Bell state measurement to project *his* photon into the same state as the photon Alice originally wanted to project. Since photons are identical particles this is completely equivalent to teleporting the original photon from Alice to Bob. Anton Zeilinger's group at Innsbruk and others demonstrated quantum teleportation in the laboratory in the 1990s.

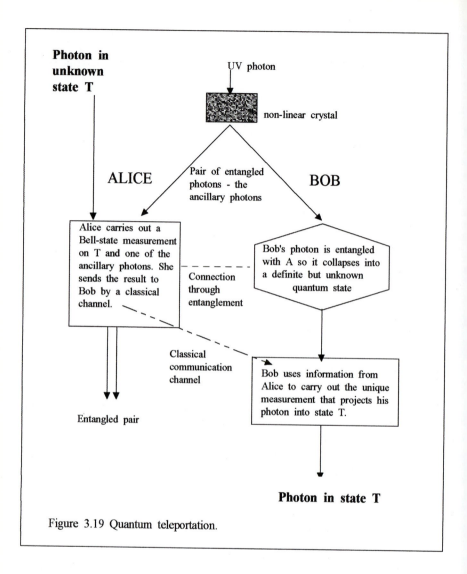

Figure 3.19 Quantum teleportation.

4
QED

4.1 TOWARD A QUANTUM FIELD THEORY

4.1.1 Introduction

" ... what I would like to tell you about today are the sequence of events, really the sequence of ideas, which occurred, and by which I finally came out the other end with an unsolved problem for which I received a prize."

(Richard Feynman, Nobel Lecture, December 11, 1965)

" ... there are two kinds of field reactions in the case of the relativistic electron and electromagnetic field. One of them ought to be called 'of mass type' and the other 'of vacuum polarization type'. The field reaction of mass type changes the apparent electronic mass from its original value ... On the other hand, the field reaction of vacuum polarization type changes the apparent electronic charge from its original value."

(Sin-Itiro Tomonoga, Nobel Lecture, May 6, 1966)

"Indeed, relativistic quantum mechanics – the union of the complementarity principle of Bohr with the relativity principle of Einstein – is quantum field theory."

(Julian Schwinger, Nobel Lecture, December 11, 1965)

The quantum theory of Bohr, Schrödinger and Heisenberg was a non-relativistic theory constructed from analogies with classical mechanics and electromagnetism. The Schrödinger equation itself described the evolution of wavefunctions in absolute time and space, concepts that had been discredited since Einstein's work on relativity in the early years of the century. Dirac was the first to launch a serious attack on this problem, and it led him, eventually, to the Dirac equation.

This gave a beautiful relativistic description of the electron, explaining its spin and predicting the existence of antimatter, as we have seen. However, we only know that electrons are there because they interact with light, and Dirac knew that a relativistic quantum field theory would be needed to describe the interactions of light and matter. He tried to construct such a theory by developing the Hamiltonian method to include interaction and field terms, but his work was inconclusive. The solution came by discarding this method and approaching the problem from a completely new direction (although, to be fair, Dirac himself had considered a related approach).

The new theory, quantum electrodynamics (or QED) was developed independently by three physicists, Julian Schwinger and Richard Feynman in America and Sin-Itiro Tomonoga in Japan. They shared the Nobel Prize for Physics in 1965 *"for their fundamental work in quantum electrodynamics, with deep-ploughing consequences for the physics of elementary particles"*. The way each of these physicists attacked the problem was different, and each method has its merits, but Feynman's approach is the simplest to describe and the most intuitive to apply so we shall base most of what we say on his description of the underlying processes. However, it should be borne in mind that, as happened with matrix mechanics and wave mechanics, Feynman's techniques are equivalent to those of Schwinger and Tomonoga and may be seen as a short cut to the mathematical formalism rather than an alternative theory.

4.1.2 Infinities

Dirac's attempts to write a Hamiltonian for the electromagnetic field ran into serious problems. Unlike the electron, which can be described by giving its position as a function of time, a field requires an infinite number of position co-ordinates. This creates a major problem – what times correspond to each position co-ordinate? According to relativity simultaneous events at separate positions in one frame of reference will not be simultaneous for another observer, so a simple theory in which the wavefunction representing the entire field evolves in a deterministic way seems impossible. For a while Dirac advocated a 'many-times' theory in which each small volume of space had its own time co-ordinate attached, but Feynamn had a different idea. He wanted to get rid of the field altogether!

The idea of doing away with electromagnetic fields had intrigued Feynman since his time as an undergraduate student at MIT. There he had come across a problem in classical physics that was to plague theorists as they tried to construct a relativistic quantum field theory for electromagnetism. The problem itself went back to the early electron theory of Lorentz and Abraham at the end of the nineteenth century. According to classical electromagnetism electromagnetic fields are set up by charges and also act back on charges. This is fine, it explains how an electron here can scatter an electron there, but it also leads to a difficult problem. If an electron sets up an electromagnetic field then it ought to act on itself. This self-interaction is a headache because it predicts an infinite self-energy if the electron is a point charge (the classical self-energy is finite if the electron is an extended body, but this leads to other clashes with relativity). The problem was complicated still further because one part of the self-interaction gave a correct expression for the 'radiation reaction', the drag on an electron when it accelerates and radiates electromagnetic waves. Feynman was also worried about the field. If it is treated as a collection of oscillators then it has an infinite number of degrees of freedom and so an infinite zero point energy. These two problems, infinite self-energy and infinite zero point energy would both disappear if the field could be removed!

4.1.3 The Absorber Theory

The electromagnetic field is not directly observable. What it does is allow us to side-step the issue of action-at-a-distance. An electron A here affects an electron B there because A sets up a field and this field changes when A moves. First the bit of field immediately adjacent to A changes and then the bit next to that and so on,

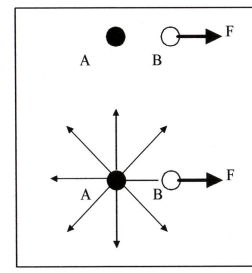

Figure 4.1 Action-at-a-distance versus field theory. In the former, charge A exerts a remote force on B with no intermediary mechanism. (B also acts back on A). In the latter, A creates a field and the field acts *locally* on B. Changes in the field take time to travel through it. (B also creates a field that acts back on A). Field theory raises the thorny problem of self-interaction: are A or B affected in any way by their own fields?

a causal influence spreading outward at the speed of light until it passes B and influences B's motion. Of course, the observable parts are the effects on A and B. Feynman's idea was to describe the interaction between A and B *directly*, without introducing the intermediary fields. To do this he needed A to have a direct (albeit delayed) effect on B, and for B to act back on A in a similar way. The clue to how this might be achieved was suggested by John Wheeler (Feynman's supervisor) and based on work done by Paul Dirac, who had already tried using both advanced and retarded solutions to Maxwell's equations to tackle the interaction of light and electrons. The retarded solutions are the familiar delayed effects described above – if A is moved then a disturbance spreads to B at the speed of light, and B moves

Figure 4.2 **Richard Phillips Feynman** 1911-1988 made his mark when he worked on the atomic bomb project under Robert Oppenheimer at Los Alamos from 1943 to 1945. After that he moved to Cornell where he attempted to formulate a quantum theory of the interaction of electrons and atoms that would be free of infinities. He succeeded and shared the 1965 Nobel Prize for Physics with Tomonaga and Schwinger for QED. He made many other major contributions to theoretical physics including work on liquid helium, nuclear forces and the quark structure of matter. In 1986 he served on a presidential commission to investigate the disaster of the Challenger space shuttle, and realised the explosion was caused by the failure of rubber O-rings on the booster rockets.   Artwork by Nick Adams

some time later. However, it was well-known that Maxwell's equations have two solutions: in addition to the retarded solutions there are time-symmetric *advanced solutions* which describe waves emitted by B prior to A's movement which then converge onto A at the moment it begins to move. Most physicists had ignored advanced solutions in much the same way that a school student might discard the negative solution to a quadratic equation as unphysical. Dirac and Feynman suspected that the advanced solutions might hold a clue to a correct theory of quantum fields.

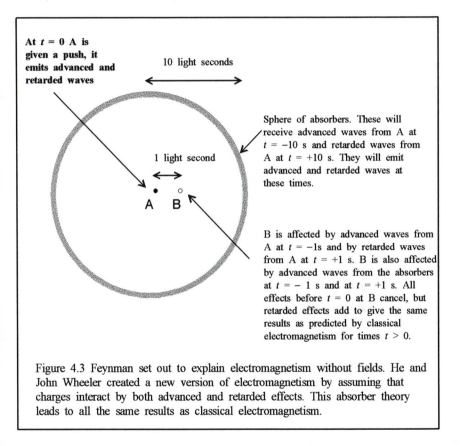

At $t = 0$ A is given a push, it emits advanced and retarded waves

10 light seconds

1 light second

A   B

Sphere of absorbers. These will receive advanced waves from A at $t = -10$ s and retarded waves from A at $t = +10$ s. They will emit advanced and retarded waves at these times.

B is affected by advanced waves from A at $t = -1$s and by retarded waves from A at $t = +1$ s. B is also affected by advanced waves from the absorbers at $t = -1$ s and at $t = +1$ s. All effects before $t = 0$ at B cancel, but retarded effects add to give the same results as predicted by classical electromagnetism for times $t > 0$.

Figure 4.3 Feynman set out to explain electromagnetism without fields. He and John Wheeler created a new version of electromagnetism by assuming that charges interact by both advanced and retarded effects. This absorber theory leads to all the same results as classical electromagnetism.

The mathematical detail of Feynman's work need not concern us here, but the central idea was this: A affects B directly, by both advanced and retarded effects, and B acts back on A in a similar way. In addition to this both A and B interact with all other charges elsewhere by the same mechanism. However neither A nor B act on themselves. Surprisingly this approach reproduced all the results of classical electromagnetism and the mathematical formulation of the theory is completely equivalent to Maxwell's equations. To see how the theory works, imagine that A is placed at the centre of a sphere of radius 10 light seconds and that B is displaced 1 light second to the right of A. The sphere itself represents the rest of the universe as a uniform distribution of charged particles which will also interact with A and B. Imagine A is given a push. It affects B twice, at $t = 1$ s and $t = -1$ s through retarded and advanced influences. It will affect the sphere of surrounding charge at $t = \pm 10$ s and these charges will act back on A at $t = 20$ s and $t = 0$ s. The surrounding charges (the *absorbers*) will also act back on B at times $t = -1$ s and $t = +1$ s (via advanced effects from either side of the sphere). Feynman showed that the combination of advanced effects from A and advanced effects from the absorbing sphere that occur at $t = -1$ s cancel (destructive interference) whilst those influences at B at $t = +1$ s add (constructive interference) to exactly the effect expected from the retarded waves from A alone. Furthermore, the advanced waves from the absorbers act back on A with just the right force to account for radiation resistance.

"So all of classical electrodynamics was contained in this very simple form. It looked good, and therefore, it was undoubtedly true, at least to the beginner. It automatically gave half-advanced and half-retarded effects and it was without fields."

(Richard Feynman, Nobel Lecture, op.cit)

Feynman derived equations for the classical 'action' for electromagnetism, allowing him to recast the theory as a principle of least action. The idea of 'least action' is related to Fermat's theorem in classical optics – the assertion that light travels by the shortest path between two points. It was applied to mechanics by Maupertuis, Lagrange, Hamilton and others and effectively says that the way a mechanical system changes with time is such that the 'action integral' has a stationary value. Feynman had found a way to write down the action integral for interacting charges without introducing electromagnetic fields (although there is an alternative interpretation that reintroduces them). This was an important step toward formulating a correct quantum field theory for electromagnetism but no-one knew how to go from these general expressions for the action to an acceptable quantum model.

4.1.4 Path Integrals

Reformulating classical electromagnetism was only the first step. Feynman's real aim was to rid classical electrodynamics of the infinities that arise from self-interaction and the infinite degrees of freedom of the field and then to convert the new theory into a well-behaved quantum electrodynamics. But now he hit a big problem. The conventional way to convert classical theories to quantum theories was to go via the Hamiltonian and convert momentum and energy to mathematical operators using a standard recipe. The trouble was, Feynman's version was not conventional. As soon as advanced and retarded effects become involved it is not clear how to write down a Hamiltonian version that evolves from moment to moment and can be represented by a simple differential equation like the Schrödinger equation.

Maths box: Hamiltonians and Lagrangians

In the eighteenth and nineteenth centuries Newtonian mechanics was generalised by Lagrange and Hamilton. Both methods provide a way to derive the results of Newtonian mechanics directly from potential functions rather than from Newton's laws. Lagrange showed that the equations of motion for of multi-particle systems could be expressed interms of derviatives of a single function, the Lagrangian:

$$L = T - V$$

where T is kinetic energy, V potential energy and L the 'Lagrangian'. The Lagrange equations are:

$$\frac{d}{dt}\left(\frac{\partial L}{\partial \dot{q}_i}\right) - \frac{\partial L}{\partial q_i} = F_i$$

where q_i is a generalised position variable and $\dot{q}$ is its associated generalised velocity. F_i is a generalised force. The idea of using generalised co-ordinates allows the principle to be applied extremely widely (and beyond the limits of simple mechanical systems). The action integral is found by integrating the Lagrangian between two times:

$$I = \int_{t_1}^{t_2} L \, dt .$$

The idea that a mechanical system evolves according to a principle of least action is called 'Hamilton's Principle' after the great nineteenth century Irish mathematician, Sir William Rowan Hamilton. Hamilton also showed that the equations of motion can be derived from a new Hamiltonian function H which is in certain cases equal to the total energy of the system:

$$H = T + V$$

Hamilton's equations have an advantage over Lagrange's in being first order:

$$\frac{dq_i}{dt} = \frac{\partial H}{\partial p_i}$$

$$\frac{dp_i}{dt} = -\frac{\partial H}{\partial x_i}$$

where the p_i are generalised momenta.

The Hamiltonian method forms a bridge between classical mechanics and Schrödinger's version of quantum mechanics. The Lagrangian method, which is more easily adapted to a relativistic treatment, is linked to Feynman's path integral approach to quantum mechanics and to quantum field theories.

In the early 1940s Feynman realised that he might be able to construct a quantum model using the classical Lagrangian approach, a method of theoretical mechanics closely related to the action integral. The Lagrangian approach had the added advantage that it can be expressed in a relativistically invariant way whereas the specialised role of time in the Hamiltonian formulation makes it essentially non-relativistic.

Once again it was the influence of Dirac that helped guide Feynman to a solution. Dirac had already tried to incorporate the Lagrangian into quantum theory and had suggested a way it could be used to calculate how the wavefunction changes from one moment to the next. Feynman developed this into the idea of path amplitude, something he incorporated into his 'sum-over-histories' approach to quantum theory.

The essential idea behind Feynman's approach is quite simple. Electrons and photons are described in the same way, using amplitudes that are calculated along the paths they may or may not follow. These amplitudes behave a bit like waves, their phase changes as the quantum object moves along its path and the amplitudes for different possible paths superpose and interfere. This interference between alternative paths (or 'histories') may lead to reinforcement larger amplitude) or cancellation. The amplitude itself is related to probability in the same way that the amplitude of a wavefunction is related to probability – through its magnitude-squared.

The essential link between the Lagrangian approach in the principle of least action and quantum theory is through the phase of the quantum amplitudes. In a short time δt the phase changes by an amount $\delta \phi$ where:

$$\delta \phi = \frac{2\pi L}{h} \delta t$$

and L is the Lagrangian.

How is this used in practice? Feynman summarised it very simply:

"So now, I present to you the three basic actions, from which all the phenomena of light and electrons arise.

- *ACTION 1: A photon goes from place to place.*
- *ACTION 2: An electron goes from place to place.*
- *ACTION 3: An electron emits or absorbs a photon.*

Each of these actions has an amplitude ... that can be calculated from certain rules."

(Richard Feynman, *QED*, Princeton University Press, 1985)

We can illustrate the technique using a few simple examples. In each case we must consider all the different ways in which a particular event may occur and then add the amplitudes for each way (taking into account their magnitudes *and* phases). The probability of any particular outcome is given by the square of the resultant amplitude for that outcome. Feynman reduced this to three general principles, which are paraphrased below.

1   **The probability that a system evolves from state A to state B is represented by the absolute square of a complex number called *a probability amplitude.*** For example, the probability that a particle leaves a source (A) and arrives at some distant point (B) can be written:

Probability amplitude to go from A to B $= \langle B | A \rangle$

Probability to go from A to B $= \left| \langle B | A \rangle \right|^2$

(using Dirac's notation)

2   **If there are two or more ways the system can change from A to B then the total probability amplitude for the process is the sum of the probability amplitudes for all the possible ways.** For example, if a particle can travel from a source (A) to a point on a screen (B) via two distinct routes then:

Probability amplitude for A to B $\langle B | A \rangle_{total} = \langle B | A \rangle_{route\,1} + \langle B | A \rangle_{route\,2}$

3   **If the transition from state A to state B goes via an intermediate state I then the probability amplitude from A to B is the product of amplitudes from A to I and I to B:**

Probability amplitude from A to B via I $= \langle B | I \rangle \langle I | A \rangle$

For example, if a photon can travel to a point on a screen via either of two small holes in a barrier then (as in Young's double slit experiment) then the probability amplitude for arriving at point B is given by:

$$\langle B | A \rangle_{via\,1\,or\,2} = \langle B | 1 \rangle \langle 1 | A \rangle + \langle B | 2 \rangle \langle 2 | A \rangle$$

Since the phase of the probability amplitude changes as the photon moves along its path the sum of these probability amplitudes could lead to constructive or destructive interference effects depending on the path difference from A to B via hole 1 or hole 2. Imagine that the path difference is quite small so that amplitudes have roughly equal magnitudes via each hole. This will lead to a total amplitude with both holes open that is double the amplitude from a single hole when they add in phase and zero when they add 180° out of phase. The probabilities, which are calculated from the absolute squares of these values, will therefore vary periodically between four times the single hole probability $(2 \times \text{amplitude})^2$ and zero as point B moves along the screen. This is exactly the result expected for the double slit experiment. Four times as many photons arrive at the maxima with both holes open as arrive at the same point when just one of the holes is open.

Figure 4.4 Feynman's path integral approach to quantum theory. The arrow on the rotating wheel above represents a probabiity amplitude for a particle to travel from A to B along the path shown. As the wheel rotates the arrow points in different directions. This change of phase will be important when it superposes with other probability amplitudes at B. To work out the chance of going from A to B we consider all possible routes and add all the arrows at B to find the resultant amplitude. The probability is related to the magnitude of this resultant probability amplitude.

However, the key to Feynman's method is in knowing how to calculate the amplitudes themselves. If we are dealing with a simple situation, such as Young's slits, where the particles travel from source to screen via two paths then the particles are effectively free – there is no potential energy term in the Lagrangian and the phase of the probability amplitude changes steadily along each path. In fact a perfect analogy to the double slit experiment is to imagine two people rolling a wheel from some starting point via two different points to meet again some time later. They have covered different distances so their wheels have turned through different angles. If there are arrows painted on each wheel then these behave like the probability amplitudes, keeping the same magnitude but changing phase continuously. When the two people meet the two arrows will usually be pointing in different directions, but a resultant amplitude can be constructed by placing them end to end and adding them like vectors. If they end up pointing in the same direction this is like constructive interference, and if they end up pointing in opposite directions it is like destructive interference.

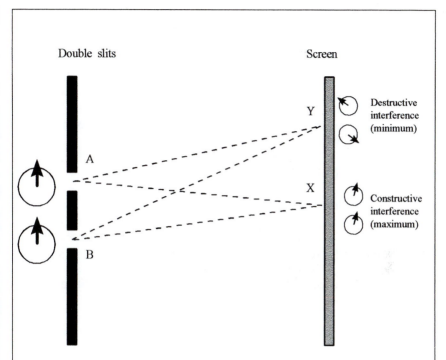

Figure 4.5 The double slit experiment. The narrow slits are treated like coherent point sources. The paths AX and BX are equal so the rotating arrows (which start in phase) rotate the same amount on their journey and add in phase to create a maximum amplitude and intensity at X. On the other hand, the path BY is longer than AY by just enough that the arrow from B completes an extra half rotation and ends up pointing in the opposite direction to the one from A. They cancel to create a minimum. A repeating pattern of maxima and minima (Young's fringes) appears on the screen.

Let us make this more explicit. Imagine Young's experiment with light. The Lagrangian is:

$$L = E = pc$$

where p is the photon momentum and c the speed of light.
The phase change during a time δt is then:

$$\delta\phi = \frac{2\pi pc\delta t}{h} = \frac{2\pi p\delta x}{h}$$

where δx is the distance the photon moves along its path in the time δt.
The probability amplitude to go from the source S to slit 1 is then given by:

$$\langle 1 | S \rangle = \frac{ke^{\frac{2\pi i p x}{h}}}{x}$$

where k is a constant, and x is the distance from the source to hole 1. The factor of $1/x$ comes in because light from the source obeys an inverse square law (and the intensity will depend on the square of the probability amplitude). The same equation can be applied to electrons or other particles by substituting the appropriate relativistic or non-relativistic momentum.

4.2 QUANTUM ELECTRODYNAMICS

4.2.1 The Lamb Shift

"Beginning with Willis Lamb's report on the first morning it was clear to all that that a new chapter in physics was upon us."

(Abraham Pais, recollecting the first Shelter Island conference on the foundations of quantum mechanics, in *Inward Bound*, p.451, OUP, 1986)

The Dirac theory predicted energy levels for electrons in atoms and the strength of interaction between electrons and external electromagnetic fields. However, in the 1930s these calculations ignored effects due to vacuum polarisation and self-energy and it was not immediately obvious whether the theory had to modified to get better agreement with experiment or if it was just a matter of adding a few extra terms to the perturbation series that were currently employed. The reasons for this uncertainty are twofold, no one was really sure how to include the extra effects and there was not any clear experimental evidence to suggest serious problems with the way things were already being done. It is true that one or two experiments had raised questions about the exact energy levels leading to the fine structure of the hydrogen spectrum and there seemed to be an anomaly with the magnetic moment of the electron. As the Second World War approached most physicists were aware of the problem, but had no clear idea what to do about it.

During the late 1930s and early 1940s many physicists became deeply involved in war work and while some progress was made in refining experimental techniques and thinking about alternative approaches to the problem, the development of quantum electrodynamics was effectively on hold. Almost immediately the war ended physicists were delighted to return to a real problem eating at the heart of their subject and by 1950 a finite and incredibly powerful version of quantum electrodynamics was in place.

The crucial historical event was the Shelter Island Conference of 1947. This was a small meeting arranged at the Ram's Head Inn on Shelter Island, Long Island, New York from 2 to 4 June 1947. The aim of the meeting was to bring leading theoreticians and experimentalists together for a brainstorming session which might lead to significant progress in some area of fundamental physics. According to Richard Feynman, one of the younger physicists present:

"There have been many conferences in the world since, but I've never felt any to be as important as this."

(Quoted in *The Second Creation*, R.P. Crease, C.P. Mann, Macmillan, 1986)

Among other participants were Robert Oppenheimer, Willis Lamb, Hans Bethe, and Julian Schwinger, all of whom had made or were about to make significant contributions to the new theory.

In April of 1947 Lamb succeeded in carrying out a very accurate experimental measurement of the fine structure in the hydrogen spectrum. He was particularly interested in the energy difference between the $2^2P_{1/2}$ and $2^2S_{1/2}$ levels which in the Dirac theory should be degenerate (i.e. have the same energy). Lamb joined forces with Robert Retherford, a graduate student at Columbia and together they carried out the experiment that would set the agenda at the Shelter Island Conference.

The idea was quite simple. They fired an electron beam at a jet of hydrogen. Each electron in the beam had just enough energy to split the hydrogen molecules into individual atoms and excite the atoms to the $2^2S_{1/2}$ state. This is a metastable state. One way to detect the presence of such a state is to pass the atoms close to a charged metal plate – if the field is strong enough an electron is ejected from the atom and strikes the plate. With many such atoms in the beam a small current can be measured. Lamb's idea was to fire microwaves at the beam of metastable atoms and promote them into the $2^2P_{3/2}$ state. If this worked then the current should suddenly fall and the microwave frequency would be a measure of the energy difference between the two levels. From this the energy of the $2^2S_{1/2}$ level could be calculated (assuming the $2^2P_{3/2}$ level is not displaced – since it is farther from the nucleus and so in a much weaker field).

They set the microwave generator to the frequency that would be needed if the Dirac theory were correct. Nothing happened, the detector current remained steady. However, when the frequency was reduced by about 10% the current suddenly dropped. This indicated that the energy difference was smaller than expected, so the $2^2S_{1/2}$ level must be at a slightly higher energy than predicted.

Where did the extra energy come from? The most likely sources had to be self-energy and vacuum polarisation, but how could these be incorporated into Dirac's theory? Hans Bethe decided that a complete relativistic treatment would be tricky and very time consuming, so he tried using a non-relativistic approach as he rode the train back from the conference. To his surprise it accounted for 95% of the required correction, but more importantly, it showed that taking a count of self-interactions could solve the problems of quantum electrodynamics. What he had done was to include the self-energies for electrons in both states and then derive the difference between them. The beauty of this approach is that the infinities traditionally associated with self-interactions cancel to leave a finite energy difference. Feynman described this as the greatest discovery leading to quantum electrodynamics. Bethe had a preliminary draft paper ready just five days after the close of the conference. Now the problem was how to get the rest of the correction from a fully relativistic approach.

A sequel to the Shelter Island Conference was held at Pocono Manor in Pennsylvania from 30 March to 2 April 1948. Here Julian Schwinger, a mathematical prodigy, and Richard Feynman presented their alternative versions of quantum electrodynamics. Neither was particularly well-received or well-understood. Schwinger spent 5 hours over a highly technical and mathematically brilliant exposition that left many of the assembled physicists reeling. It also left them with a very serious worry – the theory was very impressive, but who could use it? It seemed too difficult to apply to run-of-the-mill problems in real physics. On the other hand Feynman's presentation relied on intuitive leaps and pictorial analogies that also baffled his audience, although, once again everyone was well aware that they were seeing new and probably revolutionary ideas and techniques. At one point Bohr objected to Feynman's use of diagrams claiming that the Uncertainty Principle had destroyed the concept of well-defined trajectories and so the diagrams could not be taken seriously. Feynman was especially disappointed by the reception for his ideas; after all, he was sure he was right. Fortunately he had plenty of opportunity to discuss physics with Schwinger and when they compared notes it was clear that they were both solving the same problem in essentially the same way.

Shortly after the Poconos meeting Oppenheimer received a letter from Tomonoga in Japan. Tomonoga had learnt of Lamb's results from an article in Newsweek and realised he had been working independently but in parallel with the Americans. In fact Tomonoga had produced a similar solution before Schwinger, and in some respects his results were even better, but the isolation of Japan from the rest of world science following the war meant he had been unaware of their progress or interest.

The central idea in solving both the Lamb shift and magnetic moment problems was to include the possibility that the electron might emit and re-absorb a virtual photon. Since this emission and re-absorption could occur at any pair of points in space-time, and because the energy and lifetime of the virtual photon could take any value, this approach involved the inclusion of a very large number of possibilities. The advantage of Feynman's method was that it provided an intuitive, pictorial way to keep track of all the terms involved in the calculation.

The pictures, called Feynman diagrams, correspond to the alternative ways in which a particular process can occur. Think of an electron in free space going from one point A to another B. We have already seen how to calculate the probability amplitude for this. It includes a sum of all the amplitudes for going from A to B via a third point C that could be anywhere at all. A similar calculation can be done for electron paths in the coulomb field of a hydrogen nucleus. The solution to this problem gives the probability amplitudes for finding electrons in various orbitals, and this leads to the energy levels for the hydrogen atom and the energy differences between those levels. If the electron is allowed to emit and re-absorb a virtual photon there are a new set of probability amplitudes which must be summed to give the total amplitude for this second order process. Then the amplitudes for the first and second order processes must be summed to give the total probability amplitude. It is quite clear from this description that this is just

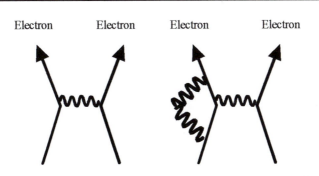

Electron   Electron   Electron   Electron

Figure 4.6 Feynman diagrams are a simple way to represent the many different ways in which a particular process can occur. The diagrams above show two ways in which electrons might scatter from one another. In the first, they exchange a single virtual photon. In the second, one electron also emits and reabsorbs a photon. This secondary process is less likely and so contributes less to the sum. In general there will be an infinite series of such diagrams including ever more exotic and less probable processes. The more of these that are included in a calculation, the more accurate the result.

the start of an infinite series of terms corresponding to progressively more complex processes – electrons emitting and re-absorbing two or more virtual photons, virtual photons converting to virtual pairs of electrons and positrons which annihilate one another to recreate the virtual photons, etc. Fortunately the terms themselves tend to get smaller by a factor of approximately 1/137 each time. This means that higher order terms contribute progressively smaller corrections to the final probability amplitude and hence to calculated energies. That is reassuring, as the number and complexity of calculations increases rapidly as higher and higher order processes are considered. The factor 1/137 is called the fine structure constant and usually denoted by the symbol α. This is a dimensionless constant relating several of the fundamental constants together:

$$\alpha = \frac{e^2}{\hbar c} = \frac{2\pi e^2}{hc} = \frac{1}{137.03604}$$

This mysterious number connects fundamental constants in relativity, quantum theory, electromagnetism and geometry and has been a curiosity which many theorists have tried (and failed) to derive from first principles. It remains as mysterious today as it was in the 1930s and 1940s; perhaps it will drop naturally out of twenty-first century physics.

Freeman Dyson proved that the apparent differences between the methods employed by Feynman, Schwinger and Tomonoga were not differences of principle, the three approaches were mathematically equivalent. However, they were not equally easy to use, and Feynman's intuitive approach quickly established itself as the most transparent and popular approach to quantum electrodynamics.

4.2.2 The Magnetic Moment of the Electron

Julian Schwinger was more concerned with the magnetic moment of the electron than with the lamb shift, but it was clear that the two anomalies were likely to be linked. Electrons interact with magnetic fields because of their spin and orbital angular momenta. By 1947 there was growing evidence that the strength of the magnetic interaction of the electron (its magnetic moment) was not predicted correctly by Dirac's theory:

*"That was much more shocking," Schwinger said. The Lamb effect, as Bethe showed, could be accounted for almost entirely without the use of relativity. "The magnetic moment of the electron, which came from Dirac's relativistic theory, was something that **no** nonrelativistic theory could describe correctly. It was a fundamentally relativistic phenomenon, and to be told (a) that the physical answer was not what Dirac's theory gave; and (b) that there was no simpleminded way of thinking about it, that was the real challenge. That's the one I jumped on."*

(Julian Schwinger, quoted by Crease and Mann, p.132, op. cit.)

Dirac's calculation of the magnetic moment of the electron depended on the exchange of a single photon between the field and the electron. This is represented on a Feynman diagram by a single photon intersecting the world line of an electron and scattering it. This first order approximation gives a value μ_{dirac} for the magnetic moment. Schwinger's calculation, presented to the Poconos meeting in 1948 included second order corrections by considering interactions in which the electron emits and reabsorbs a virtual photon and interacts with a photon from the magnet. The total amplitudes for each alternative process interfere to produce a resultant probability amplitude from which the electron's magnetic moment is calculated. This led to a slightly stronger coupling with the field giving a value for the magnetic moment that is 1.00116 times greater than Dirac's and closer to the experimental values. As experimental measurements improved still further, more complex interactions have to be included. These involve the emission and reabsorption of two virtual photons and the creation and annihilation of virtual electron positron pairs by the emitted virtual photons. The calculations themselves become frighteningly complex and take years to carry out even using high-powered computers. Experimental and theoretical values from the early 1990s are compared below:

$$\mu_e(\text{expt.}) = 1.00115965246 \pm 0.00000000020 \ \mu_{\text{dirac}}$$
$$\mu_e(\text{theory}) = 1.00115965221 \pm 0.00000000004 \ \mu_{\text{dirac}}$$

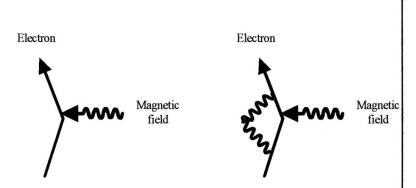

Figure 4.7 The magnetic moment of an electron depends on its interaction with virtual photons in the magnetic field. Dirac's first-order approximation included the exchange of a single virtual photon. Schwinger took this further by including the effect of the second-order process in which the electron also emits and reabsorbs a photon.

Feynman pointed out that this is equivalent to measuring the distance from Los Angeles to New York (over 3000 miles) to within the width of a human hair! QED is amazing, the most accurate theory ever, and the one that applies most generally to the world around us – think about it, if we miss out gravity and the nuclear forces, everything in our experience derives from electrons interacting with photons.

It would be easy to get carried away by the accuracy and universality of QED, but there are good reasons to retain a little caution. One is that the infinities have not been explained away, they have been cancelled, by an ingenious mathematical trick called renormalisation. QED works with the experimental values of mass and charge and loses infinite self-energies by considering only differences between infinite quantities. In this respect the ghost of classical infinities still haunts the foundations of quantum field theory – and leaves something for twenty-first century physicists to work on. One other point that should be mentioned is that the higher order terms in the perturbation series used in QED calculations allows any processes that do not violate conservation laws to contribute. This means, for example, that there is a probability amplitude for each virtual photon to create a particle anti-particle pair of any type of particles whatsoever. The problem here is that whilst QED is brilliant when applied to a world containing only electrons and photons, we do not have such an accurate and well-developed theory for interactions involving quarks and colour forces, so it is not guaranteed that the expansion will always converge. In the 1990s high energy scattering experiments have revealed these photon-to-virtual-quark-pair-to-photon effects – making photons probed at high energies appear like a composite particle!

4.3 BACKWARDS AND FORWARDS IN TIME

4.3.1 Symmetry

Feynman's programme had been to remove the infinities from classical electromagnetism and use the new theory as a foundation for quantum electrodynamics. He and Wheeler succeeded in the first step; the absorber theory was a successful reformulation of classical electromagnetism. However, the routes to QED were rather different and involved the controversial technique of remormalisation, in which infinities arising from self-interaction were set against one another to leave a finite residue. However, in grappling with the ideas of half-advanced and half-retarded solutions to Maxwell's equations, Feynman began to take the idea of interactions taking place in both time directions quite seriously. This idea was retained in QED where the sum over all possiblilities includes time-reversed effects. In particular, he realised that positrons moving forwards in time could be treated as if they were simply electrons travelling backward in time. For example, a scattering between an electron and a positron both moving forward in time is exactly the same as if the electron moving forward in time had scattered from an electron moving backwards in time (i.e. from future to past). This symmetry appealed to Feynman's intuition and could be illustrated very simply on a Feynman diagram by reversing the arrows on positrons and changing their sign.

This also suggested an alternative to Dirac's hole model of antimatter, perhaps all antimatter is just ordinary matter travelling backward in time? If this is the case we should expect all particle *reactions* involving matter to have counterparts obtained by exchanging all particles for their antiparticles. This operation is called 'charge conjugation' and is denoted by the letter C. In other words, if we and our surroundings were made entirely from antimatter then all our experiments would yield exactly the same results as they do in our matter world. Surprisingly this turned out *not* to be the case. Experiments in the 1950s involving the decay of positive kaons and the emission of electrons from cobalt-60 nuclei showed that in some cases antimatter reactions give slightly different results to matter reactions – the symmetry is not perfect. In fact it turned out that charge conjugation symmetry is one of a set of three fundamental symmetries involved in particle interactions:

C   charge conjugation (swap matter and antimatter);
P   parity (reflect all co-ordinates in the origin);
T   time reversal.

Feynman and Murray Gell-Mann proposed that although C and P symmetries may

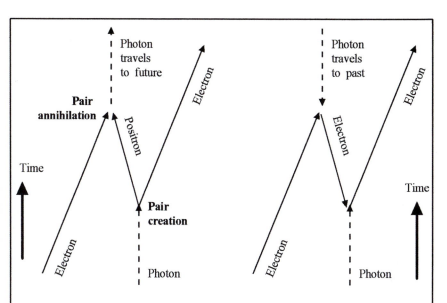

Figure 4.8 Time reversal. These two diagrams give alternative interpretations of the same events. On the left a gamma-ray photon decays to form an electron positron pair. The positron later annihilates with another electron to create another gamma-ray photon. On the right an electron interacts with a gamma-ray photon coming from the future which scatters it back into the past. It then scatters from another gamma-ray photon and ends up moving toward the future. In this interpretation what we see as a positron is simply an electron moving backward in time.

be separately violated the combined CP symmetry (i.e. swap matter for antimatter *and* reflect in the origin) might lead to viable reactions in all cases. This too turned out to be incorrect – in 1964 James Christenson, James Cronin, Val Fitch and Rene Turlay at Brookhaven discovered that neutral kaon decays violate CP symmetry. However, it is thought that the combination of all three symmetry operations, CPT (matter for antimatter plus reflection in the origin plus time reversal) *always* leads to viable reactions. So far no violation of CPT symmetry has ever been observed.

Meanwhile experiments continue to investigate CP violation, which has only ever been seen in decays of neutral kaons. The outcome of such experiments is very important because symmetry principles are built into mathematical theories at a very fundamental level and are closely related to conservation laws.

4.3.2 The Transactional Interpretation of Quantum Mechanics

Time reversal lives on in another recent idea. John Cramer at the University of Washington published a new interpretation of quantum theory in the 1980s. He

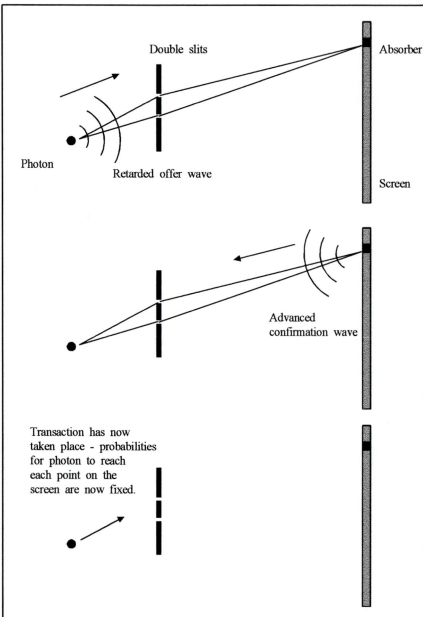

Figure 4.9 Cramer's 'Transactional Interpretation'. This gets round the traditional paradoxes by introducing a modified version of the Wheeler-Feynman absorber theory. A retarded offer wave is sent out. Absorbers respond with an advanced confirmation wave. This arrives at the source at the instant it emits its 'offer'. The transaction determines all probabilities for future paths. This explains how photons passing through one slit seem to 'know' whether the other slit is open.

effectively adapted the Feynman-Wheeler absorber theory and applied it to the Schrödinger equation. Once you get used to the idea that time is reversible the mechanism is quite simple. Imagine an experiment in which a photon can go via either of two holes to a screen. The mystery in conventional quantum theory is how the probability of reaching the screen depends on the hole through which the photon does not pass. In Cramer's interpretation (which leads to all the same results as conventional theory) a retarded wave (the 'offer wave') travels via both holes to the 'absorbers' on the screen. The absorbers send out advanced waves ('confirmation waves') which travel back through the holes and back in time to arrive at the source at the instant the photon is emitted. These advanced waves 'inform' the photon that the barrier it is about to approach has two holes in it so the mystery disappears. Interference of advanced and retarded waves along all possible paths results in an interference pattern leading to the same probability distribution on the screen as in any other version of quantum theory. The essential difference here is that the advanced and retarded interactions allow *the entire experimental configuration* to affect the photon at the instant it leaves the source, and the problems of accounting for effects due to paths it did not take go away. In one sense Cramer's transactional interpretation has something in common with the Copenhagen interpretation and complementarity – it implies that mutually exclusive experimental arrangements will lead to complementary sets of observations. The difference is that Cramer's version explains why this is the case. Where Copenhagen forces the *post hoc* collapse of the wavefunction, Cramer's interpretation collapses it from the instant the photon is emitted. This gets around the kind of logical problems that arise in thought experiments like Schrödinger's cat and Wheeler's delayed choice.

Part 2
EXPLAINING MATTER

5 Atoms and Nuclei
6 The Standard Model
7 Particle Detectors
8 Particle Accelerators
9 Toward a Theory of Everything

5 *Research into cathode rays shows that atoms are not indivisible –
all atoms contain electrons.
Alpha-particle scattering reveals the nucleus.
Rutherford proposes the nuclear 'planetary model' of the atom.
Chadwick discovers the neutron.
Rutherford explains nuclear transformations and radioactivity.
Yukawa proposes a model for the strong nuclear force binding nucleons.
The nucleus is modelled as a liquid drop and as orbiting nucleons.*

6 *Research into cosmic rays, radioactivity and particle collsions reveal
many new particles that fit into distinct families:*

*leptons – electron-like particles and neutrinos;
hadrons – baryons and mesons – affected by the strong nuclear force;
hadron patterns suggest a deeper layer – the quarks.*

Gauge bosons mediate particle interactions.

7
8 *Research in high-energy physics uses two essential tools:*

Accelerators – from the Van der Graaf to the LHC.

Detectors – from the electroscope to ALEPH.

9 *Unification as a grand aim of physics.
Electromagnetism as the first great unification.
Electroweak unification.
Symmetry to supersymmetry.
QED to QCD and the problem of quantum gravity.
String theory.*

5

ATOMS AND NUCLEI

5.1 THE ATOM

5.1.1 Why Atoms?

The idea that all matter is made from a combination of a small number of simple things goes back to the beginning of recorded history. It is an attempt to unify a diverse range of experiences in a universal theory. The ancient Greeks proposed several unifying ideas, one of which was a fairly well-developed atomic theory, attributed to the pre-Socratic philosophers Leucippus and Democritus and described by the poet Lucretius (c. 94-55 BC) in his poem 'On the Nature of Things':

"No rest is granted to the atoms throughout the profound void, but rather driven by incessant and varied motions, some after being pressed together then leap back with wide intervals, some again after the blow are tossed about within a narrow compass. And those being held in combination more closely condensed collide with and leap back through tiny intervals ... "

(Quoted from *Greek Science in Antiquity*, Marshall Claggett, Collier, 1955)

This visionary theory pre-empts nineteenth century kinetic theory and endows atoms with properties of mass, size and motion. It also embeds them in a 'void' an empty space which acts as a reference frame, allowing them to move and recombine in new arrangements. The name atom derives from *atomos*, meaning 'uncuttable' a reflection of the Greek belief that all matter consists of an arrangement of a few kinds of fundamental, indivisible particles.

  In the nineteenth century the atomic hypothesis developed into a powerful explanatory system, particularly with the work of Maxwell, Boltzmann and Clausius. They showed that much of macroscopic thermodynamics, and in particular gas theory, could be derived from the assumption that matter on the smallest scale consists of tiny hard spheres flying about in the void. Whilst the arguments used to derive these results were often subtle and involved, the underlying theory is little different from that of Democritus and Leucippus. However, discoveries in spectroscopy, chemistry and electrochemistry, and the development of electromagnetic theory, revealed an unexpected complexity associated with the elements and raised questions about the nature and possible structure of atoms.

Some physicists took the atomic theory very seriously and assumed that atoms themselves are real, as William Thomson (Lord Kelvin) put it, an atoms is:

"a piece of matter with shape, motion and laws of actions, intelligible subjects for scientific investigation."

(Inaugural lecture to the British Association, 1871, quoted from *Inward Bound*, Abraham Pais, OUP, 1986)

It is worth pointing out that some influential scientists and philosophers, Ernst Mach for example, maintained that the atom itself was no more than a mathematical construct, useful for explaining things but not corresponding to any objective aspect of physical reality. Their opposition to atomic theory faded with the advent of the twentieth century, but was an important foil for the atomists. It acted as an incentive to those who believed in atoms to provide some kind of empirical evidence for their existence. It may even have contributed to the frustration and depression that led to Boltzmann's suicide in 1906, a year after Einstein published a paper on Brownian motion that effectively provided a mathematical link between experimental results and atomic theory.

5.1.2 Spectroscopy

It would be hard to overemphasise the importance of spectroscopy to atomic theory (or, for that matter, to astronomy and cosmology). The beauty of a spectrum is that it carries precise numerical information in the wavelengths and intensities of the various spectral lines. This information cried out for an explanation in terms of the emitting bodies and led to the idea that the characteristic structure of spectra should be related at a fundamental level to the structure of the emitting atoms. This idea had even occurred to Isaac Newton two hundred years earlier:

"Do not all fixed bodies, when heated beyond a certain degree, emit light and shine; and is not this emission performed by the vibrating motions of their parts?"

(Isaac Newton, *Opticks*, Query 8)

It is echoed in 1852 by Sir George Stokes, who used 'molecules' where we might substitute the word 'atom' (perhaps because the idea of atomic structure runs counter to the idea of indivisibility):

"In all probability .... the molecular vibrations by which .... light is produced are not vibrations in which the molecules move among one another, but vibrations among the constituent parts of the molecules themselves, performed by virtue of the internal forces which hold the parts of the molecules together."

(Quoted from *Inward Bound*, Abraham Pais, p.175, OUP, 1986)

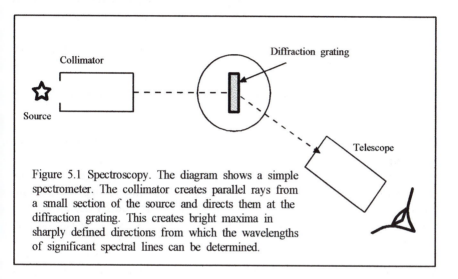

Figure 5.1 Spectroscopy. The diagram shows a simple spectrometer. The collimator creates parallel rays from a small section of the source and directs them at the diffraction grating. This creates bright maxima in sharply defined directions from which the wavelengths of significant spectral lines can be determined.

The hydrogen spectrum was first measured by Anders Jonas Ångstrom in 1853, but the beginnings of systemmatic spectral analysis derive from the work of Gustav Robert Kirchoff (1824–1887) and Robert Bunsen (1811–1899) in the 1860s. They used the Bunsen burner, which had the great advantage over previous heaters that it provided a non-luminous flame and so allowed them to see the spectra of heated objects much more clearly. They concluded that:

- Spectra can be measured with incredible accuracy and the spectrum of an element labels that element in unique way.
- Spectral analysis can be used to identify undiscovered elements.
- The spectra of light from stars may reveal the elements and compounds present in their atmospheres.

Maxwell's equations of electromagnetism were published in 1864 and included the idea that electromagnetic waves are emitted by accelerated charges. This suggested that the atom itself might contain charges and that their motion inside the atom could account for the spectra. However, it also brought problems. Any internal degrees of freedom should, according to classical thermodynamics, contribute to the specific heat of the element. That is, if the material is heated an internal degree of freedom is yet another mode of vibration that can absorb some energy, so the amount of energy required to increase the temperature by one degree is greater. The number of internal degrees of freedom needed to explain the complex structures of atomic spectra would make theoretical specific heats far too large. Maxwell himself saw this as a major problem for the kinetic theory of gases:

"The spectroscope tells us that some molecules can execute a great many different kinds of vibrations. They must therefore be systems of a very considerable degree of complexity, having far more than six variables [the number characteristic of a rigid body] ... every additional variable increases the specific heat ... every additional degree of complexity which we attribute to the molecule can only increase the difficulty of reconciling the observed with the calculated value of the specific heat. I have now put before you what I consider the greatest difficulty yet encountered by the molecular theory."

(James Clerk Maxwell, 1875, quoted in *Inward Bound*, p.175)

Faraday's experiments on electrolysis showed that matter and electricity are intimately connected and led to the first estimate, in 1874, of the fundamental charge, e, a *unit* called the 'electron' by Stoney in 1891. Faraday realised that an electrical atom could be used to explain chemical bonding as well as atomic structure:

"Although we know nothing of what an atom is, yet we cannot resist forming some idea of a small particle, which represents it to the mind ... there is an immensity of facts which justify us in believing that the atoms of matter are in some way endowed with electrical powers, to which they owe their most striking qualities, and amongst them their chemical affinity."

(Michael Faraday, *Experimental Researches into Electricity*, London, 1839, quoted in *Inward Bound*, p.80)

By the end of the nineteenth century the situation was ready for a major breakthrough. The atomic theory had achieved some impressive successes, particularly in explaining the behaviour of gases and chemical combinations, but was still treated with suspicion by many scientists. Spectroscopy had provided a large amount of detailed numerical information that hinted at an ordered internal atomic structure. Both spectroscopy and electrolysis seemed to require that atoms contain charges with various degrees of freedom and that these charges are discrete multiples of an electronic unit. However, there was no clear model of the atom and although the electron had been named, the name did not refer to a particle, only a smallest possible unit of charge. Various models were proposed for the atom including William Thomson's vortex rings in the ether and James Thomson's 'plum-pudding' model in which negative charges were embedded in a diffuse positive material. However, two key discoveries led to the more familiar

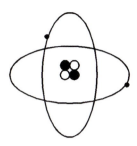

1. Thomson's plum-pudding model. Negative charges embedded in a positive medium.

2. Rutherford's nuclear model. Electrons orbit the nucleus like planets orbit the Sun.

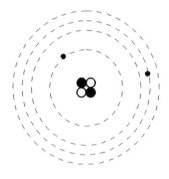

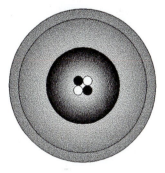

3. Bohr's quantum atom. Electron orbits are quantised. Only discrete energies and radii are allowed.

4. Schrodinger's wave-mechanical atom. Electron waves form stationary patterns around the nucleus.

Figure 5.2 The evolution of atomic theory.

planetary model – J.J. Thomson's discovery of the electron in 1897 and Geiger, Marsden and Rutherford's discovery of the atomic nucleus in 1909.

From 1895 to 1925 the physicist's model of the atom evolved through:

• Rutherford's nuclear atom;
• Bohr's quantised atom;
• Schrödinger's wave-mechanical atom.

The development of quantum theory, and the explanation of atomic spectra is dealt with in Chapter 2 so here we shall concentrate on the discoveries that revealed the structure and content of the atom and nucleus and the beginnings of particle physics.

5.1.3 The Electron

Figure 5.3 **Joseph John Thomson** 1856-1940 won the 1906 Nobel Prize for Physics for the discovery of the electron. His later research with ion beams was developed by Aston into the mass spectrometer.

Artwork by Nick Adams

"For Thomson, his cathode ray work was one step along the way to establishing a coherent theory of gaseous discharge. It was for this, rather than the cathode ray work, that he won the Nobel Prize in 1906. Neither cathode rays, corpuscles nor electrons were mentioned in the citation. Indeed, for many at the time Thomson was not the clear-cut 'discoverer of the electron'. Alternative accounts, in which Thomson was only of minor importance, viewed developments by Lorentz, Larmor, Zeeman or Wiechert as the significant steps which established the existence of electrons.

Viewed with hindsight, though, it was Thomson who made the nineteenth century electron 'real'. Arriving at the theoretical idea of an electron was not much of a problem in 1897. But Thomson pinpointed an experimental phenomenon in which the electron could be identified, manipulated and experimented upon. He did this most clearly in the e/m experiment, showing how electrons could be deflected magnetically and electrically, how measurements could be made upon them and how to attach meaning to those measurements. Through the e/m experiment electron theory changed from an abstract mathematical hypothesis to an empirical reality, expanding its meaning in the process."

(Isobel Falconer, J.J. Thomson and the Discovery of the Electron, in Physics Education, IOP, July 1997)

The electron was the first subatomic particle to be discovered. It was discovered in a particle accelerator (Crooke's tube) and the experimental work that confirmed its existence went a long way to persuade scientists to take the invisible microscopic world seriously as an objective reality containing particles with distinct properties. This naive realism was essential for the developments that followed (the nucleus, protons, neutrons etc.) but ran into deep trouble when the full implications of quantum theory began to be appreciated in the 1920s. Thomson was not the first to speculate that electrons were particles, and that electricity is effectively quantised, Hermann von Helmholtz states as much quite clearly in his Faraday lecture of 1881:

"If we accept the hypothesis that the elementary substances are composed of atoms, we cannot avoid concluding that electricity also, positive as well as negative, is divided into definite elementary portions which behave like atoms of electricity."

(Quoted in Physics Education , July, 1997, p.221)

Thomson did not use the term 'electron', he called the negative particles in cathode rays 'corpuscles'. His achievement was to convince everyone that the corpuscles were real, that the electron is a particle and can be pushed around with electric and magnetic fields and can take its place inside the atom as a part of the atomic structure. This provided an understanding of ionisation and chemical bonding and, whilst Thomson's own model (the 'plum-pudding' model) turned out to be mistaken, electrons took their place as satellites of the positive nucleus in Rutherford's model.

Thomson's measurement of the specific charge (charge to mass ratio, e/m) for the electron gave a value very much larger than the specific charge on a hydrogen ion, calculated from electrolysis experiments. There were several reasons for thinking that the charge carried by an electron would not be significantly different from that on an ion, so Thomson took this as evidence that electrons have a much smaller mass and size than hydrogen atoms (as seemed to be borne out by their ability to penetrate thin metal foils). A modern value for the specific charge is $e/m = -1.758\,820 \times 10^{11}$ Ckg^{-1} (about 1800 times greater than q/m for a hydrogen ion).

The method used to determine e/m involves several of the most important techniques found in all modern particle accelerators. Thomson's series of experiments can be seen, together with the Geiger and Marsden alpha scattering experiments, as the beginning of experimental particle physics – the 'big-science' of high energy physics that has dominated so much of twentieth century research:

- The action takes place in a high vacuum inside a Crooke's tube. It has to be a vacuum otherwise the particles would collide with gas molecules and scatter. The vacuums needed in modern particle accelerators are much lower than those in Thomson's experiments because the beam energies used in modern experiments are much greater and the paths taken by the accelerated particles are much longer.
- The charged particles are accelerated between electrodes by an electric field.

In a modern synchrotron this is likely to be achieved at several accelerating points around the circular beam tube. Electrodes are driven by radio frequency A.C. voltages that are tuned to the rotation of the particles so that they have the correct polarity to accelerate them as they pass through.

- The beams were deflected using electric and magnetic fields. The amount of deflection is related to the momentum and charge of the particles. Even the large layered detectors at facilities like CERN still use this method.
- The deflected beam was revealed by fluorescence – electrons transfer their kinetic energy to atoms in a target material and the atoms emit light. Many related processes are used in modern detectors to measure particle energy, and the same technique is used in television sets to produce a picture.

Box: Measuring *e/m* (see Fig 5.3)

This method is similar to J.J. Thomson's second method for *e/m* used in July 1897. An electron beam is passed through region in which perpendicular magnetic and electric fields are orientated so the forces they exert on the beam produce similar deflections (they could be set up so that the forces they exert on the electrons oppose one another and the beam is undeflected). This means they are exerting equal forces on the electrons so:

$$\text{Electric force} = Ee$$

$$\text{Magnetic force} = Bev$$

$$Ee = Bev$$

$$E = Bv \tag{1}$$

E and B can be calculated from measurements made during the experiment and v is related to the kinetic energy of the electrons, which is in turn linked to the accelerating voltage V:

Loss of electrical P.E. as electron passes from cathode to anode : $\Delta PE = -eV$

Gain in K.E. between cathode and anode : $\Delta KE = \dfrac{1}{2}mv^2$

$$\frac{1}{2}mv^2 = eV \tag{2}$$

Combining (1) and (2) to eliminate v gives :

$$\frac{e}{m} = \frac{E^2}{2B^2V}$$

In 1899 Thomson showed that the negative particles emitted in the photoelectric effect have the same ratio *e/m* as electrons. Several experimenters, in particular Walter Kaufmann, showed that (at least for small velocities) *e/m* for beta rays was the same as for electrons.

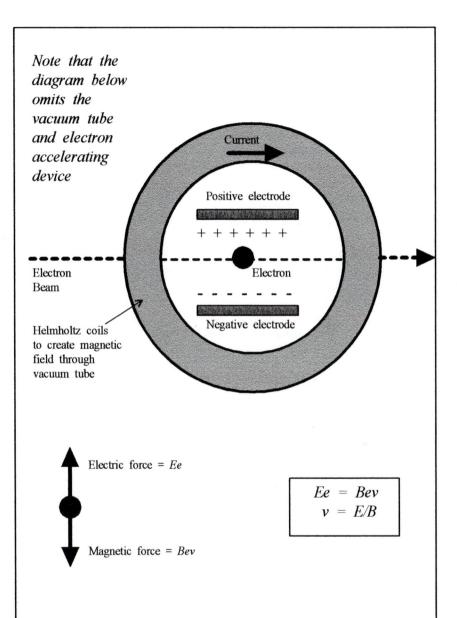

Note that the diagram below omits the vacuum tube and electron accelerating device

Current

Positive electrode

+ + + + + +

Electron Beam

Electron

Helmholtz coils to create magnetic field through vacuum tube

- - - - - - -

Negative electrode

Electric force = *Ee*

Magnetic force = *Bev*

$$Ee = Bev$$
$$v = E/B$$

Figure 5.4 Measuring *e/m*. The electrons will be undeflected when the electric and magnetic fields exert equal and opposite forces on them. For electrons travelling at a particular speed there is only one ratio of field strengths for whcih this is the case.

5.2 THE NUCLEUS

5.2.1 The 'Rutherford Scattering Experiment'

"It was quite the most incredible event that has ever happened to me in my life. It was almost as incredible as if you fired a 15-inch shell at a piece of tissue paper and it came back and hit you."

(Rutherford, quoted in *Inward Bound*, Abraham Pais, OUP, 1986)

Henri Becquerel discovered radioactivity in 1896 and in 1898 Rutherford distinguished strongly ionising alpha rays from the more penetrating but less strongly ionising beta rays. In 1909 (the year following Rutherford's Nobel Prize for Chemistry!) Rutherford suggested that two of his research students, Geiger and Marsden, should investigate what happens when a thin beam of alpha particles is fired at a thin gold foil. To their great surprise about 1 in 8000 of the incident alpha particles bounced off the foil.

Why was this such a startling result? By this time alpha particles were known to carry a positive charge twice as large as the negative charge on an electron, and to have a mass about 8000 times as large as the electron's. They were also known to travel at about 5% of the speed of light and were able to penetrate thin metal foils. Putting all this together suggested that the alpha particles were tiny (compared to atoms) highly charged particles with a great deal of energy and momentum. It was inconceivable that a diffuse mass of positive charge with a few low mass electrons embedded in it could do more than cause slight deflections of the incoming alpha particles. But they did, and it took Rutherford two years to explain why.

In 1911 Rutherford proposed his nuclear atom:

- Most of the mass (>99.9%) and all the positive charge of the atom is concentrated in a tiny central region (less than 1/10 000 of the diameter of the atom). This is the atomic nucleus.
- Electrons orbit the nucleus at relatively great distances and the total negative charge of the electrons exactly balances the positive charge on the nucleus.

This was not just qualitative speculation. Rutherford calculated how incoming alpha particles scatter from the nucleus and explained the scattering pattern measured by Geiger and Marsden. In this respect he was lucky. The alpha particles had energies of about 5.0 MeV, large enough for scattering from orbiting electrons to be ignored but not so large that the assumption of coulomb scattering breaks down. Higher energy alpha particles make a closer approach to target nuclei and come under the influence of the strong nuclear force (unsuspected in 1911) as would 5.0 MeV alpha particles if directed at less massive targets. That said, Rutherford's scattering formula worked very well, and it was soon used to work back from the scattering pattern to properties (like charge) of the target

Figure 5.5 **Ernest Rutherford** 1871-1937 was a visionary experimental physicist whose work in atomic and nuclear physics laid the foundation for the investigation of matter throughout the twentieth century. His early work on radioactivity clarified the nature of the emissions and in 1910 he directed Geiger and Marsden in the alpha scattering experiments which identified the atomic nucleus. With Soddy he explained the mechanisms of nuclear transformation and in 1919 he split the atom producing artificial radioactivity for the first time. Surprisingly he did not win the Nobel Prize for Physics, but he did win it for Chemistry, in 1908.   Artwork by Nick Adams

nucleus. This helped confirm the hierarchy of atomic structures underlying the periodic table and the patterns of nuclear transformation in radioactive decay.

5.2.2 Nucleons

Although the nucleus was identified in 1911 the proton-neutron model was not generally accepted until the 1930s. It was clear very early on that nuclei cannot be simple combinations of hydrogen nuclei (given the name 'protons' by Rutherford as late as 1920) since the atomic mass increases more rapidly than the nuclear charge:

Element	Nuclear charge /e	Atomic number Z	Atomic mass/g	Atomic mass number A
Hydrogen	1	1	1.008	1
Helium-4	2	2	4.003	4
Iron-56	26	26	55.847	56
Uranium-238	92	92	238.029	238

Rutherford assumed that helium nuclei (alpha-particles) must consist of four hydrogen nuclei (protons) and two electrons. This would account for both the mass and charge of the nucleus in terms of particles well known to physics. Furthermore, beta particles are ejected from nuclei, and they are electrons, so it seemed obvious at that time that electrons must be nuclear constituents. Gradually, however, it was realised that the proton-electron model could not be correct:

- The classical radius of the electron, found by equating its electrostatic self-energy to its mass-energy, gave it a size comparable to that of a nucleus, so it seemed hard to imagine many electrons and protons crammed together in such a small space if they were to be bound by electrostatic forces.
- The uncertainty principle states that the uncertainty in particle momentum increases as the uncertainty in its position gets smaller. If electrons were contained in the nucleus the uncertainty in their position would be very small indeed. This would imply a wildly uncertain momentum with its associated kinetic energy. In fact the energy would be much greater than the electrostatic binding energy so any electron venturing into the nucleus would soon be ejected. The uncertainty principle effectively sets a minimum size for atoms.
- Electron spin was also problematic. The proton-electron model involved a large number of spinning particles, so the nucleus itself should have a magnetic moment (it does) which would affect orbiting electron energy levels and increase the size of the hyperfine splitting of spectral lines above that observed experimentally.
- Another spin-related problem concerned the conservation laws. In the proton-electron model a cadmium nucleus contains an even number of spin-1/2

particles, but one isotope has a measured nuclear spin of 1/2, this is impossible to achieve with an even number of spin-1/2 particles.

- We have already mentioned that fermions have different statistics from bosons. A nucleus containing an odd number of spin-1/2 particles should behave like a fermion and one containing an even number should behave like a boson. By 1929 two nuclei, nitrogen-14 and lithium-6 were known to have the 'wrong' statistics. Both behave like bosons whereas the proton-electron model would make them fermions (14 + 7 particles in the nitrogen and 6 + 3 particles in the lithium).

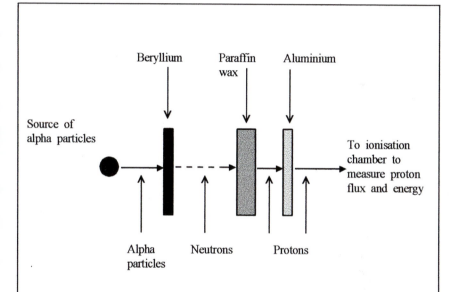

Figure 5.6 Discovering the neutron. When alpha particles hit beryllium they knock neutrons out of the metal. These are difficult to detect but have almost the same mass as protons, which they knock out of paraffin wax. Aluminium absorbers were used by Chadwick to measure the energy spectrum of the emitted protons and infer the existence of the neutrons.

The idea that nuclei might contain a neutral particle was considered by Rutherford (who else?) in the early 1920s. Rutherford's neutron was a tightly bound system consisting of one proton and one electron (rather like his model of an alpha particle with four protons and two electrons) and the term 'neutron' was introduced in the late 1920s. Rutherford suggested many ways in which the neutron might be formed or detected including the idea that hydrogen atoms might collapse to neutrons as electrons fall into the nucleus, but the crucial observations that led to the discovery of the neutron were made by Walther Bothe and Herbert Becker in 1930 and the Joliot-Curies in 1932:

- Bothe and Becker noted that beryllium emits a very penetrating nuclear radiation when alpha particles are fired at it.
- The Joliot-Curies discovered that this radiation could eject protons from paraffin (a hydrocarbon with many hydrogen atoms bound to carbon atoms).

Neither of these groups correctly interpreted their observations, but Chadwick was working with Rutherford who had likened finding the neutron to finding an invisible man in Piccadilly Circus. The invisible man would give himself away by recoil; you could tell where he was by the trail of people recoiling from him. The neutron too would give itself away by recoil. Chadwick explained the observations like this:

- When alpha particles hit beryllium they convert it to carbon-12 and emit a neutron:

$$_2^4He + {}_4^9Be \rightarrow {}_6^{12}C + {}_0^1n$$

- The neutrons collide with protons in paraffin and, having approximately equal mass, eject them.

He supported his explanation by showing that the recoil velocities of the protons were consistent with the conservation laws if they had been struck by a particle of similar mass, and that the idea of gamma rays being emitted (an alternative explanation) could not explain the results without violating these laws. It is worth pointing out that the neutron was the last subatomic particle to be discovered in a small-scale bench-top experiment (Chadwick's original apparatus was just 15 cm long and 4 cm wide).

Once the neutron was confirmed as a nuclear particle the proton-neutron model became established and the proton-electron model faded away. Protons and neutrons are called 'nucleons'.

The nucleus of an atom is described by two numbers:

- Z = proton number (atomic number = position in Periodic Table);
- A = nucleon number (atomic mass number - not to be confused with atomic mass).

A particular species of nucleus (nuclide) is labelled: $_Z^AX$, e.g. $_6^{12}C$. A third number, the neutron number N is also sometimes used: $N = A - Z$.

Isotopes of a particular element have the same proton number but different numbers of neutrons (same Z but different A). The value of Z also determines the number of orbital electrons and so the electronic configuration and chemical properties. This is why isotopes of the same element have the same chemistry and can be very difficult to separate; this was the main problem facing the allied physicists developing the atom bomb during WWII, separating U-235 from U-238.

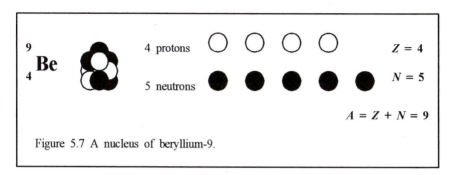

Figure 5.7 A nucleus of beryllium-9.

Of course, discovering the neutron leaves a number of questions unanswered. To solve the problems with spin, the neutron must be a spin-1/2 particle, not a simple bound state of an electron and proton as Rutherford had thought. On the other hand, beta decay (see Fig. 5.6 below) requires that, in some circumstances, the neutron can decay to a proton and an electron, a process that apparently violates the law of conservation of angular momentum (since a spin-half particle apparently turns into two spin-half particles). The early model of the proton-neutron nucleus was given a theoretical treatment by Heisenberg, the decay of the nucleus via a new weak nuclear force was investigated by Fermi, and the apparent violation of conservation laws was solved by Pauli when he proposed the existence of a new particle (now called an anti-neutrino).

5.2.3 Radioactivity and Nuclear Transformation

"We may then conclude, from these experiments that the phosphorescent substance in question emits radiations which penetrate paper opaque to light and reduces the salts of silver."

(Henri Becquerel, On the Radiation Emitted in Phosphorescence, Comptes rendus de l'Académie des Sciences, Paris, February, 1896.)

Henri Becquerel (1852-1908), like many physicists in early 1896, was intrigued by Röntgen's discovery of X-rays. He wondered whether fluorescent and phosphorescent salts, i.e. those that emit light after absorbing it, would also emit X-rays, and began a series of experiments to find out. His experiments showed that the radiation persisted whether or not he excited the salt in any way, and was not diminished when the salt was melted or dissolved (conditions known to stop phosphorescence). He also showed that the radiation produced ionisation in the air that could discharge an electroscope. The phosphorescent materials he used were uranium salts and he concluded that it was not the phosphorescence that was essential to the emission of radiation, but the uranium.

In 1898 Rutherford studied the penetrating and ionising power of the radiation emitted by uranium and discovered that it consisted of two parts, highly ionising rays with little penetration, which he called 'alpha rays' and less strongly ionising but more penetrating rays, which he called 'beta rays'. Gamma rays were discovered in 1900 by Paul Villard (1860-1934) as particularly penetrating radiations emitted by radium He also showed that they were not deflected by electric or magnetic fields.

Figure 5.8 **Marie Sklodowska Curie** 1867-1934 discovered radium and polonium and showed that radioactive decay must be an atomic process. Marie and Pierre Curie shared the 1903 Nobel Prize for Physics with Henri Becquerel.
Artwork by Nick Adams

- Kaufmann identified beta rays with electrons in 1902 by deflecting them in a magnetic field and measuring their specific charge (e/m).
- Rutherford and Royds identified alpha particles as helium nuclei in 1908. They stopped alpha particles in a vacuum tube, and collecting enough of the gas produced to show it was helium (the alpha particles simply gain two electrons when they slow down). They did this by analysing the spectrum emitted when a spark was discharge through the gas.
- Rutherford and Andrade confirmed that gamma rays and X-rays were similar kinds of radiation in 1912. They observed the reflections of gamma rays from crystal surfaces and compared these with the expected reflections for X-rays.

Radiation	Nature	Ionising power	Stopped by
Alpha	Helium nucleus ^4_2He or $^4_2\alpha$	Great	Card / skin Few cm of air
Beta	Electron $^0_{-1}e$ or $^0_{-1}\beta$	Moderate	Few mm of Al Few metres of air
Gamma	High frequency photon $^0_0\gamma$	Low	Few cm of lead

Two other important developments took place in the same period:

- The Curies showed that thorium emits radiation in a similar way to uranium (but is more active) and isolated the new elements polonium and radium. It was their work which really showed that radioactivity is an atomic process. They also introduced the term 'radioactivity'.
- Rutherford took up work begun by R.B. Owens (1870-1940) who had noticed that the activity of thorium salts varied unpredictably in the presence of air currents. Rutherford showed that this was because it releases a new radioactive element, an alpha-emitting gas, radon (then called radium-emanation) into the air. Furthermore, if the radon is captured and monitored, its activity falls in a regular way, halving every 55.5 s.

The Curie's work, which laid the foundations for radiochemistry, earned them a share in the 1903 Nobel Prize for Physics. Rutherford, meanwhile, continued to explore the properties of thorium compounds and, with Soddy, began to formulate a theory about the processes of radioactive transformation. They published this in 1902 and developed the details during the following five years. At that time the dominant atomic model was Thomson's 'plum pudding'. However we have already discussed Rutherford's later nuclear model of the atom, so it makes sense to lose the thread of history at this point and discuss the mechanism of radioactive decay as a process of nuclear transformation.

Radioactive decay is a nuclear process. It involves the decay of nuclei from one element and their transformation into nuclei of another element. Spontaneous decay occurs when the transformation results in a release of energy. This can be calculated using Einstein's mass-energy equation:

$$E = mc^2$$

If the total mass of the products is less than the total mass of the original particles then the decay can occur (assuming other conservation laws are not violated). For example, in the alpha decay of uranium-238 to thorium-234:

$$^{238}_{92}U \rightarrow \, ^{234}_{90}Th + \, ^{4}_{2}He$$

The total mass of a helium-4 nucleus plus a thorium-234 nucleus is less than the mass of a uranium-238 nucleus. There is a mass defect Δm. The energy-equivalent of this mass ($c^2\Delta m$) is released as kinetic energy of the two product nuclei. It is worth pointing out that total mass (including the mass of the energy released) is conserved, as is total energy (including all rest energies). Alpha decay is common in heavy nuclei with too many protons.

On the face of it beta decay seems easier to explain using a proton-electron model than a proton-neutron model. The latter contains no electrons, so how can electrons be emitted? Beta decay caused major problems for theoretical physicists during the first third of the twentieth century, leading some (including Niels Bohr) to suggest that the law of conservation of energy might need to be abandoned inside the nucleus! The main problems were these:

- Where do the electrons come from?
- The emitted electrons have a continuous energy spectrum, what happens to the 'missing energy'?
- Angular momentum does not add up. How can a nucleus containing an even number of spin-1/2 particles (e.g. carbon-14) emit another spin-1/2 particle? The spins of the odd number of particles after the decay cannot be combined to equal the even spin of the original nucleus.

The solution to these problems needed new physics. This was supplied by Wolfgang Pauli, who proposed that another particle (now known as an anti-neutrino) was emitted along with the beta particle in the decay, and by Enrico Fermi who introduced the idea of a weak nuclear force mediating beta decays.

Pauli proposed the new particle in 1930. The extract below is from a letter he sent to a meeting of experts on radioactivity taking place in Tubingen on 4 December:

"Dear radioactive ladies and genntlemen,
I have come upon a desperate way out regarding the 'wrong' statistics of the N-and Li 6-nuclei, as well as the continuous beta-spectrum ... To wit, the possibility that there could exist in the nucleus electrically neutral particles, which I shall call neutrons [these are the anti-neutrinos, neutrons had not been discovered in 1930] *which have spin 1/2 and satisfy the exclusion principle ...*
For the time being I dare not publish anything about this idea and address myself confidentially first to you, dear radioactive ones, with the question how it would be with the experimental proof of such a neutron [anti-neutrino]."

(Quoted in *Inward Bound*, Abraham Pais, OUP, 1986, p.315)

The particle we now call the neutron was difficult to find, only revealing itself by recoil, but at least it packed a punch large enough to eject protons. The particle predicted by Pauli was a very different thing – it had little or no mass, no charge and interacted only by the weak nuclear force (as opposed to the strong and electromagnetic interactions of neutrons). It is not surprising that neutrinos were not found experimentally until 1956, and their interactions remain controversial. At the time of writing, results from the Japanese neutrino experiments at Kamiokande seem to have shown that neutrinos do indeed have mass, a result (if confirmed) that has enormous significance for particle physics and cosmology. But more of that later – how does the anti-neutrino solve the beta decay problems and how can neutrinos or anti-neutrinos be detected?

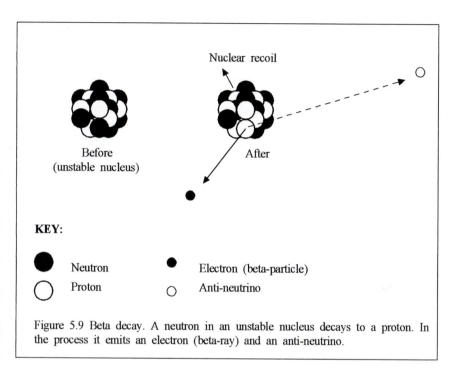

KEY:

Neutron   Electron (beta-particle)

Proton   Anti-neutrino

Figure 5.9 Beta decay. A neutron in an unstable nucleus decays to a proton. In the process it emits an electron (beta-ray) and an anti-neutrino.

The essential process underlying beta decay is the transformation of a neutron to a proton in the nucleus. The beta particle (electron) ensures that charge is conserved, but cannot by itself conserve energy and angular momentum. This is where the anti-neutrino comes in. It is also a spin-1/2 particle created along with the electron at the instant of decay so that the decay of, say carbon-14, is actually:

$$^{14}_{6}C \rightarrow {}^{14}_{7}N + {}^{0}_{-1}\beta + {}^{0-}_{0}\overline{\nu}$$

where the top numbers are nucleon numbers and the bottom numbers are charge. The bar over the neutrino symbol indicates an anti-particle. It must be emphasized that the electron and anti-neutrino are created in the decay and did not exist (despite Pauli's comments above) in the nucleus before the decay. The underlying nucleon transformation is:

$$^{1}_{0}n \rightarrow {}^{1}_{1}p + {}^{0}_{-1}\beta + {}^{0-}_{0}\overline{\nu}$$

This reaction is spontaneous for free neutrons (since neutron mass is greater than the sum of masses of a proton and an electron) which decay to protons with a half-life of about 11.7 minutes. A few comments:

- The creation of a pair of spin-1/2 particles means that angular momentum can be conserved. For example, if a nucleus X decays to a nucleus Y by beta decay and X and Y have the same spin then the electron and anti-neutrino emitted must have anti-parallel spins. If X and Y have spins differing by 1 unit then the electron and anti-electron must have parallel spins.

- The emission of two particles means that the energy released, which goes almost entirely to these two light particles, can be shared in any ratio between them. This explains the continuous energy spectrum of the beta rays. They can have any energy from near zero to almost the total amount of energy released in the decay (because the anti-neutrino rest energy is very small indeed).

- The emission of two particles means that the recoil of the nucleus is not directly opposite the direction in which the emitted beta particle leaves. This was observed in bubble chamber photographs of beta decay.

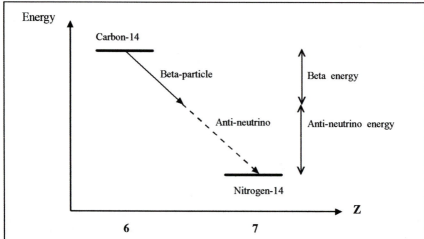

Figure 5.10 Energy is conserved in beta-decay, but it is shared randomly between the electron and an undetected anti-neutrino. This explains why experiments to measure beta-particle energies reveal a continuous spectrum. A very small amount of the total energy goes into the new nucleus as it recoils.

Neutrinos interact so weakly with normal matter that for them the Earth itself differs little from empty space. The vast majority of neutrinos reaching the Earth (from the Sun and elsewhere) pass straight through it without interacting (in the time you took to read this sentence hundreds of billions of solar neutrinos passed through you). Detecting neutrinos is a bit like trying to win a lottery: you can increase your chances by investing in lots of tickets. Cowan and Reines did this using a 10 tonne water-filled detector in which they hoped anti-neutrinos from the Savannah River Nuclear Reactor would collide with protons in water and induce an inverse-beta process, changing the protons to neutrons and emitting an anti-electron (positron): $\quad {}^{0}_{0}\bar{\nu} + {}^{1}_{1}p \rightarrow {}^{1}_{0}n + {}^{0}_{-1}\beta^{+}$

The water contained a solution of cadmium chloride. Cadmium is a good neutron absorber and was included to catch the neutrons created in the inverse process. The water tanks were sandwiched between liquid scintillators used to detect the gamma ray photons emitted by the positrons as they annihilated with ordinary matter and then, 5.5 microseconds later, more gamma ray photons emitted by cadmium nuclei as they absorb the recoiling neutrons. This double signature revealed the existence of anti-neutrinos in the summer of 1956 and on 14 June they sent a telegram to Pauli to tell him they had found the particle he predicted 26 years before. Reines shared the 1995 Nobel Prize *"for the detection of the neutrino"*.

Gamma rays are high frequency photons. They are emitted after an alpha or beta-decay when the newly formed nucleus is left in an excited state. They take away energy but do not affect A or Z; they simply allow the nucleus to de-excite. The existence of discrete gamma ray spectra led to the idea that nuclei, like atoms, can exist in a set of quantised energy levels.

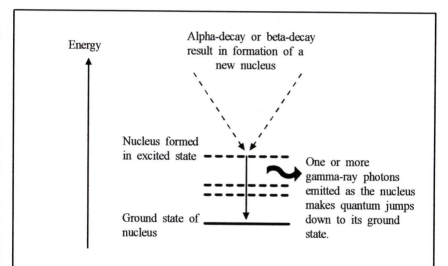

Figure 5.11 Gamma-ray emission often follows alpha or beta-decay. It occurs when the newly formed nucleus is in an excited state. The emission of a gamma-ray photon removes excess energy but does not affect the proton or neutron numbers.

5.2.4 Nuclear Forces

In 1919 Rutherford investigated collisions between alpha particles and light nuclei, including hydrogen. Low energy alpha particles scattered in agreement with the scattering formula he had derived to explain the Geiger and Marsden results. However, higher energy alpha particles showed significant deviations from this formula, and for hydrogen, the lightest nucleus, the scattering of 5 MeV alpha particles was nothing like the theoretical prediction.

Higher energy alpha particles make a closer approach to the target nucleus, and with hydrogen this could be very close indeed. The Rutherford scattering formula seemed to break down when the closest approach was about 3.5×10^{-15} m. It was clear that one of the assumptions used to derive the formula must fail at close range. He got two research students, Chadwick and Beiler, to investigate in more detail. They concluded that:

"The present experiments do not seem to throw any light on the nature of the law of variation of the forces at the seat of an electric charge, but merely show that the forces are of very great intensity ... It is our task to find some field of force which will reproduce these effects."

(J. Chadwick and E.S. Beiler, Philosophical Magazine, 42, 923, 1921 quoted in *Inward Bound*, Abraham Pais, OUP, 1986, p.240)

Early attempts to understand this new force field suffered from the confusion over what particles actually constitute the nucleus. Even after Chadwick's discovery of the neutron many physicists continued to think of it as a composite particle, a kind of collapsed hydrogen atom containing an electron and proton in a tightly bound state (even if this did mean that some of the electron's intrinsic properties must be suppressed when they are inside a nucleus). The idea that nucleons are bound by a new strong nuclear force developed through several distinct theoretical stages and incorporated a growing body of data gathered from nuclear experiments:

- In January 1932 while Chadwick was writing up his discovery of the neutron, Harold Urey at Columbia discovered deuterium, an isotope of hydrogen with $A = 2$, having a proton and a neutron in the nucleus. This simple nucleus played much the same part in the development of nuclear theory as the hydrogen atom did in the development of atomic theory.
- High-energy physics started properly with John Cockroft and Ernest Walton's accelerator (see Chapter 8), a voltage-multiplying device that could produce high-energy (approaching 1 MeV) protons to penetrate the coulomb barrier around nuclei and induce transformations. Proton-proton and proton-neutron scattering provided a great deal of information about the nature of the Strong Force. Cockroft and Walton shared the 1951 Nobel Prize for Physics *"for their pioneering work on the transmutation of atomic nuclei by artificially accelerated atomic particles"*. Later in 1932 Ernest Lawrence at Berkeley completed building the first working cyclotron and accelerated protons to 5 MeV. Lawrence was awarded the 1939 Nobel Prize for Physics *"for the invention and development of the cyclotron and for results obtained with it, especially with regard to artificial radioactive elements"*.
- Measurements of size and binding energy showed that nuclei containing over

about 40 nucleons had a constant density. This follows from the relation between nuclear radius and nucleon number:

$$R = r_0 A^{1/3} \quad \text{with } r_0 \approx 1.2\,\text{fm}$$

This indicated that the nuclear force saturates, i.e. unlike the long-range forces of electromagnetism and gravitation, the nuclear force 'switches off' beyond a certain maximum range. It also implies that it must become a repulsive force at very short range (to prevent collapse).

- In 1932 Heisenberg published a three-part paper on nuclear forces in which he developed a theory of exchange forces. The essential idea in this theory is that neutrons and protons, neutrons and neutrons and protons and protons bind to one another by exchanging properties such as charge and position. In Heisenberg's model the neutron and proton are effectively different states of the same particle, a nucleon. This implied that the strong nuclear force is charge-independent so that the interaction between any pair of nucleons is the same (after correcting for the coulomb force).

- The inter-nucleon forces do depend on spin. This was shown by the fact that the deuteron contains a neutron and a proton with parallel spins but there is no state in which their spins align anti-parallel.

- In 1934 Irène Joliot-Curie and Frédéric Joliot discovered β^+ decay, the emission of a positron from a proton-rich nucleus. The underlying process here must be the transformation of a nuclear proton to a neutron, by creating a positron/neutrino pair. This emphasised the symmetry between processes involving protons and neutrons and further undermined the view that neutrons are a bound state of a proton and a neutron.

- In 1933 Enrico Fermi constructed a theory for beta decay using Pauli's neutrino hypothesis and a variation on Dirac's quantum field theories. Fermi's theory is another exchange theory, but this time involving a pair of particles, the electron and its anti-neutrino. There is an analogy between atomic transitions in which a photon is emitted and the nucleon transformations in which a lepton pair is emitted. Being an interaction between nucleons this too was considered as a candidate for nuclear binding. Unfortunately its effects were too small by many orders of magnitude, but it provided the first theoretical model of another new nuclear interaction, the weak force, involved in beta-decay.

- Hideki Yukawa was working in relative isolation in Japan when he began to develop his meson theory of the strong nuclear force and it did not become well known in the west until the late 1930s. He decided to start from the properties of the nuclear force and work back to a suitable exchange particle. This idea made use of the Uncertainty Principle – for two nucleons to interact they must exchange a quantum of the new field and the more massive this quantum is, the smaller its range. Knowing the range, the mass could be predicted.

- In 1937 a particle of the right mass turned up in cosmic rays, but this later proved to be a different particle altogether – a kind of 'heavy electron' now called a muon. However, it served to draw attention to Yukawa's ideas.

- Yukawa's initial idea was developed by two other Japanese physicists, Sakata and Inoue into an exchange force in which the field quanta are charged and neutral particles, now called pions. The characteristic two-step decay of a pi-plus was first observed using sensitive photographic emulsions by Cecil Powell at Bristol in 1947. Yukawa's work changed the way we think about forces and he won the

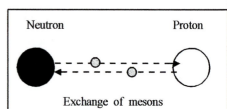

Neutron   Proton

Exchange of mesons

Figure 5.12 Yukawa's theory of the strong nuclear force. Nucleons exchange mesons. This binds them together. The force is limited in range because the mesons are massive particles and the energy needed to create them can only be borrowed for a short time.

1949 Nobel Prize for Physics *"for his prediction of the existence of mesons on the basis of theoretical work on nuclear forces"*. Powell also won the Physics Nobel Prize, in 1950, *"for his development of the photographic method of studying nuclear processes and his discoveries regarding mesons made with this method"*.

Maths Box: Yukawa's Pion

In Yukawa's theory the force that binds a neutron and proton together is analogous to the force between a proton and a hydrogen atom in the H^+ ion. In the ion the force arises from the exchange of an electron, for the proton-neutron force it is a pion that is exchanged. However, the reaction:

$$\,^1_1p \rightarrow \,^1_0n + \,^1_1\pi^+$$

is forbidden as a real process by energy conservation (the mass of a pion plus a neutron being greater than the mass of a proton). The exchange can take place, however, as a virtual process as long as the pion radiated by the proton is absorbed within a time limited by the energy-time uncertainty relation:

$$\Delta E \Delta t \approx \frac{h}{2\pi}$$

During this time the meson can travel a maximum distance $c\Delta t$. Yukawa set this equal to the range of the nuclear force as revealed by nucleon scattering experiments, $R \approx 2 \times 10^{-15}$ m and used it to predict the mass of the meson:

$$\Delta t = \frac{R}{c} \approx \frac{h}{2\pi\Delta E} = \frac{h}{2\pi mc^2} \qquad \text{where } m \text{ is the mass of the meson radiated.}$$

$$m \approx \frac{h}{2\pi Rc}$$

$$m \approx 2 \times 10^{-28} \text{kg} \approx 100 \text{ MeV}/c^2$$

This corresponds to about 200 times the mass of an electron and one-tenth the mass of a nucleon. Hence the name 'meson' because of the intermediate mass (actually there are other mesons more massive than nucleons, so the name can be misleading). The pi-plus and pi-minus particles turned out to have a mass of about 140 MeV/c^2 and the neutral pion has a mass of 135 MeV/c^2.

5.2.5 Models of the Nucleus

There have been two distinct approaches to modeling the atomic nucleus based on conflicting ideas about the way nucleons must interact inside it:

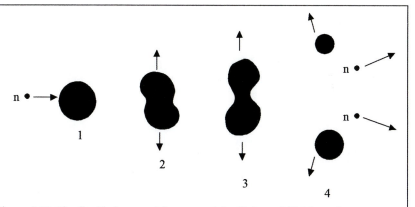

Figure 5.13 The liquid drop model, proposed by Bohr and Kalckar gives a simple picture for induced nuclear fission:

1. A neutron is absorbed by a heavy nucleus (e.g. U-235)
2. The nucleus becomes excited and, unstable, it oscillates violently.
3. When it distorts far enough two 'droplets' are able to break free because of mutual electrostatic repulsion.
4. The new nuclei are neutron-rich so a few neutrons may also be emitted, raising the possibility of a chain reaction.

- The collective model, proposed by Niels Bohr and Jörgen Kalckar started from the assumption that the nuclear forces are so strong that nucleons cannot be treated as individual particles and the nucleus as a whole can distort, vibrate and rotate like a charged liquid drop. This model led to a semi-empirical formula for nuclear masses and binding energies and was used with great success, by Bohr and Wheeler, to model nuclear fission.
- The shell model was put forward by Maria Goeppert-Mayer (then Maria Goeppert) in 1948 to explain the existence of particularly stable nuclei with 'magic numbers' of protons or neutrons (2, 8, 14, 20, 28, 50, 82, 126). The approach was by analogy with the problem of electrons in atoms – she assumed that each nucleon moved independently in the same central field. The 'magic numbers' should then arise like the closed shells in atoms. (The fact that the magic numbers differ from the numbers of electrons in closed atomic shells is because the spin and orbital angular momenta of nucleons interact strongly whereas those of the electrons and nuclei do not).

The binding energy B of a nucleus is the amount of energy needed to separate all the nucleons to a great distance from one another. The binding energy per nucleon is simply B/A and is often the best quantity to use to compare the stability of one nucleus with another. The total binding energy of a nucleus is given (in joules) by:

$$B = \left\{ m_{nucl} - Zm_p - (A-Z)m_n \right\} c^2$$

and most stable nuclei above about $A=16$ have approximately the same binding energy per nucleon (between 7.5 and 8.5 MeV per nucleon). However, a detailed survey of nuclear binding energies and stability shows up some important trends:

- The binding energy per nucleon rises rapidly for the light nuclei and has significant local maxima at $A = 4, 8, 12, 16, 24$ suggesting that the alpha particle itself is a particularly stable arrangement of nucleons. This provides evidence that the nucleons tend to pair up in their orbits.
- The predominance of stable nuclei with even numbers of both protons and neutrons and the small number of stable even/odd or odd/even nuclei emphasizes nucleon pairing.
- The maximum binding energy per nucleon occurs for iron-56, for $A > 56$ the binding energy per nucleon falls. The main reason for this is the long-range repulsion of the positive charges on protons.
- Light stable nuclei have approximately equal numbers of protons and neutrons, heavy stable nuclei have more neutrons than protons.

When these considerations are fed into the liquid drop model it is possible to construct a semi-empirical formula for the mass and binding energy of a nucleus. It is semi-empirical because the formula contains a number of constants that must be put in to make the formula fit the nuclear data. The semi-empirical mass formula is written below together with a key to the terms:

$$m(A,Z) = Zm_p + (A - Z)m_n - a_1 A + a_2 A^{2/3} + \frac{a_3}{A}(A - 2Z)^2 + \frac{a_4 Z^2}{A^{1/3}} + \delta(A,Z)$$

where:

1. $Zm_p + (A - Z)m_n$ is the total mass of the nucleons as free particles.
2. $-a_1 A$ is a reduction in mass proportional to the number of nucleons present. This arises from the average binding energy per nucleon in continuous nucleonic matter, about 14 MeV per nucleon).
3. $+a_2 A^{2/3}$ is an increase in mass because surface nucleons are not bound so tightly as those inside the nucleus which are completely surrounded by other nucleons. The factor $A^{2/3}$ arises because nuclear radius is proportional to $A^{1/3}$ and surface area to radius-squared.
4. $+\frac{a_3}{A}(A - 2Z)^2$ accounts for the fact that nuclei gain stability by having equal numbers of protons and neutrons. If $A = 2Z$ this term is zero. If $A \neq Z$ the term adds to the total mass destabilising the nucleus.
5. $+\frac{a_4 Z^2}{A^{1/3}}$ is the coulomb energy, the electrostatic potential energy of all the protons squeezed together in the nucleus. It is this term that grows as Z increases effectively limiting the number of possible stable nuclei.
6. $+\delta(A,Z)$ is a term that increases the mass if A and Z are not both even.

The coefficients needed to complete the equation can be put in by hand or estimated using theoretical ideas. Fermi suggested the following values:

$$
\begin{aligned}
a_1 &= 14.0 && \text{MeV} \\
a_2 &= 13.0 && \text{MeV} \\
a_3 &= 19.3 && \text{MeV} \\
a_4 &= 0.58 && \text{MeV} \\
\delta &= 33.5 A^{-3/4} && \text{MeV}
\end{aligned}
$$

The charged liquid drop model has been surprisingly effective in describing nuclear properties.

The existence of 'magic numbers' prompted Mayer and Haxel, Jensen and Suess to construct a model in which nucleons in the nucleus move more or less independently in a spherically symmetric electrostatic potential well. Energy levels for nucleon orbits can then be calculated using the Schrödinger equation in much the same way that electron orbits are calculated in the hydrogen atom. A simple approach did give closed shells, but not at the magic numbers. A non-central force component had to be added due to spin-orbit coupling. Once this had been done the magic numbers were explained.

One of the most fascinating aspects of these nuclear models is that two apparently contradictory assumptions, that the nucleons act independently of one another or that they act collectively, both lead to models that are in good general agreement both with each other and with the behaviour of real nuclei. It is also interesting that it was Aage Bohr, Niels Bohr's son, working with Ben Moffelson who did most to explain this connection. Aaage Bohr, Ben Moffelson and James Rainwater shared the 1975 Nobel Prize1975 for Physics *"for the discovery of the connection between collective motion and particle motion in atomic nuclei and the development of the theory of the atomic nucleus based on this connection"*.

5.2.6 Extending the Periodic Table

In 1934 the Italian physicist, Enrico Fermi suggested that new elements might be synthesised by firing neutrons at the nuclei of known elements. Adding an extra neutron to an existing nucleus does not create a new element because the atomic number is not changed. However, neutron-rich nuclei are often beta-emitters and the underlying process in beta-decay is the transformation of a neutron to a proton. The first element to be produced by this method was neptunium ($Z = 93$) at Berkeley by Edwin McMillan and Philip Anderson in 1940. The rapid advances in nuclear science and technology made in the 1940s and 1950s as a result of wartime research into atomic weapons led to the creation of elements 94 through to 100 (plutonium $Z = 94$, americium $Z = 95$, curium $Z = 96$, berkelium $Z = 97$, californium $Z = 98$, einsteinium $Z = 99$ and fermium $Z = 100$, respectively). It is appropriate that fermium turns out to be the last element that can be created in this way; for larger atomic numbers beta-decay does not occur.

However, the production of new elements does not end with fermium. Larger nuclei can be created by bombarding nuclei of transuranic elements with light nuclei such as helium, carbon, nitrogen or oxygen. This method led to the synthesis of new elements up to element 106. They are: mendelevium $Z = 101$, nobelium $Z = 102$, lawrencium $Z = 103$, rutherfordium $Z = 104$, dubnium $Z = 105$, and seaborgium $Z = 106$. Beyond $Z = 106$ this 'hot fusion' technique ceased to be effective.

In 1974 Yuri Oganessian and Alexander Demin, researchers at the Joint Institute for Nuclear Research at Dubna in Russia (where several of the transuranic elements had been discovered), managed to create fermium by lower energy collisions between argon ($Z = 18$) and lead ($Z = 82$). They realised that the collision itself transferred much less energy to the new nucleus than a collision with a light nucleus. This increases the probability of a heavy nucleus surviving the impact. This technique was taken up by a team at the Institute for Heavy Ion Research at Darmstadt in Germany. They created bhorium ($Z = 107$), hassium ($Z = 108$) and meitnerium ($Z = 109$) in the 1980s and elements 110, 111 and 112 in the 1990s.

Most of the superheavy elements are extremely short-lived unstable nuclei that decay almost as soon as they are formed. However, one of the reasons for continuing to create new super-heavy elements is to test the idea that there will be islands of stability at particular 'magic numbers' of nucleons corresponding to closed nuclear shells. This was predicted by the nuclear shell model developed by Maria Goeppert-Mayer and Hans Jensen in the 1950s. Theory suggests that there should be a magic nucleus around $Z = 114$ and it may have a half-life long enough to stockpile a significant amount of it and perhaps lead to a new chemistry of the superheavy elements.

In January 1999 scientists from the Dubna facility announced that they had created an isotope of element 114 (with atomic mass 289). Furthermore, its half-life is half a minute, far in excess of other nearby superheavies. Theoreticians have suggested that an isotope of this element with atomic number 298 might have an extremely long half-life, perhaps billions of years. Watch this space! Enrico Fermi won the Nobel Prize in Physics in 1938 *"for his demonstrations of the existence of new radioactive elements produced by neutron irradiation, and for his related discovery of nuclear reactions brought about by slow neutrons"*.

5.2.7 Fission, Fusion, Bombs and Reactors

When Einstein published his mass-energy relation in 1905 he suggested that it might account for the enormous amount of energy released during the lifetime of a radioactive substance. He was correct. Rutherford 'split the atom' in 1919, but in 1933 he said that *"The energy produced by the breaking down of the atom is a very poor kind of thing. Anyone who expects a source of power from the transformation of these atoms is talking moonshine"*. He was wrong.

In 1939, as Europe descended into war, Otto Hahn and Fritz Strassman in Berlin confirmed that elements such as barium ($Z = 56$) and lanthanum ($Z = 57$) are formed when uranium ($Z = 92$) is bombarded by neutrons. Lise Meitner in Stockholm and her nephew Otto Frisch in Copenhagen correctly interpreted this as nuclear fission, the splitting of a heavy nucleus into two lighter nuclei. Bohr announced the discovery to the American Physical Society and the main characteristics of nuclear fission were soon established in a series of simple experiments carried out at various laboratories around the world. Later in 1939 Bohr and Wheeler published a paper which explained the process using the charged liquid drop model. When fission occurs:

- The uranium nucleus can split in many different ways producing a wide range of daughter nuclei.
- Daughter nuclei are neutron-rich so decay by a sequence of beta emissions.
- The recoiling daughter nuclei have energies of around 75 MeV, giving them a range of about 3 cm in air.
- A few fast neutrons are ejected with the daughter nuclei, so there is the possibility of a chain reaction.
- Natural uranium is 99.3% uranium-238 which is a neutron-absorber. The

other 0.7% is almost entirely uranium-235, which can undergo fission when it absorbs a slow neutron.

A typical example of an induced nuclear fission is:

$$\,_0^1n + \,_{92}^{235}U \rightarrow \,_{57}^{148}La + \,_{35}^{85}Br + 3\,_0^1n$$

This reaction has a mass defect of about 0.2 u (1 u = 1.6611 $\times$ 10^{-27} kg – approximately equal to the mass of a proton) and releases about 200 MeV, mainly in the kinetic energy of the daughter nuclei. For this to lead to a self-sustaining chain reaction at least one neutron per fission (on average) must collide with another U-235 nucleus and induce a fission reaction. This reaction was used in the 'atom bomb' dropped at Hiroshima in 1945, and is also the reaction used in nuclear reactors to generate electrical energy.

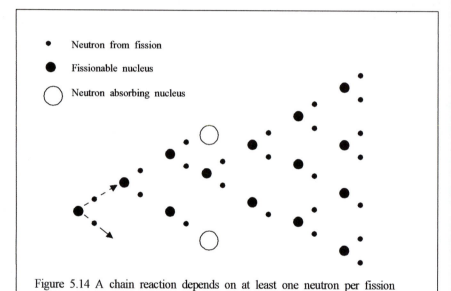

Figure 5.14 A chain reaction depends on at least one neutron per fission inducing fission in another nucleus.

The bomb dropped at Nagasaki was also a fission weapon, but its fissile material was the transuranic element plutonium (Z = 94) which is created inside nuclear reactors as uranium-238 absorbs neutrons and decays:

$$\,_0^1n + \,_{92}^{238}U \rightarrow \,_{92}^{239}U$$

$$\,_{92}^{239}U \rightarrow \,_{93}^{239}Np + \,_{-1}^{0}\beta + \,_{0}^{0-}\nu$$

$$\,_{93}^{239}Np \rightarrow \,_{94}^{239}Pu + \,_{-1}^{0}\beta + \,_{0}^{0-}\nu$$

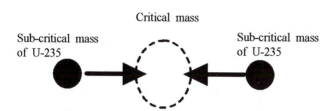

(i) Two-sub-critical masses of U-235 (or plutonium) are blasted together using conventional explosives. They form a critical mass for a short time. However, the generation time for a chain reaction is so short that most of the fissionable material is consumed releasing an enormous amount of energy.

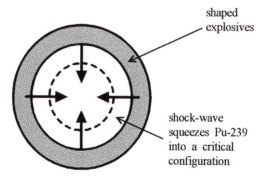

(ii) Another way to form a critical mass is to take a sub-critical mass and compress it using shock-waves from shaped explosives. This reduces the surface area to volume ratio so that fewer neutrons are lost from the surface. This tips the balance in favour of a chain reaction.

Figure 5.15 An atom bomb induces a chain reaction in fissionable material. The diagrams above show two ways in which this can be achieved. The only nuclear weapons used in warfare were dropped on Japan in 1945. The Hiroshima bomb used uranium and method (i). The Nagasaki bomb used plutonium and method (ii) Both destroyed most of the city on which they fell, killing more than one hundred thousand people. They each released the energy equivalent of about ten thousand tonnes of TNT. However, large fusion weapons (H-bombs) release up to a thousand times this amount of energy.

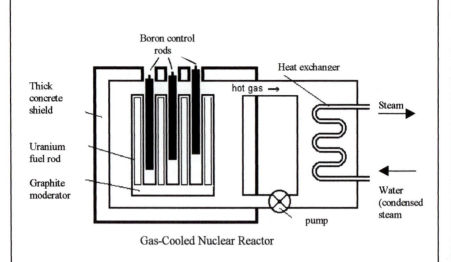

Gas-Cooled Nuclear Reactor

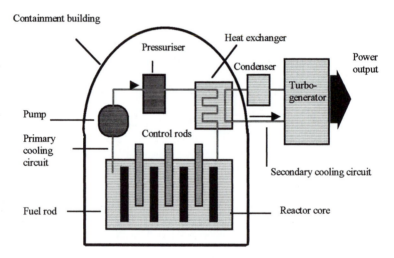

Pressurised water reactor (PWR)

Figure 5.16 Commercial nuclear reactors generate energy from the fission of uranium-235. For this to work neutrons released in fission reactions must be slowed down by collisions inside a moderating material. This increases their chance of inducing further fission reactions and maintaining a chain reaction. Heat is removed by a suitable coolant pumped through the reactor core. In the gas-cooled reactor above the moderator is graphite and the coolant carbon dioxide. In the pressurised water reactor (PWR) below water acts as both moderator and coolant.

Plutonium is fissionable and has a half-life of 24 360 years. It can be retrieved from nuclear reactors by reprocessing the spent fuel rods, a process that is much simpler and more productive than trying to separate U-235 from U-238. This is why there is such a close link between the development of civil nuclear power programs and nuclear weapons.

In the 1920s Eddington and Atkinson and Houtermans suggested that the steady energy release of the Sun could be explained by nuclear fusion reactions taking place in its core. In the 1930s Cockcroft and Walton used their early accelerator to induce fusion reactions in light elements, and confirmed the details of the process. In 1940 Hans Bethe worked out how the Sun converts hydrogen to helium in the stepwise 'proton-proton' reaction whose net effect is to convert four protons to one helium nucleus, two positrons, two neutrinos and some gamma rays (see Chapter 14). In 1941 Fermi and Teller discussed the idea that a fission explosion might generate temperatures high enough to ignite fusion reactions in deuterium and a project to develop a fusion weapon, the 'Super', began alongside the drive for fission weapons at Los Alamos. Following the bombing of Hiroshima and Nagasaki and the end of the Second World War many physicists opposed the continuation of the 'Super' project. However, in 1950 President Truman was presented with evidence that the Russians had their own nuclear fission weapons and he gave the green light to the fusion bomb project at Los Alamos. The leading theorists involved in the project were Edward Teller and Stanislaw Ulam. By this time they had discovered that fusion would be more likely if the explosive material was a mixture of deuterium and tritium and if the fission explosion was set up so that it caused intense compression as well as intense heating of the fusion fuel. In 1952 the 'Mike' tests were successfully carried out and the explosion yielded about 500 times the energy of the Nagasaki fission bomb.

From the 1950s there has been a great deal of research and experimentation aimed at constructing a safe, economical fusion reactor to generate electricity. The main obstacle to this is the difficulty of containing matter at the incredibly high temperatures (tens of millions of kelvin) needed to initiate and sustain fusion. There have been two main strands to this research, the conventional attempts to confine the reacting plasma using magnetic fields and a more recent approach based on imploding fuel pellets using focussed lasers ('inertial confinement'). The most notable successes so far have been achieved using Tokamaks, toroidal magnetic confinement reactors based on a design proposed by the Russian physicists, Andrey Sakharov and Igor Tamm in 1950.

Some fusion reactions that may be candidates for a reactor are listed below:

$$^2_1H + ^2_1H \rightarrow ^1_1H + ^3_1H + 4.04\,\text{MeV}$$

$$^2_1H + ^2_1H \rightarrow ^1_0n + ^3_2He + 3.27\,\text{MeV}$$

$$^2_1H + ^3_1H \rightarrow ^1_0n + ^4_2He + 17.6\,\text{MeV}$$

$$^2_1H + ^3_2He \rightarrow ^1_1H + ^4_2He + 18.3\,\text{MeV}$$

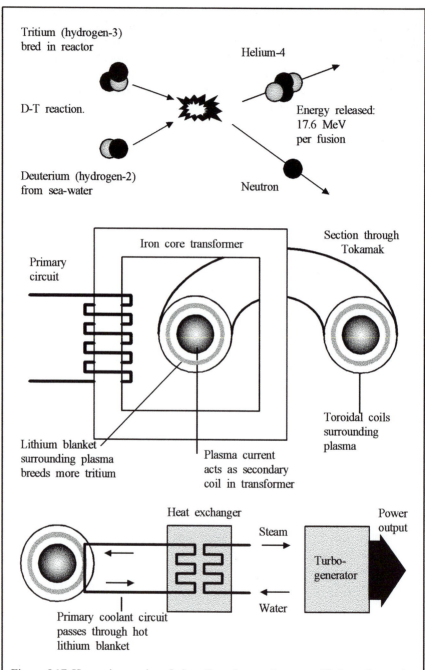

Tritium (hydrogen-3)
bred in reactor

Helium-4

D-T reaction.

Energy released:
17.6 MeV
per fusion

Deuterium (hydrogen-2)
from sea-water

Neutron

Iron core transformer

Section through
Tokamak

Primary
circuit

Lithium blanket
surrounding plasma
breeds more tritium

Plasma current
acts as secondary
coil in transformer

Toroidal coils
surrounding
plasma

Heat exchanger

Power
output

Steam

Turbo-
generator

Water

Primary coolant circuit
passes through hot
lithium blanket

Figure 5.17 Harnessing nuclear fusion. Top: the reaction most likely to be used
in a commercial reactor (D-T reaction). Middle: magnetic confinement in a
Tokamak. Bottom: heat extraction and power generation.

The last two reactions release considerably more energy per nucleon than nuclear fission reactions, and the reaction between deuterium and tritium can be initiated at a lower temperature than deuterium-deuterium fusion reactions. This makes it the best candidate for fuel in a fusion reactor. Deuterium is present as 0.015% of natural hydrogen, but can be retrieved fairly easily by fractional distillation of water. Tritium is only present in trace quantities on Earth but can be created by bombarding lithium, which is plentiful, with neutrons:

$$\begin{array}{l} {}^{6}_{3}\text{Li} + {}^{1}_{0}\text{n} \rightarrow {}^{3}_{1}\text{H} + {}^{4}_{2}\text{He} \end{array}$$

This could be carried out in the reactor itself by surrounding the core with a blanket containing lithium. Neutrons released in fusion reactions pass into the blanket and convert lithium to tritium.

Reactions involving nuclei with larger Z are unsuitable because their large coulomb repulsion requires a very high temperature for fusion to occur and also results in large energy losses due to X-rays as the nuclei scatter from one another. Hydrogen itself is unsuitable because much of the energy released in its fusion reactions end up as gamma rays and it is difficult to transfer this to kinetic energy of protons, which is needed to produce further fusion reactions.

6

THE STANDARD MODEL

6.1 PARTICLE PATTERNS

"The ... reductionist tendency is visible within physics. This is not a matter of how we carry on the practice of physics, but how we view nature itself. There are many fascinating problems that await solution, some like turbulence, left over from the past, and others recently encountered, like high-temperature superconductivity. These problems have to be addressed in their own terms, not by reduction to elementary particle physics. But when these problems are solved, the solution will take the form of a deduction of the phenomenon from known physical principles, such as the equations of hydrodynamics or of electrodynamics, and when we ask why these equations are what they are, we trace the answers through many intermediate steps to the same source: the Standard Model of elementary particles ... But the Standard Model is clearly not the end of the story. We do not know why it obeys certain symmetries and not others, or why it contains six types of quarks, and not more or less. Beyond this, appearing in the Standard Model there are about 18 numerical parameters (like ratios of quark masses) that must be adjusted by hand to make the predictions of the model agree with experiment."

(Steven Weinberg, Nature Itself, 1995, in *Twentieth Century Physics*, IOP, 1995)

The Periodic Table of the elements and its subsequent explanation in terms of atomic structure is a wonderful example of the rewards that a reductionist approach can provide. It is hardly surprising that twentieth century particle physics follows a similar path. As more and more particles were discovered patterns began to emerge. The Standard Model is our best attempt to explain these patterns in terms of a small number of fundamental particles and forces. The amazing success of the Standard Model has given physicists enormous confidence in the general principles used to construct it. It is now recognised that symmetry principles and conservation laws are intimately linked and the underlying assumption that nature is understandable and (in a mathematical sense) both simple and beautiful seems justified.

Attempts to understand atomic spectra, radioactivity and nuclear stability gradually widened the scope of particle physics from the proton, electron (1897) and photon (1905) to include the neutron (1932), neutrino and anti-neutrino (predicted 1932, discovered 1956), positron (1932), muon and anti-muon (1936) and pions (1947-50) by about 1950. Increased variety and sensitivity of particle detectors revealed a large number of strongly interacting particles produced by cosmic rays and the ever-higher energy particle accelerators in collision events during the next two decades. Slowly a new pattern began to emerge. Some the key ideas and discoveries are listed below.

- There are two main types of particle: fermions (with half-integer spin) and bosons (with integer spin).
- All particles have corresponding anti-particles, although in some cases the anti-particle is indistinguishable from the particle (e.g. photons).
- Matter is made from fermions (e.g. electrons, protons and neutrons).
- Interactions take place by the exchange of bosons (e.g. photons and mesons).
- There are four fundamental interactions: the strong nuclear force (actually a consequence of the colour force between quarks), the weak nuclear force (involved in beta decay), electromagnetism and gravitation.
- Particles can also be classified by their interactions:
    hadrons interact by the strong force. Half-integer spin hadrons (such as protons and neutrons) are called baryons and integer spin hadrons are called mesons (such as the pion).
    leptons (e.g. electrons and neutrinos) not affected by strong interactions.
- Whilst leptons *appear* to be structureless point-like particles the hadrons are all made from more fundamental units called quarks.
- There are three generations of leptons and three generations of quarks. In each generation there is a pair of particles. This 'quark-lepton symmetry' remains unexplained but hints at a deeper connection yet to be discovered.

The tables below summarise the Standard Model as it stands at the end of the twentieth century.

	Leptons		**Quarks**		**Found in**
First generation	Electron	Electron-neutrino	Up	Down	Ordinary matter
Second generation	Muon	Muon-neutrino	Charm	Strangeness	Only found in cosmic
Third generation	Tau	Tau-neutrino	Top	Bottom	rays and accelerators

Force	Gauge bosons	Acts on	Conservation laws	Typical lifetime	Typical range
Gravity	Graviton zero rest mass	Everything	Q, B, S, L	No decays	Infinite inverse square law
EM	Photon zero rest mass	Quarks and charged leptons	Q, B, S, L	10^{-18} s	Infinite inverse square law
Weak	W^+ W^- 83 GeV/c^2 Z^0 93 GeV/c^2	Quarks and leptons	Q, B, L but not S	10^{-10} s	10^{-18} m
Strong	8 gluons zero rest mass?	Quarks	Q, B, S	10^{-23} s	10^{-15} m

The conservation laws in the table above are: Q charge, B baryon number, S strangeness, L lepton number.

6.2 LEPTONS

6.2.1 'Heavy Electrons'

We have already discussed J.J. Thomson's discovery of the electron. C.D. Anderson and S.H. Nedermeyer discovered the muon in cosmic ray cloud chamber tracks in 1936. No one expected to discover a 'heavy electron' and the muon was intially mistaken for Yukawa's meson. However, it was soon realised that muons do not take part in the strong interaction but do behave just like electrons would if their mass was increased 207 times, except that they are unstable. When they decay they do so by creating an electron and an anti-neutrino, but this anti-neutrino is *not* the same as the anti-neutrino emitted in beta-decay. It seems that the muon and muon-neutrino form a distinct pair of leptons, just like the electron and electron-neutrino. These are the first two generations of leptons. In 1974 Martin Perl discovered an even heavier lepton, the tau particle in the SPEAR (Stanford Positron Electron Asymmetric Rings) electron-postron collider at Stanford. Once again there was a short period of confusion concerning the identity of the unexpected particle, but it was soon confirmed that there is a third generation of leptons, the tau and tau-neutrino.

All leptons have a 'lepton number' L, which is +1 for the particles and −1 for their anti-particles. All known interactions conserve total lepton number, this is why beta decay must create an anti-neutrino ($L= -1$) along with the beta-particle, an electron ($L = +1$). Individual generations are individually conserved, so in addition to the conservation of lepton number there are separate conservation laws for electron- muon- and tau-lepton numbers.

G	Particle	Mass (rel)	Charge (rel)	Spin	L	L_e	L_μ	L_τ
1	Electron	1	−1	$^1/_2$	1	1	0	0
1	Electron-neutrino	$\approx 10^{-6}$	0	$^1/_2$	1	1	0	0
1	Positron	1	+1	$^1/_2$	−1	−1	0	0
1	Anti-electron neutrino	$\approx 10^{-6}$	0	$^1/_2$	−1	−1	0	0
2	Muon	207	−1	$^1/_2$	1	0	1	0
2	Muon-neutrino	$\approx 10^{-6}$	0	$^1/_2$	1	0	1	0
2	Anti-muon	207	+1	$^1/_2$	−1	0	−1	0
2	Anti-muon neutrino	$\approx 10^{-6}$	0	$^1/_2$	−1	0	−1	0
3	Tau	3490	−1	$^1/_2$	1	0	0	1
3	Tau-neutrino	$\approx 10^{-6}$	0	$^1/_2$	1	0	0	1
3	Anti-tau	3490	+1	$^1/_2$	−1	0	0	−1
3	Anti-tau neutrino	$\approx 10^{-6}$	0	$^1/_2$	−1	0	0	−1

G in the table above stands for 'generation'.

The 1988 Nobel Prize for Physics was awarded jointly to Leon Lederman, Melvin Schwartz and Jack Steinberger "*for the neutrino beam method and the demonstration of the doublet structure of the leptons through the discovery of the muon-neutrino*". In 1995 the Prize was awarded "*for pioneering experimental contributions to lepton physics*" to Martin Perl "*for the discovery of the tau lepton*" and (at last) Frederick Reines "*for the detection of the neutrino*".

6.2.2 Neutrino Experiments

"*Neutrinos they are very small.*
They have no charge and have no mass
And do not interact at all.
The earth is just a silly ball
To them, through which they simply pass
Like dustmaids down a drafty hall."

(John Updike, from *Cosmic Gall*)

The problem with neutrinos is that they only interact with matter through the weak force, so they are extremely difficult to detect. A 1 MeV neutrino is likely to interact only once as it passes through a lead sheet one light year thick! This is why it took a quarter of a century from Pauli's prediction to the discovery of the electron anti-neutrino by Cowan and Reines in 1956. We have already seen that fission products are necessarily neutron-rich and so are beta-emitters. This means that the core of a working reactor is an intense source of anti-neutrinos. Cowan and Reines set up an experiment at the Savannah River Reactor to look for the inverse-beta process:

$$_0^0\bar{\nu}_e + {}_1^1p \rightarrow {}_0^1n + {}_1^0e^+$$

The reaction was confirmed by looking for the gamma rays emitted by postiron-electron annihilation followed a short time after by gamma ray pulses from the neutrons as they are absorbed in cadmium nuclei (see Chapter 5).

The existence of the muon-neutrino was suspected because no one had observed the decay of a muon to an electron and a gamma ray (with no neutrinos). This should occur by the following virtual step if muon and electron neutrinos are identical:

$$_{-1}^0\mu^- \rightarrow {}_{-1}^0e^- + {}_0^0\bar{\nu} + {}_0^0\nu \rightarrow {}_{-1}^0e^- + {}_0^0\gamma$$

In other words the virtual neutrino created by the decay of the muon should annihilate with the virtual anti-neutrino created along with the electron. This is never observed. If the two neutrinos are distinct from one another this would make sense, but how could this be shown to be the case? In 1962 Schwartz's group at Columbia created high-energy beams of muon-neutrinos and fired them at protons to induce inverse-beta transformations to neutrons:

$$_0^0\nu_\mu + {}_1^1p \rightarrow {}_0^1n + ?$$

Lepton number must be conserved, so the unknown particle could be a muon. If there is no difference between muon and electron neutrinos then the unknown particle could also be an electron. In the first eight months of the experiment they detected 29 muons and no electrons. This confirmed that the muon-neutrino is a separate particle from the electron-neutrino and that muon- and electron-lepton numbers are separately conserved. The correct equation for the muon decay above should therefore be written:

$$_{-1}^0\mu^- \rightarrow {}_{-1}^0e^- + {}_0^0\bar{\nu}_e + {}_0^0\nu_\mu$$

This decay conserves lepton number L ($1 = 1 - 1 + 1$), electron lepton number L_e ($0 = 1 - 1 + 0$) and muon lepton number L_μ ($1 = 0 + 0 + 1$).

This was the first time neutrino beams had been used as an experimental tool. The beams themselves were generated by colliding protons into nuclei in a fixed target and then using magnetic fields to select the pions created in the collision.

Pions decay to a beam of muons and muon-neutrinos. This beam was aimed at an iron wall tens of metres thick (made from iron plates scavenged from the scrapped battleship, Missouri). The iron filtered out everything except the muon-neutrinos. The cross-section for interaction is much greater for high-energy neutrinos, so the probability of an interaction in the detectors placed beyond the iron wall was considerably higher than it would have been for the lower energy electron neutrinos emitted from a reactor.

Fusion reactions taking place in the core of the Sun result in a large flux of neutrinos, about 2×10^{38} per second. This gives astronomers a way to 'see' into the core of the Sun if they can detect any of the 10^{14} neutrinos per square metre per second that reach the Earth. The first group to attempt this was led by Raymond Davis at the Brookhaven National Laboratories in 1964. They used an induced beta process once again, but their target was the isotope chlorine-37 which converts to argon-37 when an incident high-energy neutrino converts a neutron to a proton in its nucleus. Chlorine-37 is 25% of natural chlorine, so they used a huge pool (400 000 litres) of C_2Cl_4 (cleaning fluid!) as the detector. This was set-up 1500 m underground in an old gold mine. The plan was to flush helium through the fluid and measure how much argon came out with the helium. Argon-37 is unstable and decays by K-capture so it can be identified by the gamma rays it emits. The activity is therefore a measure of how many solar neutrinos have been captured. The theoretical prediction was that roughly one neutrino should be captured per day. The measured rate was only about a third of this value.

This discrepancy between theoretical and experimental values for the flux of solar neutrinos is known as the 'solar neutrino problem'. It has been confirmed by more recent measurements at the Japanese Kamiokande II detector. However, Kamiokande, like the Brookhaven experiment, detects high-energy electrons from a rare side reaction of the main proton-proton chain in the Sun:

$$^{7}_{4}\text{Be} + ^{1}_{1}\text{p} \rightarrow ^{8}_{5}\text{B} + ^{0}_{0}\gamma$$

$$^{8}_{5}\text{B} \rightarrow ^{8}_{4}\text{Be} + ^{0}_{1}\text{e}^+ + ^{0}_{0}\nu_e$$

$$^{8}_{4}\text{Be} \rightarrow 2^{4}_{2}\text{He}$$

Lower energy neutrinos are emitted from the first reaction of the proton-proton chain:

$$^{1}_{1}\text{p} + ^{1}_{1}\text{p} \rightarrow ^{2}_{1}\text{H} + ^{0}_{1}\text{e}^+ + ^{0}_{0}\nu$$

There are far more of these lower energy neutrinos than the high-energy neutrinos from the beryllium decays, but they are far more difficult to detect. Gallium is more sensitive to the low energy neutrinos and results from the GALLEX detector were reported in 1992; they too gave a low result for the neutrino flux. There are two ways the solar neutrino problem could be solved:

- Our model of the sun may be incorrect.
- Our understanding of neutrino physics could be incorrect.

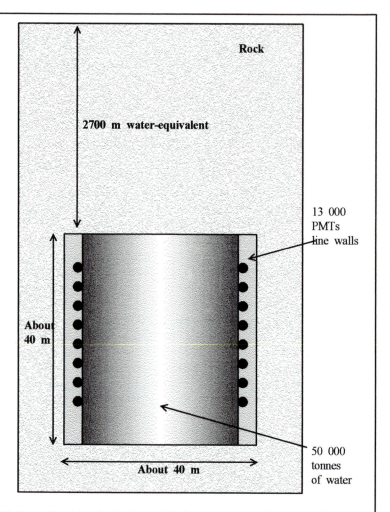

Figure 6.1 Super-Kamiokande is the world's largest neutrino detector. It is a huge cylinder of water 1.7 km inside the Kamioka mine in the Japan Alps about 300 km NW of Tokyo. The walls of the detector are lined with several thousand photomultiplier tubes (PMTs) designed to detect Cerenkov radiation from fast electrons knocked out of atoms by high-energy neutrinos. These electrons emit Cerenkov radiation in a characteristic cone of bluish light whose axis is parallel to the electron's path. The surrounding rock acts as a shield against most cosmic rays. This is important because the weakly interacting neutrinos are incredibly difficult to detect. Super-Kamiokande catches about 30 per day from the Sun. The experiment was designed to study neutrinos from supernovae explosions and the Sun and in 1998 it collected the first strong evidence for neutrino oscillations – the idea that electron, muon and tau neutrinos can change their nature whilst in flight. This is the key to solving the solar neutrino problem and also implies that neutrinos must have mass.

Both possibilities have been considered, but the general feeling is that the problem is in neutrino physics rather than solar physics. Recent work seems to confirm this. In 1998 a collaboration of US and Japanese physicists working at the SuperKamiokande experiment in Japan measured the rate of neutrino production by cosmic rays as they hit the Earth's atmosphere. The SuperKamiokande experiment is a Cerenkov detector containing 50 000 tonnes of pure water 1 km below ground in the Kamioka mine. This detector is sensitive to both electron and muon neutrinos and the pattern of scintillation can be used to indicate the arrival direction of the incoming neutrinos. Since cosmic rays are known to arrive pretty evenly from all directions, and since the Earth is more or less transparent to neutrinos, the experimenters expected to see equal numbers of neutrinos arriving from above and below the detector (those from below having been produced by cosmic rays hitting the atmosphere on the far side of the Earth). What they actually observed was about twice as many coming from above as from below. This would make sense if the muon-neutrinos had changed into tau-neutrinos *en route* from the far side of the world. The detector is not sensitive to the tau-neutrinos, so this would result in a 50% reduction in the total neutrino flux from that direction. Neutrinos created in the atmosphere above the detector would not have time to change. Some unified field theories suggest that the three types of neutrino are each actually quantum mechanical mixtures of several different mass states. As time goes by the relative phase of these mass components change and the nature of the neutrino changes too, oscillating between electron, muon and tau types. This would explain the solar neutrino problem quite neatly; the missing neutrinos have oscillated into other undetected forms during their journey to the Earth (e.g. if they start off as electron neutrinos they might end up as 1/3 electron, 1/3 muon and 1/3 tau neutrinos on arrival). If the neutrino oscillations do take place it would also imply that neutrinos must have mass. This has important cosmological consequences because there are so many neutrinos in the Universe. They may account for some or all of the 'missing mass' (for example, if neutrino mass is about 1 eV then neutrinos account for more mass in the Universe than all the neutrons and protons combined).

6.3 HADRONS

6.3.1 Baryons and Mesons

We have already met two distinct types of hadrons (strongly interacting particles):

- spin-1/2 hadrons such as the proton and neutron, called **baryons**;
- integer spin hadrons like the pion, called **mesons**.

These are also distinguished by their decay modes. Baryon decay always results in a proton and some other particles since the proton is the lightest hadron and probably stable. Baryons cannot decay to mesons, so protons do not decay to pions. Mesons decay to leptons. The equations below are for the decay of a baryon (neutron) and a meson (pi-plus):

$$_0^1 n \rightarrow {}_1^1 p + {}_{-1}^0 e^- + {}_0^0 \bar{\nu}_e$$

$$_1^0 \pi \rightarrow {}_1^0 \mu^+ + {}_0^0 \nu_\mu$$

The observation that baryons do not decay to mesons can be explained by introducing a new property called baryon number (the term 'baryon' was introduced in 1954 to include the nucleons and related heavy particles then known as hyperons). Baryons have baryon number +1, anti-baryons −1 and all mesons have baryon number 0. No decay has ever been observed that violates this principle of conservation of baryon number, although some unified field theories predict that it may happen and protons might decay (experiments looking for proton decay have put a lower limit on the proton lifetime greater than 10^{32} years). The way particle equations have been written in this book includes baryon number as the top number attached to each symbol – for example, in the equations above this shows 1 for the proton and neutron and 0 for all the mesons and leptons.

6.3.2 Strangeness

Rochester and Butler discovered a new heavy meson, the kaon, in 1946. They observed a 'V' shaped track made by a pi-plus pi-minus pair as the kaon decayed. Soon both charged and neutral kaons were observed. They are created in strong interactions when pions collide with protons or protons collide with nuclei, so it was expected that they would also decay by the strong interaction. Strong decays have a characteristic lifetime of about 10^{-23} s but kaons were observed to have lifetimes about 10^{15} times longer than this, around 10 ns, a time comparable to that of the weak decays. This suggested that kaons are formed by the strong interaction and yet decay by the weak interaction. It was clear that something odd was happening and kaons were called 'strange particles' because of this. A clue to the new physics needed to understand the kaon decays came from their mode of production. Whenever kaons were created in collisions between pions and protons the kaon was always accompanied by another strange particle (as in the reaction below), it never came with a pion (a reaction which should occur more frequently because the pion has less mass than any of the strange particles); for example:

$$_{-1}^0 \pi + {}_1^1 p \rightarrow {}_{-1}^0 K^- + {}_1^1 \Sigma^+ \quad \text{where the } \Sigma^+ \text{ (sigma - plus) is a strange baryon}$$

This suggested that the pair of strange particles were needed to balance out some new particle property not possessed by protons or pions. This property was called '*strangeness*', and a new quantum number S was invented. It was assumed that strangeness is associated with particles such as kaons and sigmas and is conserved in all strong interactions but not in weak interactions. This scheme was introduced by Murray Gell-Mann and Kazuhiko Nishijima in about 1953.

It works like this: if the K^+ and K^0 are given $S = 1$ and their anti-particles are given $S = -1$ then the strangeness of all other particles can be worked out from strong reactions that link them to kaons. For example, in the reaction quoted above, the Σ^- must have $S = +1$ because it is created with a K^- ($S = -1$). Kaons cannot decay by the strong interaction because they must conserve strangeness and there are no lighter strange particles. When they decay to pions kaons are violating the conservation of strangeness, so this has to be a weak interaction. In the following years a large number of strange mesons and baryons were discovered. They all fitted into this scheme and that gave physicists confidence that the new quantum number really does correspond to some inherent particle property. The equation below shows why a sigma-plus could not decay to a proton and a pi-minus, a reaction that conserves charge, lepton number and baryon number, but not strangeness:

$$ {}^{1}_{0}\Sigma^0 \rightarrow {}^{1}_{1}p + {}^{0}_{-1}\pi $$

$$
\begin{array}{lllll}
Q & 0 & = & +1 & - & 1 \\
L & 0 & = & 0 & + & 0 \\
B & +1 & = & +1 & + & 0 \\
S & +1 & \neq & 0 & + & 0
\end{array}
$$

Of course, very heavy strange particles can decay to lighter strange particles by the strong interaction, and these decays proceed in a time characteristic of the strong interactions (i.e. lifetimes of about 10^{-23} s).

6.3.3 The Eightfold Way

More and more heavy baryons and mesons were discovered as accelerator energies gradually increased. By the late 1950s there were about 15 and many more were on their way. Murray Gell-Mann and Yuval Ne'eman realised that the particles formed suggestive patterns when they were grouped according to their strangeness and charge. This was very similar to the way Mendeleev grouped the elements according to their chemical properties. He had been able to predict the existence and likely properties of undiscovered elements simply by looking at the gaps in the pattern. In particular, Gell-Mann and Ne'eman found a pattern of 8 baryons that seemed to form a complete pattern (Fig. 6.2) and a similar pattern with 7 mesons. In 1961 an eighth meson, the eta meson, was discovered. It had all the properties required to make the meson pattern identical to the baryon pattern. Gell-Mann called this scheme the 'eightfold way'. The existence of these patterns suggests some kind of underlying hadron structure. On the strength of this Gell-Mann and Ne'eman expected that still heavier baryons should form a group of 10 (fig 6.3). Nine of these particles were known, the four delta particles, the three sigma-stars and two xi-stars. When plotted in the same way as before they form a triangle with its apex missing.

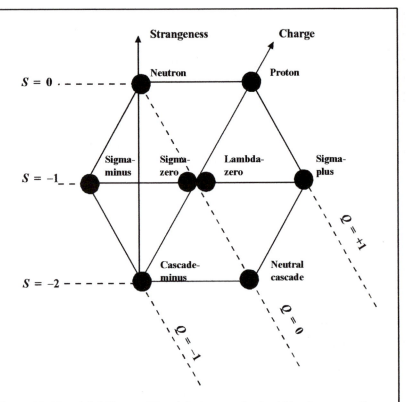

Figure 6.2 The eightfold way. The eight baryons in the table above are all spin-half particles. The simple pattern suggested that they might all be made from a set of simpler particles in much the same way that the periodicity of the Periodic Table derives from an underlying atomic structure.

Reading from the axes (of Fig. 6.3) it is clear that this particle should have strangeness −3 and charge −1. Gell-Mann also predicted its mass. He noticed that each extra unit of strangeness increased particle mass by about 150 MeV/c^2, so he predicted the existence of a particle with a mass 150 MeV/c^2 greater than that of the Xi-stars (which have strangeness −2). He named the particle 'omega-minus' and it was discovered independently at Brookhaven and at CERN in 1963. Its properties fit the predictions almost perfectly. Murray Gell-Mann won the Nobel Prize for physics in 1969 *"for his contributions and discoveries concerning the classification of elementary particles and their interactions"*.

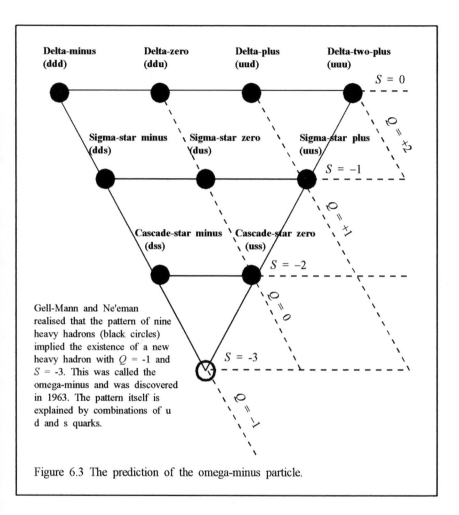

Figure 6.3 The prediction of the omega-minus particle.

"Gell-Mann borrowed a term from Buddhism and called this symmetry principle the eightfold way because the better known particles fell into families with eight members, like the neutron, proton and their six siblings. Not all families were then complete; a new particle was needed to complete a family of ten particles that are similar to neutrons and protons and hyperons but have three times higher spin. It was one of the great successes of the new SU(3) symmetry that this predicted particle was subsequently discovered in 1964 at Brookehaven and turned out to have the masss estimated by Gell-Mann."

(Steven Weinberg, in *Dreams of a Final Theory*, Radius, 1992)

The table below shows the properties of hadrons included in the eightfold way and the subsequent meson nonet and baryon decuplet. Q is charge (in multiples of e), M is mass relative to the proton, B is baryon number, S is strangeness.

Particle	Q	M (rel)	B	S	Lifetime	Spin
Proton (p^+)	1	1	1	0	stable (?)	$\frac{1}{2}$
Neutron (n^0)	0	1	1	0	15 min	$\frac{1}{2}$
Lambda (Λ^0)	0	1.1	1	-1	10^{-10} s	$\frac{1}{2}$
Sigma ($\Sigma^\pm$)	± 1	1.2	1	-1	10^{-10} s	$\frac{1}{2}$
Sigma (Σ^0)	0	1.2	1	-1	10^{-20} s	$\frac{1}{2}$
Delta ($\Delta^\pm$)	± 1	1.2	1	0	10^{-23} s	$\frac{3}{2}$
Delta-zero (Δ^0)	0	1.2	1	0	10^{-23} s	$\frac{3}{2}$
Delta-2-plus (Δ^{++})	2	1.2	1	0	10^{-23} s	$\frac{3}{2}$
Xi-zero (Ξ^0)	0	1.3	1	-2	10^{-10} s	$\frac{1}{2}$
Xi-minus (Ξ^-)	-1	1.3	1	-2	10^{-10} s	$\frac{1}{2}$
Sigma-star ($\Sigma^{*\pm}$)	± 1	1.4	1	-1	10^{-23} s	$\frac{3}{2}$
Sigma-zero-star (Σ^{*0})	0	1.4	1	-1	10^{-23} s	$\frac{3}{2}$
Xi-zero-star (Ξ^{*0})	0	1.5	1	-2	10^{-23} s	$\frac{3}{2}$
Xi-minus-star (Ξ^{*-})	-1	1.5	1	-2	10^{-23} s	$\frac{3}{2}$
Omega-minus (Ω^-)	-1	1.6	1	-3	10^{-10} s	$\frac{3}{2}$
Pion ($\pi^\pm$)	± 1	0.14	0	0	10^{-8} s	0
Pi-zero (π^0)	0	0.14	0	0	10^{-16} s	0
Eta (η^0, η'^0)	0	0.5	0	0	10^{-19} s	0
Kaon ($K^\pm$)	± 1	0.5	0	± 1	10^{-8} s	0
Neutral kaons: K^0	0	0.5	0	1	10^{-8} s	0
and $\overline{K}^0$	0	0.5	0	-1	10^{-10} s	0

6.4 QUARKS

6.4.1 Inside Hadrons

The idea that hadrons might be combinations of simpler more fundamental particles occurred to a number of physicists. In 1964 Gell-Mann and George Zweig suggested that the meson nonet and baryon octet and decuplet could all be derived by combining three distinct spin-$\frac{1}{2}$ particles, and their associated antiparticles. These new fundamental particles were called quarks. In 1967 Richard Feynman suggested an intuitive model of what happens when protons containing quarks (he called them partons and made no assumptions about how many there might be) collide with other particles. He was able to derive certain characteristics of the scattering that could be related back to the quark/parton model. In 1968

Jerome Freidman, Henry Kendall and Richard Taylor directed high-energy electrons at protons at SLAC. The showers of particles were analysed by James Björken in an attempt to figure out what is inside the proton. One thing stood out, the pattern of scattering was the same at all collision energies, it had 'scale-invariance'. Feynman happened to be visiting SLAC in the summer of 1968 and he and Björken discussed the results in the light of Feynman's parton theory. They disovered that they could account for the scattering by identifying the partons with quarks and introducing some additional particles (now called gluons) moving between the quarks. Björken and Emmanuel Paschos developed the theory further.

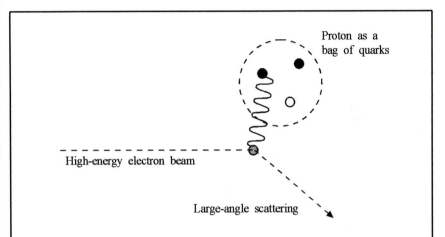

Figure 6.4 Deep inelastic scattering. The first strong evidence for the quark structure of hadrons came from experiments at SLAC in 1968. When high-energy electron beams were fired at fixed targets some of the electrons suffered large angle scatterings rather like those of the alpha particles in the original Rutherford scattering experiment. The interpretation is similar too: the electrons must be interacting with highly concentrated charged particles inside the protons. The scattering takes place via the exchange of a high-energy photon.

Three quark 'flavours' were needed to explain the hadron patterns we have discussed so far.

Quark flavour	Charge	Strangeness	Spin
Up (u)	$+\frac{2}{3}$	0	$\frac{1}{2}$
Down (d)	$-\frac{1}{3}$	0	$\frac{1}{2}$
Strange (s)	$-\frac{1}{3}$	-1	$\frac{1}{2}$
Anti-up ($\overline{u}$)	$-\frac{2}{3}$	0	$\frac{1}{2}$
Anti-down ($\overline{d}$)	$+\frac{1}{3}$	0	$\frac{1}{2}$
Anti-strange ($\overline{s}$)	$+\frac{1}{3}$	$+1$	$\frac{1}{2}$

The 1990 Nobel Prize for Physics was shared by Friedman, Kendall and Taylor *"for their pioneering investigations concerning deep inelastic scattering of electrons on protons and bound neutrons, which have been of essential importance for the development of the quark model in particle physics"*.

The quarks combine in triplets to form baryons and in pairs to form mesons. The baryon octet and decuplet are formed when the quarks combine to give spin-$\frac{1}{2}$ and spin-$\frac{3}{2}$ particles respectively. This is summarised in the table below.

Quark triplet	Strangeness	Charge	Spin-$\frac{1}{2}$	Spin-$\frac{3}{2}$
ddd	0	−1		Δ^-
udd	0	0	n^0	Δ^0
uud	0	+1	p^+	Δ^+
uuu	0	+2		Δ^{++}
dds	−1	−1	Σ^-	Σ^{*-}
uds	−1	0	Σ^0, Λ^0	Σ^{*0}
uus	−1	+1	Σ^+	Σ^{*+}
dss	−2	−1	Ξ^-	Ξ^{*-}
uss	−2	0	Ξ^0	Ξ^{*0}
sss	−3	−1		Ω^-

It is worth making a couple of comments about this table:

- The spin-$\frac{1}{2}$ baryons have two quarks with their spins aligned parallel and the third quark with its spin axis in the opposite direction. The spin-$\frac{3}{2}$ baryons have all three quark spins parallel to the same axis.
- Baryons with higher spin form if the quarks have orbital angular momentum within the baryon in addition to their intrinsic spin. This explains the fact that several hundred baryons have so far been discovered and catalogued.
- Baryons with higher spin contain more energetic quarks, so they are usually more massive.
- A particle such as the omega-minus has three similar quarks all spinning in the same direction. Quarks are fermions so they should obey the Pauli Exclusion Principle, which would rule out the possibility of finding more than one of them with the same set of quantum numbers in the same baryon. This means that quarks must have an additional property that distinguishes them inside the omega-minus. This additional property is called 'colour'. Quarks come in three 'colours', red blue and yellow (or green). The term colour is used by analogy since quarks always combine in clusters that are white or colourless. For example, in each baryon there is always one quark of each colour; it is this that distinguishes the three quarks within the omega-minus. The idea of 'colour' was first put forward by Oscar Greenberg in 1965.

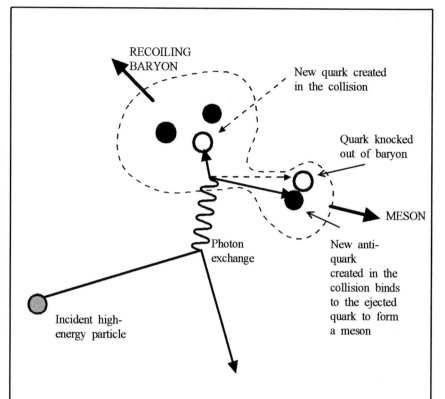

Figure 6.5 The colour force that binds quarks inside hadrons increases in strength as the quarks are moved further apart. In a violent collision the energy needed to separate the quarks becomes large enough to form quark anti-quark pairs. Any ejected quarks bind to anti-quarks to form mesons. Free quarks are never detected.

- Mesons are combinations of quark-anti-quark pairs. Anti-quarks have anti-colour, so mesons too are colourless. They can have spin-0 or spin-1 according to how their intrinsic spins combine, but orbital motion can give them higher spins too, just like the baryons. The table below shows how the meson nonet arises from combinations of up, down and strange quarks and their anti-particles.

- The idea that hadrons are clusters of quarks bound together by gluon exchange is analogous in some respects to the atomic model of electrons bound to a nucleus by photon exchange. However, there is a very important difference. Atoms can be ionised and electrons can become free. Our present theories suggest that quarks cannot become free. The colour force, which binds quarks together, has the strange property that it is very weak at close range and allows the quarks relative freedom inside the hadrons. However, as the quarks begin to separate the force increases in strength. This prevents quarks from ever becoming free particles. In a violent collision the energy needed to separate quarks is large enough to generate 'jets' of new hadrons.

The table below shows the quark structure of the mesons.

Quark pair	Strangeness	Charge	Meson (spin 0)
$d\bar{u}$	0	−1	π^-
$u\bar{u}$	0	0	$\pi^0 \, \eta^0 \, \eta'^0$
$d\bar{d}$	0	0	
$s\bar{s}$	0	0	
$u\bar{d}$	0	−1	π^-
$s\bar{d}$	−1	0	$\bar{K^0}$
$s\bar{u}$	−1	−1	K^-
$d\bar{s}$	1	0	K^0
$u\bar{s}$	1	1	K^+

Notice that the pi-zero and eta mesons do not correspond to unique quark combinations. Each of these mesons is a combination of quark states.

6.4.2 The Quark-Lepton Symmetry

Beta-decay has revealed a lot of new physics and the underlying process in which a neutron converts to a proton links quarks and leptons. The neutron is udd and the proton uud, so the weak interaction involved in beta decay has changed the flavour of a quark from d to u.

Sheldon Glashow, Steven Weinberg and Abdus Salam proposed a unified theory of weak and electromagnetic interactions in the mid-1960s. One consequence of this was a link between leptons and quarks. In particular the electron and electron-neutrino correspond to the down and up quarks. This suggested that the muon and muon-neutrino would correspond to a second generation of quarks, one of which was, presumably, the strange quark. They were able to predict the properties of the missing quark, now called the 'charmed' quark. It should have a charge $+\frac{2}{3}$ and a larger mass than any of the three known quarks (otherwise particles containing charmed quarks would already have turned up in collision experiments).

The first charmed particle, the J/ψ meson (made from a charmed quark and its anti-quark) was discovered independently by two groups, one led by Richter, Perl and Goldhaber using SPEAR (Stanford Positron Electron Asymmetric Ring) and another led by Ting at Brookhaven. The name J/ψ derives from the names given to the particle by the two groups, ψ being particularly apt because the decay of the particle leaves tracks in the shape of this character. Burton Richter and Samuel

Ting shared the 1976 Nobel Prize for Physics *"for their pioneering work in the discovery of a heavy elementary particle of a new kind"*.

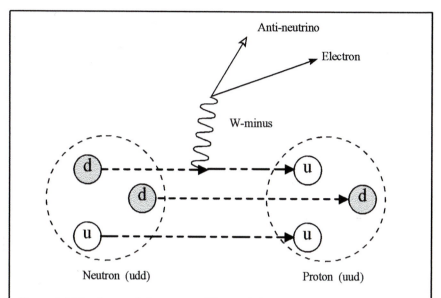

Figure 6.6 The decay of the neutron. The weak interaction is the only interaction that can change the flavour of quarks. Here a down quark radiates a W-minus boson and becomes an up quark. The W-minus rapidly decays to an electron (beta-particle) and an anti-neutrino.

We have already mentioned the discovery of the tau lepton in 1974. This, together with its tau-neutrino forms a third generation lepton doublet that is also linked to another pair of quarks, the bottom and top quarks. It might look like the quarks and leptons will take over from the hadrons in generating a never-ending sequence of new 'fundamental' particles, but this is not the case. Evidence from particle physics and cosmology fixes the number of particle generations at three:

- The Z^0 particle (a force-carrying boson required by electroweak unification) was discovered at CERN in 1983. It is a very massive particle, about 100 times the rest mass of the proton, so it should be able to decay into a large number of different lighter particles. The rate at which it is produced and its lifetime both depend on the number of particle generations (if there are more generations there will be more decay modes and so a shorter lifetime). Accurate measurements at CERN indicate that there is no fourth generation of leptons or quarks.
- The ratio of abundances of deuterium and helium in the early universe is also sensitive to the number of particle generations (if there are more generations

then the energy in the early universe would be spread more thinly and the amount of helium is less). It is possible to estimate the ratio from astronomical observations and the result is consistent with three generations only.

Generation	Leptons	Quarks
1	e^- ν_e	d u
2	μ^- ν_μ	s c
3	τ^- ν_τ	b t

The bottom quark first revealed itself in the upsilon particle discovered by Lederman at Fermilab in 1977. The top quark was also discovered at Fermilab in 1995.

Generation	Quark	Q/e	E_0/GeV	Spin/$\hbar$
1	d	$-\frac{1}{3}$	0.008	$\frac{1}{2}$
1	u	$+\frac{2}{3}$	0.004	$\frac{1}{2}$
2	s	$-\frac{1}{3}$	0.150	$\frac{1}{2}$
2	c	$+\frac{2}{3}$	1200	$\frac{1}{2}$
3	b	$-\frac{1}{3}$	4700	$\frac{1}{2}$
3	t	$+\frac{2}{3}$	93000	$\frac{1}{2}$

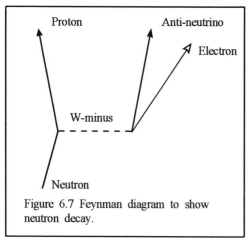

Figure 6.7 Feynman diagram to show neutron decay.

6.5 FORCE CARRIERS

6.5.1 Gauge Bosons

Quantum electro-dynamics represents electromagnetic forces by an exchange of virtual photons between charged particles. The amazing success of this theory made it into a kind of template for other quantum field theories. Fermi's theory of beta decay involved the exchange of an electron and an anti-neutrino, two particles which, if exchanged together might act like a spin-1 boson. Oscar Klein introduced the idea of a spin-1 exchange particle, the 'W boson' in 1938 and, using a Feynman diagram, the decay of the neutron is shown in Fig. 6.7.

Later attempts to unify electromagnetism and the weak interaction showed that the electroweak interaction requires *four* exchange particles, the photon, the charged W^+ and W^- particles and another massive neutral boson, the Z^0. These each mediate a characteristic electroweak interaction. Examples are shown in Feynman diagrams below.

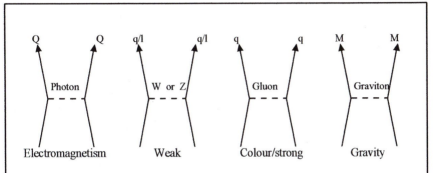

Figure 6.8 Feynman diagrams of the four fundamental forces. Electromagnetism acts on charged particles, the weak force acts on quarks and leptons, the colour force acts on quarks and gravity acts on massive particles. There is good evidence for all of the force carrying particles except the graviton.

- Photons are responsible for electromagnetic forces between charged particles, but do not affect the nature of the interacting particles.
- $W^\pm$ particles change the flavour of quarks and are responsible for beta decay processes.
- Z^0 particles allow neutral weak interactions to take place. An example would be the scattering of a neutrino from a proton, where neither particle changes its nature in the process. (Photons cannot mediate these scattering events because neutrinos do not interact via the electromagnetic interaction).
- $W^\pm$ particles and Z^0 are massive particles, 83 GeV/c^2 and 93 GeV/c^2 respectively, approaching 100 times the mass of the proton and are very short-lived, around 10^{-25} s each, so they can only be produced experimentally in very high energy collisions.

The $W^\pm$ and Z^0 particles were discovered in 1983 in proton-antiproton collisions at CERN and Carlo Rubbia and Simon van der Meer were awarded the 1984 Nobel Prize for Physics *"for their decisive contributions to the large project, which led to the discovery of the field particles W and Z, communicators of the weak interaction"*.

Yukawa's theory of the strong interaction involves the exchange of mesons between baryons. The fact that both mesons and baryons are made of quarks, which are themselves bound by gluons indicates that nucleon binding is actually a consequence of the colour force, not a fundamental interaction. The theory of the colour force is another quantum field theory, set up using QED as a guide and is called Quantum ChromoDynamics or QCD for short. The ideas are familiar: quarks are bound to one another by exchanging gauge bosons called gluons. The gluons are spin-1 bosons and come in eight varieties (each with a corresponding anti-gluon). Gluon carries a colour difference (e.g red-minus-blue) so they can change the colour of the quarks that emit and absorb them (Fig. 6.9). Colour is the source of gluon exchange just as charge is the source of photon exchange. However, whereas photons are themselves neutral and so do not interact with other photons or charges by exchanging more photons, gluons carry colour and can radiate more gluons as they are exchanged. This makes QCD far more difficult to work with than QED. Another strange property of QCD is 'asymptotic freedom'. The colour force between quarks becomes very small when the quarks are close together, as they are inside a hadron. In some respects a hadron acts like a 'bag' in which the quarks and gluons bounce about like free particles. However, as more energy is supplied and quarks separate, the colour force becomes very strong and so much energy is stored in the colour field that new quark anti-quark pairs are created. This results in jets of new particles shooting out of hadrons in high-energy collision experiments and provides the most convincing evidence for the internal structure of the hadrons.

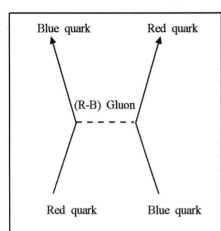

Figure 6.9 Colours are conserved. This means that gluons must carry colour differences. This means gluons can interact via the colour force, making QCD much more difficult to deal with than the electroweak forces.

The only fundamental interaction we have left out is gravity. Our best model of gravitation is Einstein's general theory of relativity in which it is a manifestation of the distortion of space-time by the presence of mass and energy. So far no one has managed to write a quantum field theory for the gravitational interaction, but if they succeed a new exchange particle, the graviton would be required. Recent work on superstring theory suggests that a ten-dimensional quantum field theory might unify all of the fundamental interactions. One of the most suggestive aspects of the theory is that Einstein's theory of gravity is automatically built in. This is clearly an area in which major advances are likely in the twenty-first century.

Figure 6.10 Summary of the Standard Model.

MATTER

All quarks and leptons are fermions or spin-1/2 particles

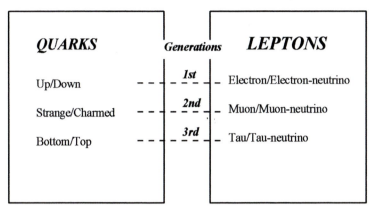

QUARKS	Generations	LEPTONS
Up/Down	1st	Electron/Electron-neutrino
Strange/Charmed	2nd	Muon/Muon-neutrino
Bottom/Top	3rd	Tau/Tau-neutrino

INTERACTIONS

All force-carriers are bosons or integer spin particles

ELECTROMAGNETISM

Force-carrier =   Photon

WEAK INTERACTION

Force-carriers = W-plus/minus
                          Z-nought

STRONG/COLOUR INTERACTION

Force-carriers = eight different gluons
(each carries a colour difference)

GRAVITATION

Force-carrier = graviton

7
PARTICLE DETECTORS

7.1 IONISING RADIATION

Particle physics in the twentieth century depended on two parallel experimental developments. Advances in accelerator physics enabled higher and higher energy collisions to take place to order in the laboratory. Increasingly sophisticated detectors enabled more rapid, sensitive and automatic observations to be made, so that the rare but significant events leading to new physics could be captured. Several Nobel Prizes were awarded to physicists who invented new kinds of detectors and others went to those who managed to use them to tease out new discoveries.

7.1.1 Catching Cosmic Rays

As we enter the twenty-first century the cost of large accelerators has become prohibitively high and the cancellation of the American superconducting supercollider (SSC) may come to symbolise the end of the era of accelerator physics. Perhaps it is reassuring to realise that, however many TeV (1 TeV = 1000 GeV = 10^{12} eV) we manage to squeeze out of existing facilities such as CERN there is still a source of high-energy particles that reaches energies we can only dream about in laboratory experiments. Cosmic rays send occasional particles with energies in excess of 10^{18} GeV! The problem is, of course, that these are rare unpredictable events, but they are observed and they could reveal new physics. In fact the majority of particles discovered during the first half of the twentieth century were as a result of cosmic ray events. Some of these are listed in the table below.

Year	Particle	Discovered by	Detector
1932	Positron	C. Anderson	Cloud chamber
1937	Muon	S. Nedemeyer & C. Anderson	Cloud chamber
1947	Charged pions	C. Powell	Photo-emulsion
1947	Charged kaons	G. Rochester & C. Butler	Cloud chamber
1947	Neutral kaon	G. Rochester & C. Butler	Cloud chamber
1951	Lambda	C. Butler	Cloud chamber
1952	Sigma	G. Tomasini	Photo-emulsion
1953	Xi-Minus	R. Amenteros	Cloud chamber

Early experiments with radioactive sources measured activity by the rate at which ionisation of the air would discharge a previously charged electroscope. Shielding the electroscope from the radioactive source could reduce this rate but never reduced it to zero. This seemed to imply the presence of another very penetrating radiation. In 1910 Father Theodor Wulf took an electroscope to the top of the Eiffel Tower assuming that the penetrating radiation would be less effective there if it originated in the Earth. In fact the radiation was still as powerful and he wondered whether it might come from an extraterrestrial source. The crucial experiments were carried out by the Austrian physicist and amateur balloonist, Victor Hess (1883-1964). In 1911 and 1912 he took electroscopes up in balloons and measured their rate of discharge up to an altitude of 5000 m. At that height the electroscopes discharged about 10 times more rapidly than at sea level. He concluded that the radiation really did come from outer space and his work was taken further by Millikan, who automated the measurements so that they could be carried out in unmanned balloons. In 1925 Millikan coined the term 'cosmic rays' for this high-energy penetrating radiation and he, like most other physicists at that time, assumed it consisted of ultra-hard gamma rays from extraterrestrial nuclear reactions.

In 1928 Hans Geiger and Walther Müller invented the 'Geiger counter' (see below) which was a much more sensitive detector that could be set up to produce an electrical pulse when an event occurred. At this time cosmic rays were thought to be high-energy gamma-rays, and these interact weakly with matter by ejecting electrons from atoms. In 1929 Walther Bothe and Werner Kolhörster in Berlin set up a pair of Geiger counters on either side of a 4.0 cm thick gold block. The outputs of the two detectors showed many coincidences. This implied that both detectors had been triggered by the same cosmic ray. However, it was extremely unlikely that one cosmic ray would happen to eject an electron in both tubes, and electrons ejected from atoms in the first detector had been prevented from reaching the second by the gold. They realised that cosmic rays must be high-energy highly penetrating charged particles, not gamma rays. Bruno Rossi in Florence took their

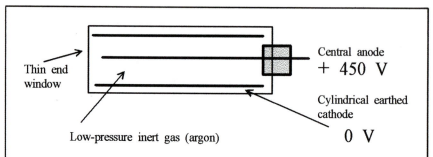

Figure 7.1 A Geiger tube. When radiation enters the tube it ionizes the low-pressure argon. Electrons accelerate toward the anode and cause further ionisation. An avalanche of charge flows and a short electrical pulse appears across the electrodes.

work further, showing that coincidences still occur with a 1.0 m thick lead absorber between the two Geiger counters – an indication of the extremely high energies involved. Rossi monitored coincidences using an electronic device, the forerunner of many modern automatic detection methods.

Experiments at different latitudes showed that the distribution of cosmic rays reaching the Earth is significantly affected by the Earth's magnetic field. It was also realised that most of the radiation received at sea level is secondary radiation generated when primary cosmic rays collide with particles near the top of the Earth's atmosphere. The flux of cosmic rays at the top of the atmosphere is about 200 000 particles per square metre per second and this falls to about 10 000 particles per square metre per second at sea level.

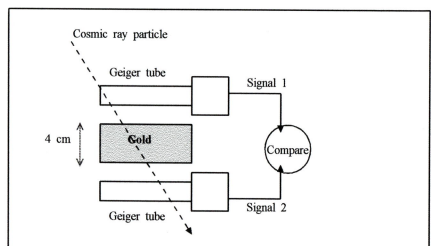

Figure 7.2 Bothe and Kolhorster showed that cosmic rays could penetrate a 4 cm gold block. They looked for simultaneous signals from Geiger tubes placed on either side of the block. This showed that the cosmic rays must have high energy.

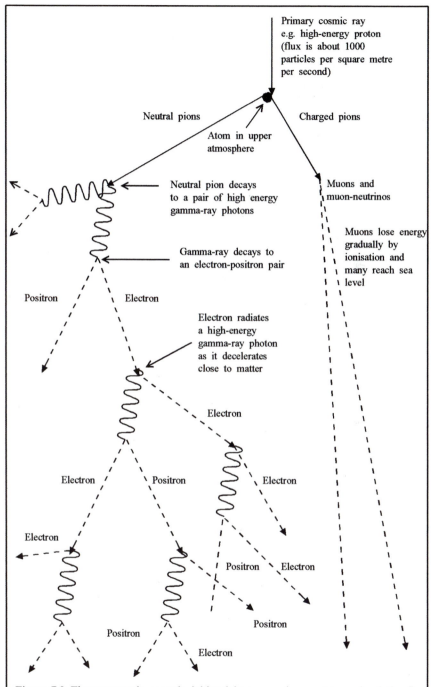

Figure 7.3 Electromagnetic cascade initiated by a cosmic ray. At sea level the flux of secondary particles is approximately 25% muons and 75% electrons/positrons.

In the 1930s Patrick Blackett and Giuseppe Occhialini used Rossi's coincidence method to trigger a cloud chamber between two Geiger counters. This led to a much higher success rate in photographing the tracks of cosmic ray particles and caught the first photographic images of 'cosmic ray showers', showing among other things that electron positron pairs are fairly common products of high energy collisions. Extremely high-energy cosmic ray particles (in excess of 10^{18} GeV) are rare, but when they hit the top of the atmosphere they generate an 'extensive air shower' which can be picked up simultaneously at detectors in laboratories many kilometres apart.

It is now known that cosmic rays are mainly high energy protons accompanied by a small proportion of helium nuclei, heavier nuclei and electrons. These primary cosmic rays collide with atoms in the atmosphere and generate showers of secondary particles. The main products from the initial high altitude collisions are charged and neutral pions. The charged pions decay to muons and muon antineutrinos whereas neutral pions decay to pairs of high-energy gamma-ray photons. These photons generate cascades of high-energy electrons and positrons, and more gamma rays as they interact with matter. Close to the ground this results in two components of cosmic rays – muons and electrons. The muons are a 'hard component' losing energy gradually as they ionise the matter through which they pass. The electrons are a 'soft component' losing energy more rapidly as they decelerate and radiate.

7.1.2 The Wilson Cloud Chamber

The cloud chamber was the first device to reveal the tracks of subatomic particles. However, it was invented quite by chance when Charles Wilson, returning from the Scottish mountains in 1894, tried to recreate atmospheric mists under laboratory conditions. His basic idea was to take a container of moist air and expand it suddenly. This lowers its temperature and the air becomes supersaturated with water vapour. Dust particles act as centres for water to condense on and mist forms. However, he also discovered, to his surprise, that a thin mist forms even if the air is completely free of dust particles. This suggested some other small particles on which the water could condense. He thought it might be ions in the air and confirmed this using an X-ray source, but did not realise the significance of his discovery for over a decade. In 1910 he placed radioactive sources in the 'cloud chamber' and observed the wispy tracks they left behind.

Blackett automated the technique by building a cloud chamber that works cyclically every few seconds and is photographed after each expansion. This allowed him to record and analyse a very large number of events including cosmic ray showers, pair production and the first direct evidence of induced nuclear transformation (an alpha particle hitting a nitrogen nucleus and forming oxygen). Charles Wilson shared the 1927 Nobel Prize for Physics *"for his method of making the paths of electrically charged particles visible by condensation of vapour"*. Lord Patrick Blackett won the Nobel Prize for Physics in 1948 *"for his development of the Wilson Cloud chamber method, and his discoveries therewith in the fields of nuclear physics and cosmic radiation".*

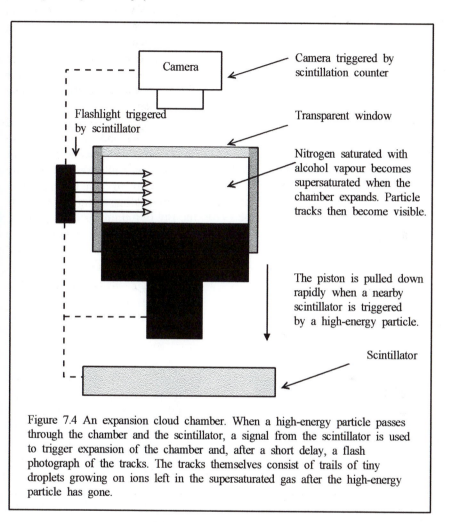

Figure 7.4 An expansion cloud chamber. When a high-energy particle passes through the chamber and the scintillator, a signal from the scintillator is used to trigger expansion of the chamber and, after a short delay, a flash photograph of the tracks. The tracks themselves consist of trails of tiny droplets growing on ions left in the supersaturated gas after the high-energy particle has gone.

The earliest observations of both radioactivity and X-rays involved a different kind of detector, photographic emulsions. The development of very sensitive emulsions in the 1940s provided another way to study both cosmic rays and the high-energy collisions brought about by accelerators. Photographic emulsions also had some advantage over cloud chambers. They were simpler to use and allowed greater accuracy in measuring particle tracks and ranges. Their crowning glory was the discovery, twelve years after its prediction, of Yukawa's pion. Occhialini had taken some photographic plates to the French observatory on the Pic du Midi in 1945 and at that high altitude he was able to see tracks of primary cosmic ray particles in the emulsion. Among these, in 1947, Powell identified the pions. Powell won the 1950 Nobel Prize for Physics *"for his development of the photographic method of studying nuclear processes and his discoveries regarding mesons made with this method"*.

7.1.3 Glaser's Bubble Chamber

The cloud chamber was an amazingly successful detector leading to many important discoveries during the first half of the twentieth century. However, as accelerators reached higher and higher energies, the chance of an important event being captured as particles shot through the tenuous gas became smaller and smaller. After 1952 no major discoveries were made using cloud chambers, but in the same year Donald Glaser at Michigan University began to develop a liquid-based detector operating on similar principles. He was encouraged by Luis Alvarez and Glaser's 2.0 cm diameter prototype soon led to much larger devices such as the 80 inch at Brookhaven and, in the 1970s, the Big European Bubble Chamber at CERN, which was 3.7 m across!

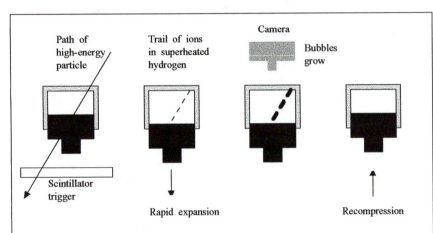

Figure 7.5 A hydrogen-filled bubble chamber. This cycle could be repeated up to 1000 times per second.

Many new particles were discovered using bubble chambers, those listed below helped confirm the eightfold way patterns that led to the quark hypothesis.

Year	Particle	Team leader	Laboratory
1956	Sigma-zero	R. Plano	Brookhaven
1959	Xi-zero	L. Alvarez	Berkeley
1964	Omega-minus	V. Barnes	Brookhaven

Bubble chambers were built using a variety of liquids, but the ideal choice was liquid hydrogen because the hydrogen nucleus is a single proton and the debris of high-energy collisions is then be simpler to analyse. The principle behind the bubble chamber is straightforward: hydrogen is kept at high pressure and low temperature in a large container with glass walls. The pressure is suddenly released and the hydrogen becomes superheated, i.e. it finds itself in a liquid state at a temperature at which it would normally (at the lower pressure) be gaseous. Bubbles of gaseous hydrogen rapidly grow on any suitable particles. If the pressure release is timed to coincide with the arrival of a pulse of particles from an accelerator then the bubbles form on the ions left behind by charged particles passing through the chamber. This is likely to include collision events in which incident particles have struck hydrogen nuclei and showers of new particles have been produced. A short time after the pressure release the chamber and its tracks are photographed and then the high pressure is restored. This process is repeated as successive pulses arrive from the accelerator. Bubble chambers have many advantages over cloud chambers:

- The liquid acts as a target as well as a detector.
- They respond more rapidly than cloud chambers and can be synchronised to expand as new particles are injected from an accelerator.
- Liquids are denser than gases, so more interactions take place per cm of track, meaning the chance of photographing an interesting but rare event is increased.
- Bubble chamber tracks resolve detail far more clearly than cloud chamber tracks.

Large bubble chambers presented a daunting technological challenge requiring complex cryogenics and powerful electromagnets (later superconducting magnets). The '72-inch' at Berkeley contained 500 litres of liquid hydrogen at -250^0 C in a magnetic field of 1.5 T. It took four years to build and cost 2.5 million dollars in the late 1950s. By the 1970s the BEBC contained 35 000 litres of liquid hydrogen. But the cost and efforts were justified. A large number of new baryons and mesons were discovered in bubble chambers and they continued to be useful in fundamental research up until about 1980. They also altered the way high energy physics is done. They could be synchronised with the cycles of the accelerators to which they were connected and generated *millions* of photographs per year. At first these were analysed by hand by teams of scientists, later the process was automated. In some cases the search for a new particle needed cooperation between research groups in several countries, just to wade through the data churned out and catalogued at the experimental site. Physics had come a long way from the simple desk-top experiments that revealed the electron and even the neutron. The heroic age where a Rutherford or Thomson could claim pretty well all the credit for a fundamental discovery had passed.

Donald Glaser won the 1960 Nobel Prize for Physics *"for the invention of the bubble chamber"*. Luis Alvarez won the Prize in 1968 *"for his decisive contributions to elementary particle physics, in particular the discovery of a large number of resonance states, made possible through his development of the technique of using hydrogen bubble chambers and data analysis"*. 'Resonances' are short-lived hadrons identified by the sharp increase in interactions at certain collision energies. The interpretation is that these energies are close to the rest mass of highly unstable hadrons which almost immediately decay to showers of lighter particles. The nature of the resonance can be implied from the properties of particles in the shower and its lifetime can be estimated from the resonance peak. This is a direct application of the Heisenberg Uncertainty Principle – the greater the uncertainty in energy at which the resonance occurs, the shorter its lifetime. Hundreds of resonances were identified from bubble chamber photographs.

7.2 RAPID RESPONSE

7.2.1 Cerenkov Detectors

It is impossible for a particle to travel faster than the speed of light in empty space. However, the speed of light in a medium is less than it is in a vacuum because the interaction of photons with electrons results in continual absorption and re-emission. This means it is possible for a high energy particle traveling through a material to move faster than the speed of light in that material. When it does so it creates an electromagnetic shock wave analogous to the sonic boom of an aircraft travelling faster than the speed of sound in air. The effect was discovered by Pavel Cerenkov in 1934 and is responsible for the blue glow that surrounds spent nuclear fuel when it is stored under water after being removed from a reactor. It has been used in a number of particle detectors to reveal the existence, path and speed of high-energy charged particles. The usual arrangement is to surround the location of interesting events with an array of photo-tubes. These react to the Cerenkov light and the pattern and order in which they detect it can be used to reconstruct the trajectory of the particle. It can also be used to filter out events in which the high-energy particles have the 'wrong' velocity. This is possible because the Cerenkov radiation is emitted at an angle to the particle direction and the angle itself depends on the particle velocity.

Cerenkov detectors were used by Hans Sègre in 1955 when he successfully identified the antiproton. Antiprotons have the same mass as a proton but opposite charge so it was essential to show that candidate particles had the same mass as protons and yet curved the other way in an applied magnetic field. The experiment was carried out at the Lawrence Berkeley Laboratory using the Bevatron, a 6 GeV proton accelerator that came into operation in 1954. The protons were fired into a fixed target where proton-proton collisions were expected to create just 1 antiproton for every 50 000 pions. In a strong magnetic field particles with similar momentum are deflected along the same path, according to the equation:

$$r = \frac{mv}{Bq}$$

The problem was that negative pions and antiprotons with the same momentum would be selected together (since both have the same B and q). However, antiprotons have larger mass and so lower velocity at the same momentum. To distinguish them from negative pions Sègre had to measure their velocity. He did this using two Cerenkov detectors, one set up to respond to the negative pions, the other to catch the slower antiprotons. The antiproton detector was a specially designed quartz counter constructed by Wiegand and Chamberlain. Quartz has a refractive index of 1.46 and a Cerenkov angle of $46.7°$. Several other materials have been used, including air ($n = 1.000283$, $\theta_c = 1.36°$), water ($n = 1.33$, $\theta_c = 41.2°$) and isobutane ($n = 1.00127$, $\theta_c = 2.89°$). The energy loss by Cerenkov radiation is small compared to ionisation processes (about 1%).

Pavel Cerenkov shared the 1958 Nobel Prize for Physics with Il'Ja Mikhailovich Frank and Igor Yevgenyevich Tamm *"for the discovery and interpretation of the Cerenkov effect"*. Emilio Gino Segrè and Owen Chamberlain shared the 1959 Nobel Prize in Physics *"for their discovery of the antiproton"*.

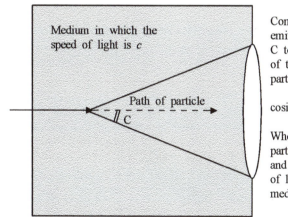

Medium in which the speed of light is c

Path of particle

C

Cone of photons emitted at angle C to the direction of the super-luminal particle:

cosine (C) - v/c

Where v is the particle velocity and c is the speed of light in the medium

Figure 7.6 Cerenkov radiation is emitted when a particle travels at a speed greater than the speed of light in a particular medium. The radiation forms a characteristic cone enabling Cerenkov detectors to be used to determine both the position of emission and the speed and direction of the emitting particle. The arriving photons are usually detected with an array of photomultiplier tubes.

7.2.2 Scintillation Counters

The earliest detectors used light emitted from fluorescent salts to indicate the arrival of charged particles. An example is the spinthariscope, used by Rutherford and his contemporaries. When alpha particles hit a screen coated with zinc sulphide small flashes of light were emitted and could be counted. This basic technique is used in a variety of ways in modern scintillators. These are multi-purpose detectors that can be used in calorimetry (calculation of total energy deposited in the material), time of flight measurements, tracking detectors (using bundles of scintillator fibres) and as triggers or vetoes for other detectors.

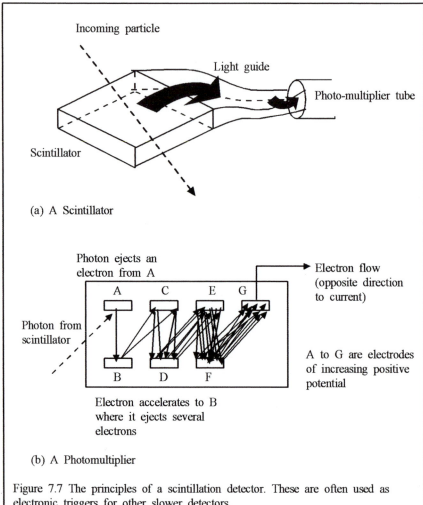

Figure 7.7 The principles of a scintillation detector. These are often used as electronic triggers for other slower detectors.

The idea is simple: energy deposited by the ionizing particle results in the emission of scintillation light (luminescence). The emitted light is then directed to a photodetector (e.g. photomultiplier tube or photodiode) where it creates an electrical signal. Various materials are used in scintillators:

- Inorganic crystalline scintillators (e.g. NaI, CsI, BaF_2, etc). The incoming particle excites electrons from the valence to conduction bands. The inevitable de-excitation emits photons either immediately (fast recombination in a few ns) or after a delay due to trapping before de-excitation can occur (perhaps after 100 ms). These scintillators have high density and are good for detecting charged particles and gamma rays.
- Liquid noble gases (e.g. Ar, Kr, Xe) are excited by the radiation and form short-lived excited or ionised molecules as they collide with other atoms in the gas. De-excitation, dissociation and recombination processes emit ultra-violet light, which can be converted to visible light using wavelength shifters.
- Organic scintillators (such as naphthalene, anthracene or liquid and plastic scintillators) contain many carbon-carbon bonds. Incoming particles excite the electrons in these bonds and de-excitation results in the emission of ultra-violet light.

The output pulse in the photodetector is proportional to the number of photons emitted in the scintillator. Since this may amount to the total energy of the incident particle, the strength of the output pulse is a measure of particle energy. This is why scintillators are used as calorimeters. Their rapid response (nanoseconds in some cases) also makes them excellent triggers.

7.3 MODERN DETECTORS

7.3.1 Multiwire Detectors

Air becomes conducting when the electric field strength exceeds about 3×10^6 Vm^{-1}. The electric field tears electrons off air molecules and accelerates them rapidly. The electrons gain enough kinetic energy between collisions to ionise more air molecules and create an 'avalanche' of charge – a spark. If the electric field between two electrodes is held just below this breakdown strength then the passage of high-energy charged particles between the electrodes is sufficient to trigger an avalanche that can be used to detect the particles. This is a spark-counter and it does effectively what a Geiger tube does, except that its electrodes are surrounded by low pressure argon. During the 1950s and 1960s the idea behind the simple spark counter was adapted by Frank Kiernan at CERN in large wire spark chambers which could record particle tracks automatically up to 1000 times per second. This was taken further by Georges Charpak, also at CERN, who invented the multiwire proportional chamber, a detector capable of extremely rapid operation and high resolution. In fact the response of these detectors is so rapid that they are often used to trigger other slower detectors just as Blackett used Geiger tubes to trigger a cloud chamber.

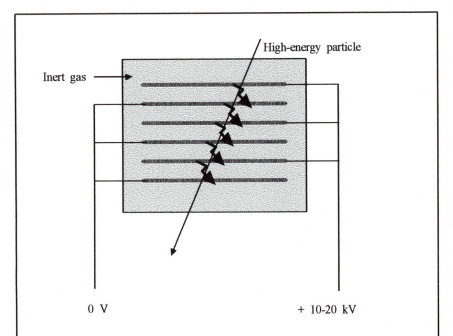

Figure 7.8 The spark chamber. High-energy particles leave a trail of ionisation in an inert gas. This is sufficient to initiate electrical breakdown between oppositely charged electrodes. A line of sparks makes the track visible. If the electrodes are replaced with fine wires, breakdown can be detected electronically by small currents that flow in the wires. These ideas led to two important modern detectors, the multi-wire proportional chamber and the time projection chamber.

Inside a multiwire proportional chamber is a lattice of fine 'sense wires' held at a positive potential with respect to planes of negative wires running through the chamber. If a charged particle passes through the chamber electrons released by ionisation rapidly move onto the nearest sense wires. The outputs of these wires are monitored electronically and a two-dimensional path can be reconstructed from the sequence of output pulses. The senses wires are just a couple of millimetres apart, so the electrons do not have far to move, giving the detector a rapid response and good resolution. Georges Charpak won the Nobel Prize for Physics in 1992 *"for his invention and development of particle detectors, in particular the multiwire proportional chamber"*.

  This is taken one step further in a 'drift chamber', which records the time at which the electrons arrive at each sense wire as well as the wire that detects them. The basic idea is very similar to the multiwire proportional chamber, but sense wires are further apart and the surrounding wires are used to set up an electric field along the length of the chamber so that electrons drift at a constant velocity toward the nearest sense wire. The advantage of knowing time of arrival is that the position of ionisation can be worked out much more accurately and paths can be resolved within about 50 μm. The disadvantage of the drift chamber is that its response is much slower (because of the drift times) than the multiwire proportional chamber.

David Nygren at SLAC invented the time projection chamber in the 1970s. This relies on the same principles as a drift chamber, with the major difference that the electrons drift up to 1 m to sense wires arranged in small sections at the end of the chamber. These wires record a 2D image of the particle path through the chamber, but they also record the time of arrival of the electrons. Knowing the arrival time makes it possible to calculate how far from the end of the chamber the electrons started drifting. Computers are used to reconstruct three-dimensional images of the particle paths. Time projection chambers are used inside some of the large layered detectors that surround the collision points in colliders such as LEP (the Large Electron positron Collider) at CERN.

All of these detectors are operated in strong magnetic fields that have to be maintained over several cubic metres. These involve the largest superconducting magnets in the world and are necessary so that the magnetic deflections of very high-energy particles are large enough to measure. There are many clues in particle tracks that can be used to work out the properties of the particles that made those tracks and in some cases to infer the existence of particles through missing tracks:

- **Curvature**: the path of a charged particle is curved because of the magnetic force exerted on it. If the curvature is measured it can be used to calculate the momentum of the particle and hence its total energy. In many early cloud chamber and bubble chamber photographs it is obvious that the curvature changes along the path, the radius of curvature reducing. This is because a particle loses energy as it interacts with the material in the chamber, and the effect can be used to determine the direction in which a particle is moving. This was important because it allowed tracks due to negatively charged particles to be distinguished from those caused by positively charged particles moving in the opposite direction.

  Magnetic force $\qquad F = Bqv$

  Radius of curvature $\qquad r = \dfrac{p}{Bq}$

  $p = mv$ (non-relativistic) and $p = \gamma mv$ (relativistic)

  $KE = \dfrac{p^2}{2m}$ (non-relativistic) and $TE = \sqrt{p^2 c^2 + m_0^2 c^4}$ (relativistic)

- **Intensity**: usually a good indication of the charge on a particle – the more charge the more intense the track. However, faster particles produce less intense tracks. Modern layered detectors use concentric rings of calorimeters laced with detectors to measure the total energy of particles stopped in them.

- **Length**: most high-energy particles produced in modern collision experiments have velocities very close to the speed of light. The length of a track is therefore a direct measure of the lifetime of the particle. On the face of it a particle with a lifetime of 10^{-12} s (typical of hadrons containing bottom quarks, e.g. B mesons) would only move 3×10^{-4} m before decaying, and it is difficult to detect a path of this length. However, time dilation means these particles survive longer in the laboratory frame and so move farther. B meson tracks can be detected close to the collision point using solid state detectors.

- **Conservation laws**: paths due to charged particles are usually reasonably easy to detect, but neutral particles and especially neutrinos are much more difficult to catch. Modern layered detectors such as ALEPH at CERN use a variety of detectors spread around the collision point and combine information from all of them, together with laws of conservation of energy, momentum, etc., to work out all the particles involved in any particular event. Neutrinos are not stopped in these detectors, but just about everything else is, so the total energy carried away by neutrinos can be calculated from the difference between the collision energy and the total energy recorded by calorimeters.

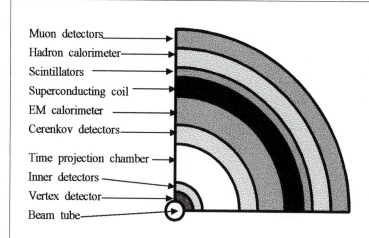

Figure 7.9 A cross-section through the DELPHI layered detector at CERN. The LEP beams collide about 30 000 times per second inside the detector, generating an enormous amount of data. A great deal of this must be rejected so the outputs from 200 000 electronic channels are analysed to identify characteristic signatures of interesting events.

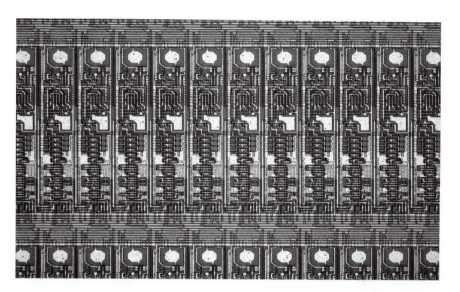

Figure 7.10 Close-up of a pixel detector readout chip. The photograph above shows an area of 1 mm × 2 mm containing 12 separate readout channels. The entire chip contains about 1000 readout channels (80 000 transistors) covering a sensitive area of 8 mm × 5 mm, The chip has been mounted on a silicon detector and high energy particles have been detected.

Photo credit: CERN

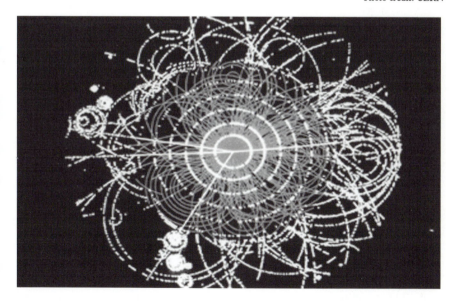

Figure 7.11 Simulation of the tracks that might be created in the ATLAS experiment by the decay of a hypothetical Higgs particle to 4 muons (strong white tracks radiating from centre of detector).

Photo credit: CERN

Figure 7.12 Super Kamiokande is designed to detect neutrinos. This image shows the inner detector half-filled with water. The walls are lined with thousands of photomultiplier tubes. Just visible on the right are some physicists in a dinghy!

Photo credit: University of Tokyo

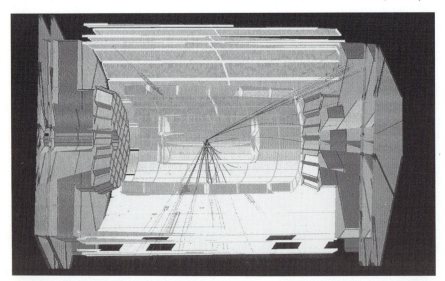

Figure 7.13 A collision between an electron and a positron at 140 GeV in which the positron emits a photon and then interacts with the electron producing a Z^0 which decays into a quark and anti-quark which each in turn produce a jet of particles. This event was observed in the DELPHI detector.

Photo credit: CERN

8
PARTICLE ACCELERATORS

8.1 ELECTROSTATIC ACCELERATORS

Rutherford used naturally occurring radioactive sources to provide the alpha particles which he used to bombard atoms and investigate the nucleus. However, this restricted both the energy of the particles and the intensity of the beam he could direct at a target. Whereas a typical alpha source might provide around 10^6 alpha particles per square centimetre per second, a positive ion beam of 1 μA would give over 10^{12} ions per square centimetre per second. It was obvious that high energy ion beams would be an extremely useful tool in nuclear research. The problem was that natural sources emit alpha particles with several MeV so the ions would need to be accelerated through extremely high voltages. In 1930 the high voltage generators needed to do this did not exist.

Two high voltage d.c. generators were developed in the next few years. At Princeton in 1931 Van de Graaf invented a high voltage electrostatic generator that he developed (at MIT) into a particle accelerator. The Van de Graaf machine provided a simple device capable of generating high voltages (eventually up to several MeV) and a later invention, the tandem generator, allowed this to be doubled resutling in electrostatic acceleration up to 14 MeV. However, the first artificial nuclear reactions inititiated in an accelerator took place at the Cavendish Laboratories in Cambridge. J.D. Cockcroft and E.T.S. Walton invented the cascade generator, a device that could multiply the peak input voltage from a high voltage transformer and use it to accelerate hydrogen ions.

Cockcroft had been inspired by a theoretical paper by George Gamow in which he had shown that the energies needed to penetrate the nucleus might not be so high as was previously thought. Classically alpha particles need enough energy to get over the coulomb potential barrier around the nucleus, usually many MeV. However, quantum theory predicts that lower energy alpha particles have a small but non-zero probability of *tunneling through* the barrier. Cockcroft's idea was to provide an intense beam of positive ions at several hundred keV and look for the effects of the few that managed to penetrate the barrier and induce artificial

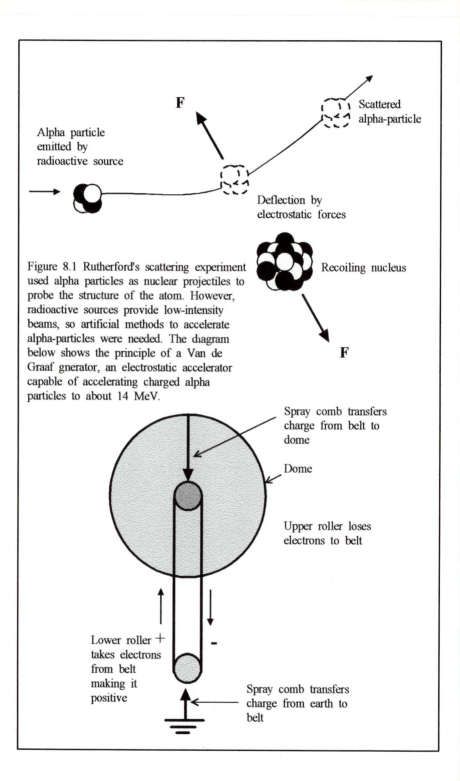

F

Alpha particle
emitted by
radioactive source

Scattered
alpha-particle

Deflection by
electrostatic forces

Figure 8.1 Rutherford's scattering experiment
used alpha particles as nuclear projectiles to
probe the structure of the atom. However,
radioactive sources provide low-intensity
beams, so artificial methods to accelerate
alpha-particles were needed. The diagram
below shows the principle of a Van de
Graaf gnerator, an electrostatic accelerator
capable of accelerating charged alpha
particles to about 14 MeV.

Recoiling nucleus

F

Spray comb transfers
charge from belt to
dome

Dome

Upper roller loses
electrons to belt

Lower roller +
takes electrons
from belt
making it
positive

−

Spray comb transfers
charge from earth to
belt

nuclear reactions (he had calculated that 1 in 1000 hydrogen ions of 300 keV should penetrate the nucleus of a boron target). The first hydrogen ion beams produced in the Cockcroft-Walton device delivered a current of 10 μA at 280 keV, an accelerating voltage that was later raised to 700 keV. Later versions provided beam currents up to 10 mA at 3 MeV.

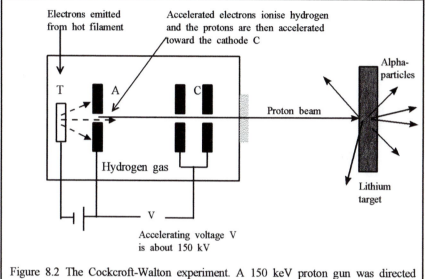

Figure 8.2 The Cockcroft-Walton experiment. A 150 keV proton gun was directed at a lithium target and 8.5 MeV alpha-particles were detected. The experiment succeeded because some of the alpha-particles tunnelled through the potential barrier and reacted with nucleons in the lithium nucleus.

In 1932 Cockcroft and Walton directed their proton beams at lithium and detected alpha particles. They had induced the first artificial nuclear transmutation:

$$\,^1_1H + \,^7_3Li \rightarrow \,^4_2He + \,^4_2He + 17.2 \text{ MeV}$$

This result was the first triumph for accelerators and heralded the start of 'big science', a strand of high energy physics that dominated much of the rest of the century. Sir John Cockcroft and Ernest Walton shared the 1951 Nobel Prize for Physics *"for their pioneering work on the transmutation of atomic nuclei by artificially accelerated atomic particles"*. Cockcroft was also a leading scientist in the development of radar and became director of the newly formed Atomic Energy Authority at Harwell, a post he kept until 1959, when he became Master of Churchill College, Cambridge.

8.2 LINEAR ACCELERATORS

The accelerators described above rely on continuous acceleration using a single large potential difference. Linear accelerators (LINACs) depend on repeated acceleration as charged particles pass through a series of axial 'drift tubes', a technique suggested by Wideröe in the late 1920s and developed by Lawrence in the 1930s. In this way a radio frequency a.c. supply provides an alternating potential difference between pairs of drift tubes and the lengths of the tubes are set so that the charges are between them only when the electric field is in the right direction to increase their acceleration. When the field reverses polarity the charges are shielded from it inside the metal tubes. Luis Alvarez took this idea further making use of the resonant behaviour of the drift cavities to accelerate particles on the crest of the travelling electromagnetic wave. He kept the wave in phase with the accelerating electrons by 'loading' the cylindrical waveguide with metal plates. This technique is more suited to the acceleration of electrons whose extreme relativistic motion means they are effectively moving at the speed of light. Electrostatic accelerators are used to inject the charged particles and LINACs may also be used as injectors for synchrotrons.

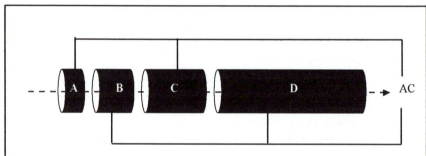

Figure 8.3 A linear accelerator or LINAC. Alternate cylinders are connected to opposite poles of an alternating high voltage supply. A positive ion is injected from the left. The inside of each cylinder is field-free. The ion is accelerated by voltages across the gaps between cylinders. These must have the correct polarity when the ion reaches them. This means the separation between gaps must increase along the accelerator.

The longest and most famous linear accelerator in the world is the Stanford linear accelerator (SLAC) near Berkeley. This came into operation in 1966 accelerating electrons through 3 km of resonant cavities up to an energy (then) of 20 GeV. At the end of these tubes electrons are directed onto fixed targets and the debris of the collisions is analysed in an experimental end station. Experiments at SLAC have led to many important discoveries, perhaps the most significant being the confirmation of Gell-Mann and Zweig's quark model in the mid-1960s and Perl's discovery of the tau particle in 1975.

Linear accelerators have a major advantage over circular accelerators such as cyclotrons and particularly their higher energy relatives, the synchrotrons. To keep a charged particle moving in a circle, it must have a centripetal acceleration. However, accelerated charges radiate and the energy loss due to this 'synchrotron radiation' becomes a very serious problem in high-energy orbital accelerators:

Power losses due to synchrotron radiation are proportional to the fourth power of the beam energy divided by the radius of curvature of the path. This is very significant for high-energy machines, about 14 MW at a beam energy of 90 GeV. This power must be replaced in the r.f. cavities where particle acceleration takes place. Such large losses suggest that electron positron colliders might be close to their limiting energy and was one of the motivating arguments for the construction of the SLC (Stanford Linear Collider).

Synchrotron radiation is more of a problem for electrons than protons or ions because the electrons have much lower mass, higher velocity and greater acceleration at the same energy.

8.3 CIRCULAR ACCELERATORS

8.3.1 Lawrence's Cyclotron

The advantage of circular accelerators over linear accelerators is that they can accelerate charged particles to high-energy using a lower voltage repeatedly rather than using a single very high voltage once only. The idea was first investigated by Lawrence and Edlefsen at Berkeley in 1930 and Lawrence constructed his first working device in 1931 – it accelerated protons to 80 keV. By 1932 (while Cockcroft and Walton were busy using a lower energy device to initiate nuclear reactions) he and a research student, Stanley Livingstone, had built a 28 cm diameter cyclotron that reached 1 MeV. By 1939 a 1.5 m cyclotron (the '60-inch') was generating 19 MeV deuterons and Lawrence had plans to go beyond this.

The basic principle of the cyclotron is to deflect moving charges into a spiral path using a perpendicular magnetic field at the same time as repeated electric pulses are used to accelerate them. The particles move in a vacuum chamber between the poles of an electromagnet. The vacuum chamber itself is a hollow cylinder made up of two D-shaped chambers connected to a radio-frequency oscillator. Inside each chamber the particles are shielded from the field between them and move in a semi-circular path. The frequency of the r.f. voltage is adjusted so that the polarity of the electric field between the 'dees' accelerates them as the particles cross the gap between them. This means they are travelling faster after crossing the gap and the radius of curvature increases. If particles are injected near the centre of the chambers they will move in circles of increasing radius until they reach the outside of the chambers. The technique relies on the fact that the period of rotation (for non-relativistic particles) is independent of their speed (otherwise they would get out of synchronisation with the accelerating field). When they reach very high velocities and relativistic corrections can no longer be ignored this assumption is no longer valid. This limits the maximum energy of a simple cyclotron to about 25 MeV for protons (at this speed they have a velocity of about 0.5 *c*). Ernest Lawrence was awarded the 1939 Nobel Prize for Physics "*for the invention and development of the cyclotron and for results obtained with it, especially with regard to artificial radioactive elements*".

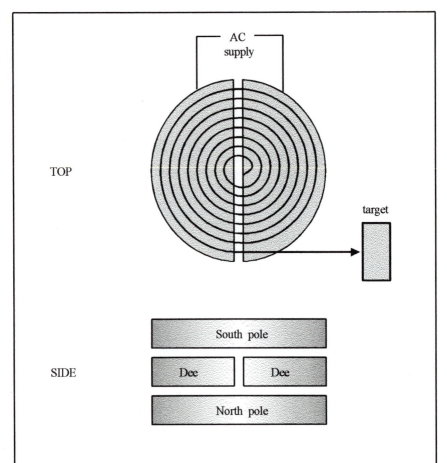

Figure 8.4 The principle of the cyclotron. Acceleration takes place as electrons move across the gap between two semi-circular dees. The dees are attached to a high-frequency high-voltage supply between the poles of a very strong magnet. The radius of electron orbits increases steadily until they emerge from the edge of the accelerator to be directed at some target.

Maths Box: Cyclotron Resonance

Consider a particle with charge q and mass m moving at velocity v perpendicular to a magnetic field of field strength B.

Magnetic force on particle $= Bqv = \dfrac{mv^2}{r}$

Radius of curvature $\qquad r = \dfrac{mv}{Bq}$

Period of rotation $\qquad T = \dfrac{2\pi r}{v} = \dfrac{2\pi mv}{Bqv} = \dfrac{2\pi m}{Bq}$

Frequency of rotation $\qquad f_c = \dfrac{Bq}{2\pi m}$ (cyclotron frequency)

This frequency is called the cyclotron frequency.

If the particles move with relativistic speeds their mass increases with velocity and the cyclotron frequency (see box) is no longer independent of velocity: as the particles speed up the frequency decreases. Ed McMillan suggested that the 25 MeV 'limit' could be overcome in a device in which the frequency of the r.f. supply was gradually reduced as the particles speeded up. This idea was put into practice after the second world war in a 4.6 m 'synchro-cyclotron' (the '184-inch') built at Berkeley to look for Yukawa's pions. The neutral pion was discovered there in 1949.

However, beating the 25 MeV limit only delayed the demise of the cyclotron as a front-runner in the high-energy particle stakes. High energy meant high velocity and this resulted in a large radius of curvature and an enormous electromagnet (the magnet used in the Berkeley 184-inch synchro-cyclotron had been scavenged from war research into the separation of uranium isotopes for the atomic bomb). A new method was needed that was not limited by the size of the magnet – the synchrotron.

8.3.2 Synchrotrons

Higher energies were needed to investigate the heavy strange particles discovered in the late 1940s. One way to solve this problem would to be keep the particles on the same radius orbit and use a series of dipole bending magnets around the circumference of the orbit rather than a single magnet covering the entire diameter. For protons this can be done by decreasing the r.f. frequency (to match the changing orbital frequency, as in the synchro-cyclotron) and gradually increasing the magnetic field strength to keep a constant radius of curvature. For electrons it is simpler. High-energy electrons are effectively travelling at the speed of light, so their orbital frequency does not change significantly as their energy increases. This means that electron synchrotrons can operate from a fixed frequency r.f. supply and only need to change the magnetic field strength (as long as the electrons are injected at a high enough energy). In addition to this electrons and positrons can be accelerated at the same time in the same machine and the particle and anti-particle beams can be brought together to collide and annihilate at selected points around the orbit inside specially designed detectors.

The development of intersecting storage rings and the use of synchrotrons as

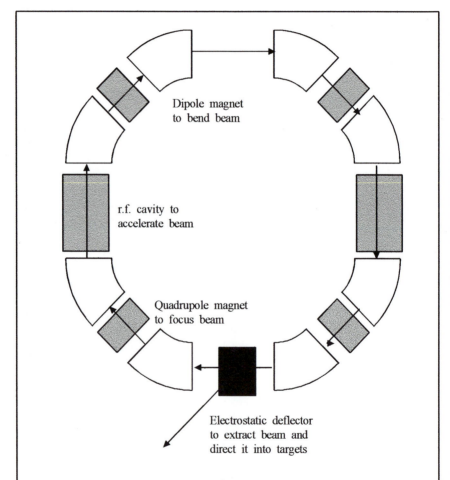

Dipole magnet
to bend beam

r.f. cavity to
accelerate beam

Quadrupole magnet
to focus beam

Electrostatic deflector
to extract beam and
direct it into targets

Figure 8.5 The principle of a synchrotron. If charged particles are injected at high energy they are already close to the speed of light so they orbit at a fixed frequency and the AC accelerating voltage can also have a fixed frequency synchronised with the beam. The particles are accelerated inside the r.f. cavities.

colliders led to a major advance in the effective energy of collisions. LINACs fire subatomic projectiles at fixed targets, usually nuclei. This means that the high incident linear momentum of the projectile must be conserved in the product particles created in the collision. In other words the products cannot be created at rest and some of the incident energy must remain as kinetic energy rather than rest energy in the particles produced. At relativistic energies this is a serious problem for fixed target machines and limits their efficiency at producing new particles. Colliding beam experiments create collisions in the centre of mass frame so in principle all the incident energy can be converted to rest energy of new particles. When LEP (the large electron-positron collider at CERN) was first operated it was 'tuned' to about 90 GeV, the rest energy of the Z^0 particle which it was designed to create. Later (1997) its energy was doubled to create $W^+ W^-$ pairs.

8.4 SOME MODERN ACCELERATORS

8.4.1 The Stanford Linear Accelerator and Collider

The 3 km linear accelerator at Stanford provides the highest energy and intensity electron beams in the world. The idea for a large linear acceleration at Stanford developed from the work of William Hansen in the 1930s and the invention with Russel and Sigurd Varian of the klystron (a powerful r.f. source that can be used to accelerate charged particles). But it was the persistence of Wolfgang Panofsky in particular who secured government support for the SLAC Project. Construction began in 1962 and experiments started in 1966, directing 20 GeV electron beams onto fixed targets. The high energy and momentum of these electrons gives them a very short de Broglie wavelength so they are ideal projectiles with which to explore the inner structure of the nucleus and the nucleons it contains.

Early experiments to investigate the inelastic scattering of electrons from protons led to the first convincing evidence for the quark model of hadrons. These experiments bore more than a passing resemblance to the famous scattering experiment of Geiger and Marsden whose results led Rutherford to propose the existence of the atomic nucleus. In effect high energy point-like electrons penetrated protons and scattered off the quarks. The existence a significant number of large angle scatterings implied point-like structures inside the proton.

This beam energy was increased to 50 GeV (with an average accelerating field strength of about 1.5×10^7 Vm^{-1}) as the LINAC was upgraded to feed electrons and positrons into the SLC (Stanford Linear Collider), an innovative accelerator that achieves 100 GeV collisions between positrons and electrons. This is an alternative to the approach at CERN where electrons and positrons orbit in opposite directions around a large storage ring (LEP). The advantage of the Stanford technique is that the acceleration is achieved while the charges are moving in a straight line thus reducing the power losses due to synchrotron radiation. The SLC also allows the electrons and positrons to be polarised, something that cannot be done at LEP. It is possible to accelerate both electrons and positron in the same direction along the same linear accelerator if the particles are injected in alternate pulses riding on successive half-cycles of the travelling r.f. wave. It is also possible to focus the beams more easily than in LEP thereby increasing the beam intensity at the points of collision. Like LEP, the SLC has been used to investigate the creation and decay of Z^0 particles – the rates of these processes are linked to the number of generations of quarks and leptons and results from SLC and LEP imply that there are no more than three generations.

Although Stanford is famous for its linear accelerators it was also the site of the

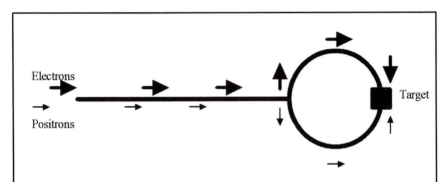

Figure 8.6 The Stanford Linear Accelerator has become the Stanford Linear Collider. Alternate pulses of electrons and positrons are accelerated along the 3 km tunnel and the emerging particles are sent in opposite directions around the collider.

first colliding beam experiments using a small device shaped like a figure of eight built on the main campus of the university. This idea was developed into SPEAR (Stanford Positron Electron Asymmetric rings) which was completed in 1972. This is an 80 m diameter ring in which electrons and positrons circulate in opposite directions with energies of 4 MeV, providing 8 GeV collisions in the centre of mass frame. Despite its relatively moderate cost the machine led to several major discoveries, including the J/Ψ particle (the first charmed quark) in 1974 and the tau particle (third generation lepton) in 1976. The J/Ψ was discovered independently at Brookhaven and Richard Burton (from Stanford) and Samuel Ting (from MIT) shared the 1976 Nobel Prize "*for their pioneering work in the discovery of a heavy elementary particle of a new kind*". Martin Perl shared the 1995 Nobel Prize for Physics "*for the discovery of the tau lepton*".

Circulating charges emit intense beams of electromagnetic radiation because of their centripetal acceleration. This is called synchrotron radiation and it has many uses in research and industry. Synchrotron radiation from SPEAR was used for research even while SPEAR itself was dedicated to collision experiments and the Stanford Synchrotron Radiation Laboratory (SSRL) was established in 1973. From 1979 the ring was dedicated to synchrotron radiation for 50% of its operational time. Since 1991 it is used solely for synchrotron radiation and has 24 experimental stations serving approximately 600 users. The high-intensity X-ray and ultraviolet beams produced by SSRL are used by biologists, chemists, geologists, materials scientists, electrical engineers, chemical engineers, physicists, astronomers and medics. Recent experiments include investigations of high-temperature superconductors, and high-resolution crystal structures, studies of progressive bone loss in live animals, and determining the trace impurities in silicon used by the semiconductor industry.

As a high-energy storage ring and collider the SLC was replaced by PEP in 1980.

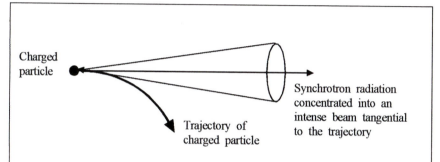

Figure 8.7 Synchrotron radiation is a major power loss from circular electron or positron accelerators. However, the radiation itself is very useful in many fields of applied physics, particularly materials science.

This 800 m diameter ring produced collisions between electrons and positrons up to 30 GeV in the centre of mass frame. This collider was used to make detailed measurements of particle lifetimes and to investigate the predictions of quantum chromodynamics (QCD). It is now being upgraded to PEP II, the 'asymmetric B factory'. This will be used to produce large numbers of B mesons in much the same way that LEP was used to produce Z^0s and LEP-2 to produce $W^+ W^-$ pairs. The term 'asymmetric' refers to the fact that electrons and positrons in PEP II will have different energies. One of the main aims for PEP II is to investigate the asymmetries between matter and antimatter.

Collider	Beam Energy/GeV	Location
DORIS	4 - 5.3	DESY (Germany)
VEPP-4	5	Novosibirsk (Russia)
CESR	8	Cornell (USA)
PEP	17	Stanford (USA)
PETRA	23	DESY (Germany)
TRISTAN	35	KEL (Japan)
LEP	50-100	CERN (Switzerland)

8.4.2 CERN

On first thought the idea of a colliding beam machine seems to make little sense. The density of particles in an accelerated beam is relatively low compared to the density of particles in a stationary target. However, if the high-energy particles are stored in counter-rotating beams they can be made to collide repeatedly and this helps to compensate for the low probability of interesting events in each individual collision. However, colliding beams have the major advantage that collisions take place in the centre of mass frame so that the total energy from both particles is available for rest energy of new particles. For example, if a 30 GeV particle collides with a fixed target the maximum available energy to create new particles is just 7.6 GeV whereas a collider using two beams each of energy 30 GeV could transfer all 60 GeV to the creation of new particles.

The construction of prototype electron-electron colliders took place at Stanford in the mid-1960s but the key development was ADA (an accumulation ring) designed and built by an Italian group led by B. Touschek. The first low intensity electron-postron annihilation events took place in ADA in 1963 and many of the major technological problems were examined and overcome using the device. The SPEAR electron positron collider at Stanford has already been mentioned and was one of the most successful of the early machines. However, the largest and most successful colllider so far is the Large Electron Positron Collider (LEP) at CERN. This machine was completed in 1989 and beam energies were set at about 50 GeV to maximise the probability of Z^0 creation (the Z^0 has a rest energy of about 90 GeV). In 1997 this energy was raised to about 90 GeV to produce $W^+ W^-$ pairs.

The LEP collider was a collaborative venture funded by many nations and between 1983 and 1989 it was the largest civil engineering project in Europe. The LEP tunnel is 26.67 km in circumference (bored to an accuracy of better than 1.0 cm) and runs underground in a specially excavated tunnel inclined at 1.4° to the horizontal between Geneva airport and the Jura Mountains. In addition to this there are four major experimental caverns, 18 pits, 3 km of secondary tunnel and more than 60 chambers and alcoves. The beam is directed by groups of magnets arranged in 31 'standard cells' around the ring. The large radius of curvature is necessary to reduce the amount of synchrotron radiation. At each of the experimental points the beam passes through a large solenoidal magnet that squeezes it to around 10 μm by 250 μm and so increases luminosity.

As with all large high-energy accelerators LEP is just final machine in a chain of injectors and pre-injectors. The particles begin their journey in two LINACs of 200 MeV and 600 MeV before they enter a 600 MeV electron-positron accumulator which transfers them to the PS (proton synchrotron) which is operated as an electron positron synchrotron at 3.5 GeV. This injects them into the SPS (super proton synchrotron) which is a 20 GeV electron-positron injector for LEP itself. (The use of SPS as an electron positron synchrotron has little or no effect on its continued use to accelerate protons since it is operated cyclically and the pulses of electrons and positrons pass through it in the dead time between proton pulses).

R.f. acceleration in the LEP ring is achieved in 128 resonant cavities powered by klystrons tuned so that they reach maximum power as the bunches of particles pass through them. Electrons and positrons orbit the 27 km ring about 11 000 times per second, but it takes typically 12 hours to fill the ring with particles for an experimental run. During this time each particle will complete about 500 million orbits, so it is essential that the vacuum in the beam tube is very good indeed (about 10^{-11} torr rising to 10^{-9} torr as a result of synchrotron radiation releasing gas molecules from the vacuum-chamber wall when there is a beam present).

From 1989 to 1995 LEP was a Z^0 factory (producing something like 30 million of them). The beam energies were varied around the 91 GeV needed to create the Z^0 and the interaction rate (related to the number of events occurring at each energy) swept out a resonance curve with its peak on 91 GeV. This was expected and the width of the curve gave physicists their first accurate measurement of the Z^0 lifetime. This was highly significant because the lifetime will be shorter the more alternative decays that are open to the particle decaying. Since the Z^0 is a relatively heavy particle (almost 100 times the rest mass of a proton) it has many possible decay mechanisms open to it. The greater the number of light particles that exist the larger the number of ways it could decay and therefore the shorter its lifetime should be. In this way the Z^0 lifetime gives information about the number of generations leptons that can exist. Three of the four LEP experiments (ALEPH, DELPHI and OPAL) have contributed to this research and confirm that there are just three types of neutrino, a result that agrees with the number needed by cosmologists to explain the abundances of atomic nuclei in the early universe.

Electron positron colliders are good for testing precise theoretical predictions (as opposed to creating new particles in the brute force collisions of proton antiproton machines) and LEP has been particularly important for testing and verifying predictions about electroweak theory. However, there is one aspect of this theory that has not yet been dealt with – why Ws and Zs have any mass at all (their companion force carrier, the photon has zero rest mass). The current theoretical explanation depends on an idea suggested by Peter Higgs and known as the Higgs mechanism. In this theory the particles aquire mass by interacting with a new field, the Higgs field. To verify this physicists hope to catch a quantum of the field, the Higgs boson, and this was one of the motivating factors in upgrading the 50 GeV LEP to a 100 GeV LEP 2. This higher energy machine will operate until after the year 2000 when the tunnel will be taken over by a higher energy proton collider, the LHC (Large Hadron Collider). The LHC is to be built in one stage and switched on in 2005.

Some facts about the LEP collider at CERN:

First experiments: 1989
Maximum beam energy: 100 GeV
Luminosity: 2.4×10^{31} cm^{-2}s^{-1}
Time between collisions: 22 μs

Bunch length:	1 cm
Beam dimensions:	0.20 mm (H) by 0.008 mm (V)
Filling time:	20 h
Acceleration period:	550 s
Injection energy:	550 GeV
R.F. frequency:	352.2 MHz
Average beam current:	55 mA (per species)
Circumference:	27.66 km
Dipole magnets in ring:	3280 +24 weaker dipoles
Peak magnetic field:	0.135 T

8.4.3 The LHC

"A recipe. Build two pipes, each about 6 centimetres wide and 27 kilometres long. Bend them both into a circle and cool to 1.9 kelvin, about 300°C below room temperature. Fill with protons – about a hundred billion of them to start with – all travelling as close to the speed of light as possible. Add a magnetic field a hundred thousand times more powerful than the Earth's to steer the particles through the pipes. And make sure the protons in each pipe move in opposite directions.
Now here's the exciting bit. Align the two pipes so that the particles collide. About twenty protons should smash into each other, creating showers of other particles. Take a good look at each one. If you spot a particle you don't recognise, shout. Finally, repeat forty million times. Each second!"

(Justin Mullin, from New Scientist, 18 June 1994, Vol. 142, No. 1930, p.34, describing the proposed plan for the LHC)

The LHC proton beams will have an energy of about 14 TeV (1.4×10^{13} eV) almost 10 times greater than the energies used at present in the Tevatron at Fermilab. The idea of slamming hadrons together at ever increasing energies is neither subtle nor new but it is certain to reveal new physics. The hope is that it will throw some light on the problem of the origin of mass (the theoretical upper limit on the mass of a Higgs particle is about 1 TeV) and the asymmetry between matter and antimatter in the universe. It may also answer questions about the origin of the three generations of quarks and leptons and the nature of 'dark matter'. Yet another possibility is that the supersymmetric particles ('sparticles') predicted in some unified field theories will be created in the high-energy collisions.

LHC will create large numbers of heavy quarks, especially B quarks (which are expected to exhibit CP violation like the lighter strange quarks) which should help to solve problems about the asymmetry between matter and antimatter in the observed universe. It will also create ever-increasing numbers of unwanted 'background particles' (since the total cross-section for proton-proton interactions increases with energy). This will necessitate a great deal of pre-selection when results begin to emerge from proposed detectors such as ATLAS (A toroidal LHC apparatus), CMS (Compact Muon Solenoid) and ALICE (designed to look at heavy ion collisions) – the rate of proton-proton interactions at 14 TeV and a beam luminosity of 10^{34} particles per cm^2 will be about 1 GHz, with every event generating about 1 Mbyte of data! To deal with this the detectors must be controlled by intelligent triggers that can automatically reject the vast majority of events when nothing 'interesting' has happened. This is done by feeding the intial data to a local computer that constructs allow resolution profile of the event. It compares this with standard patterns stored in memory representing events that should be kept and analysed further. If there is a match the data is kept, if not it is deleted. This pre-selection must be completed within a few microseconds of the original event!

Extremely high energies (>1000 TeV) will be created by accelerating and colliding heavy ions. These collisions should create bubbles of hot dense matter that mimic conditions in the early universe, just 10^{-35} s after the Big Bang.

The technological challenge of the LHC is far more daunting than for LEP even though the accelerator will be built within the existing LEP tunnel. The energy and momentum of protons and ions in LHC will be far greater than electrons in LEP so the bending magnets on the ring will have to be very much stronger (about 9 T compared to 0.135 T in LEP). This will be achieved using superconducting electromagnets wound with copper-clad niobium-titanium wires, but these only work at very low temperatures so the magnets have to be cooled to 1.9 K using superfluid liquid helium. This will require eight cryogenic plants spaced equally around the 27 km ring pumping 700 000 litres of liquid helium through 40 000 leak-proof junctions to cool 31 000 tonnes of equipment! The only parts of LHC that will not be cooled to 1.9 K are the experimental stations themselves (accounting for less than 10 % of the circumference of the ring).

The construction and financing of such a complex experimental facility depends on support from many governments, both within and beyond the European community, and in this respect the cancellation of the SSC (superconducting supercollider) in Dallas has actually helped the development of the LHC by attracting 530 million dollars from the USA.

8.4.4 Tevatron at Fermilab

Although synchrotron radiation is less of a problem for protons than electrons there are other serious difficulties when accelerating protons that tend to produce beam instabilities, so the development of hadron colliders initially lagged behind electron positron machines. The first machine of this type was the ISR (intersecting storage rings) at CERN which used 28 GeV protons from the PS (proton synchrotron). The key step forward in the ISR came from Simon van der Meer who invented a process called stochastic cooling in which the beam is sampled at points in the ring and the information on developing instabilities is fed back to correcting electrodes that return the protons to their central orbit.

Electroweak unification led to the prediction of new force-carrying particles, the W^+, W^- and Z^0 particles with rest masses around 90 GeV each. Physicists were keen to find these particles in the laboratory, but this needed very high energies, and gave a boost to the idea of a high energy hadron collider. In 1976 C. Rubbia, P. McIntyre and D. Cline proposed the conversion of an existing accelerators at Fermilab near Chicago to make a proton antiproton collider capable of creating these intermediate vector bosons. In the end the idea was put into practice at CERN using the existing SPS (super proton synchrotron) and the particles turned up in the UA1 (Underground Area 1) and UA2 detectors in 1983. C. Rubbia and S. van der Meer shared the 1984 Nobel Prize for Physics *for their decisive contributions to the large project, which led to the discovery of the field particles W and Z, communicators of the weak interaction.*

In 1985 a giant proton antiproton collider was completed at Fermilab. The beam energies were each close to 1 TeV and the machine was called the Tevatron. The accelerator itself has a diameter of 2 km and uses 2000 superconducting electromagnets to guide the protons and antiprotons inside a 10 cm diameter evacuated steel pipe. Like LEP the Tevatron itself is just the final stage in a complex sequence of accelerating machines. The protons start life in a Cockcroft-Walton generator where they are accelerated to 4% of the speed of light ($0.04c$) at an energy of 0.75 MeV, about 30 times the energy of an electron in a TV set. Next comes a 150 m LINAC by the end of which they have reached $0.55c$ and 0.2 GeV (later raised to 0.4 GeV). The LINAC is followed by a 150 m diameter rapid cycling synchrotron, the Booster, which increases their speed to $0.99c$ and their energy to 8 GeV. The Booster usually completes twelve cycles in rapid succession firing twelve bunches of protons into the Main Ring which continues the acceleration process. In the Main Ring, another proton synchrotron, their energy is increased to 150 GeV in 70 000 orbits. In 1998 this was replaced by the Main Injector which increases the collision luminosity by a factor of 5. After this the protons are transferred down into a lower ring of 1000 superconducting magnets forming the Tevatron. After a further 400 000 orbits protons have reached $0.9999995c$ with an energy of 1 TeV. During the acceleration (which lasts about 20 s) they travel a distance equal to about 7 times the Earth-Moon separation.

The Tevatron is used as both a fixed target device by extracting the particle beams and as a collider. When used as a collider antiprotons are stored in an accumulator ring before being injected into the Main Ring and passed down to the Tevatron for final acceleration prior to collision with protons travelling around the same ring with similar energy in the opposite direction.

In 1994 the top quark was discovered using the Tevatron. This involved analysing and comparing several different decay modes and resulted in a top quark mass of 199±22 GeV, in complete agreement with predictions based on the Standard Model. This discovery completed the three generations of quarks (up/down, strange/charmed, bottom/top) and helped theorists set limits on the possible masses of other particles including the Higgs boson.

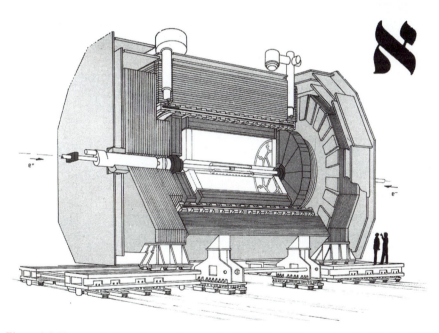

Figure 8.8 The top photograph was taken inside the 27 km LEP tunnel at CERN. The LHC will run in the same tunnel early in the new millennium. The bottom image is a cutaway view of the ALEPH layered detector. The LEP vacuum tube passes through the centre of the detector and electrons and positrons collide head-on inside it.

Photo credit: CERN

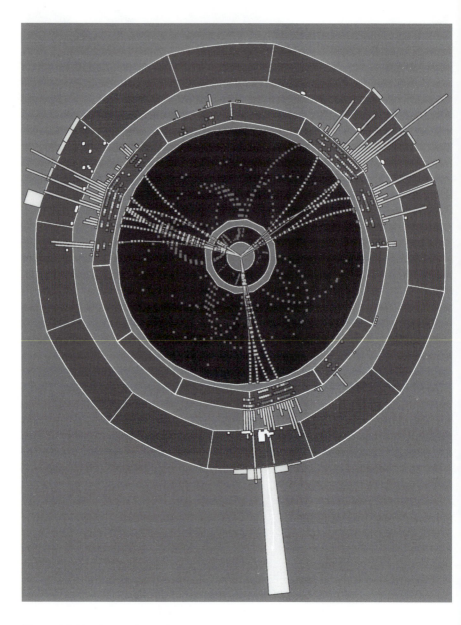

Figure 8.9 The image above shows the result of an electron positron collision inside the ALEPH detector. In this case the two particles have annihilated in the centre of the detector to create a quark anti-quark pair. One of the quarks has radiated a gluon. The two quarks and gluon create jets of new particles as they fly apart. These register in the various layers of the detector and are identified electronically.

Photo credit: CERN

9
TOWARD A THEORY OF EVERYTHING

9.1 A BRIEF HISTORY OF UNIFICATION

9.1.1 From Albert Einstein to Edward Witten

" ... it has been my greatest ambition to resolve the duality of natural laws into unity. This duality lie in the fact that physicists have hitherto been compelled to postulate two sets of laws – those which control gravitation and those which control the phenomena of electricity and of magnetism ... Many physicists have suspected that the two sets of laws must be based upon one general law, but neither experiment nor theory has, until now, succeeded in formulating this law. I believe now that I have found a proper form. I have thought out a special construction which is differentiated from that of my relativity theory, and from other theories of four-dimensional space, through certain conditions. These conditions bring under the same mathematical equations the laws which govern the electromagnetic field and those which govern the field of gravitation. The relativity theory reduced to one formula all the laws which govern space, time, and gravitation, and thus it corresponded to the demand for simplification of our physical concepts. The purpose of my work is to further this simplification, and particularly to reduce to one formula the explanation of the field of gravity and of electromagnetism. For this reason I call it a contribution to a 'unified field theory'."

(Albert Einstein, in the Daily Chronicle, 26 January 1929, quoted from *Einstein, the Life and Times*, by R.W. Clark, World Publishing Company, 1971)

Einstein established his scientific reputation with revolutionary theories about space, time and gravitation – the special and general theories of relativity. Despite his concern over the statistical nature of quantum theory he was also responsible for several of its most important early ideas and he made major contributions in many other areas of fundamental theory. However, his dream was to construct a 'unified field theory' from which the laws of electromagnetism and gravitation would derive. From the 1920s to his death in 1955 he set about doing this and his solitary research led him ever further away from the mainstream of theoretical physics. He more or less ignored quantum theory and must have assumed that the recently discovered nuclear forces would somehow drop out of the new theory. On several occasions he thought he was close to success, but each time it turned out to be a false trail and he began again, searching for the symmetries and mathematical patterns that might be a clue to the unified field theory.

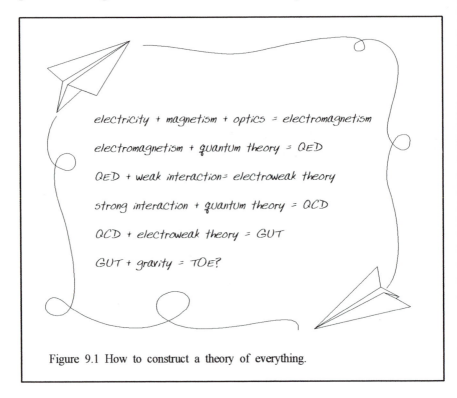

electricity + magnetism + optics = electromagnetism

electromagnetism + quantum theory = QED

QED + weak interaction = electroweak theory

strong interaction + quantum theory = QCD

QCD + electroweak theory = GUT

GUT + gravity = TOE?

Figure 9.1 How to construct a theory of everything.

For some time after his death most physicists assumed that there was no unified field theory and turned their attention to other things. After all, the development of high-energy accelerators were throwing out new particles on an almost daily basis and far from becoming simpler, the universe was looking a lot more complicated. Meanwhile Paul Dirac combined relativity and quantum theory to create the Dirac equation, a beautiful and startlingly accurate model of the electron which also predicted the existence of antimatter. Feynman, Schwinger and Tomonoga built on Dirac's work and created the first relativistic quantum field theory, QED, which explained the interaction of light and matter. But there were major problems, quantum field theories kept generating embarrassing infinities that could only be tamed using an *ad hoc* procedure called renormalisation. And gravitation theory, described by general relativity, remained a world apart, pristine, geometrical, apparently immune to quantisation or incorporation in some unified scheme.

The development of quantum field theories emphasised the importance of underlying symmetries and their links to conservation laws and unification. A theory for quark interactions (QCD) was constructed by analogy with QED. The weak interaction turned out to be very closely linked to electromagnetism and eventually the two theories were unified by Abdus Salam, Steven Weinberg and Sheldon Glashow. Electroweak theory predicted the existence of heavy force-carrying particles as partners to the massless photon, and these (the $W^\pm$ and Z^0 particles) were discovered at CERN in 1983. By combining the symmetries of the electroweak and colour interactions theorists proposed a grand unified theory (GUT) involving this 'supersymmetry'. This led to the prediction of new supersymetric particles ('sparticles') accompanying all the known particles and implied that protons would be unstable. Attempts to detect proton decay or supersymmetric particles have all so far failed, but the next generation of high-energy experiments should be able to show whether sparticles exist or not (perhaps in LEP-2 but almost certainly in the LHC).

As has happened several times in the history of physics, the proliferation of apparently diverse phenomena have begun to fall into simple (and not so simple) patterns which have led to unifying theories. Einstein's quest has been taken up by a new generation of mathematical physicists such as Ed Witten at Princeton who are attempting once again to explain everything using many dimensions, some of which have curled up or compactified so that we have no direct evidence of their existence. They noticed that the symmetries and supersymmetries of particle physics can be reproduced in a space of 26 dimensions in which the particles themselves are not points but excitations of structures called strings, waving about in this multi-dimensional space. Unfortunately there are a great many possible forms for superstring theory and no one knows how to select the one that may apply to our universe. However, one thing is certain, Einstein's theory of gravity emerges as a necessary consequence of the theory and does not have to be treated separately as it is in other versions of GUTs. This in itself is enough to persuade many physicists that string theory really is the way forward. Unfortunately the phenomena predicted by the theory operate on an energy scale so far removed from the physics of present high-energy laboratories that there seems virtually no chance that we can ever test it. If a final unique self-consistent superstring theory is ever constructed it may be that we have to take it on trust, persuaded by its elegance and simplicity rather than the empirical confirmation of its predictions. If this is the case it will mark a turning point (perhaps an end point) for fundamental physics, and maybe a beginning of a new metaphysics.

"String theory is twenty-first century physics that fell accidentally into the twentieth century."
(Ed Witten, quoted in *Hyperspace*, by Michio Kaku, OUP, 1995)

9.1.2 The First Great Unification

"If the hope of unification should prove well founded, how great and mighty and sublime in its hitherto unchangeable character is the force I am trying to deal with, and how large may be the domain of knowledge that may be opened with the mind of man."

(Michael Faraday, quoted by Abdus Salam in *Overview of Particle Physics*, in *The New Physics*, ed. Paul Davies, CUP, 1989)

Faraday was not a great mathematical physicist, but he was a visionary experimentalist whose belief that all the forces of nature spring from a common source led him to seek links between electric and magnetic forces, between magnetism and light and light and gravity. He was a prophet of unification and his ideas inspired James Clerk Maxwell to create the 'Maxwell equations' that completely describe the allowed behaviour of the classical electromagnetic field. These equations contained within themselves the confirmation of one of Faraday's hypotheses – if the fields are disturbed then this disturbance races through the field as a wave at a speed of about 300 million metres every second – exactly equal to the speed of light in a vacuum. So light is an electromagnetic effect; the apparently diverse fields of electricity, magnetism and optics had been unified in electromagnetism. The unification of diverse fields is interesting but the power of such a discovery depends on its ability to predict new and as yet unexpected phenomena. The obvious prediction was that electromagnetic waves of other frequencies must also exist. Hertz soon generated transmitted and received them using simple spark transmitters and Marconi developed the idea into a radio

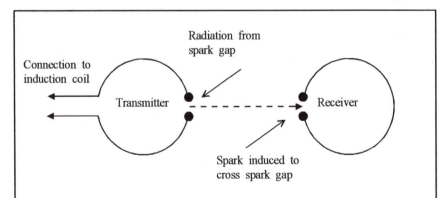

Figure 9.2 Hertz demonstrated the transmission and reception of radio waves. He also noticed that sparks jump across a spark gap more easily if the gap is illuminated with ultra-violet radiation, an observation linked to the photo-electric effect.

system able to send messages across the Atlantic Ocean from England to America. (Actually Marconi was lucky, the radio waves were repeatedly reflected from the ionosphere, otherwise they would have failed to follow the curve of the Earth and no signal would have been detected in America). Marchese Guglielmo Marconi shared the 1909 Nobel Prize for Physics with Carl Ferdinand Braun *"in recognition of their contributions to the development of wireless telegraphy"*.

Maxwell's achievement was to recognise the link between a number of electromagnetic phenomena and explain them all in a 'unified field theory'. This reduced diverse discoveries, such as Coulomb's law and Faraday's laws of electromagnetic induction, to a single coherent set of four equations describing how electric and magnetic fields depend on another and connect through space and time.

Maths Box: From Electricity and Magnetism to Electromagnetism

The main clues were scattered through the late eighteenth and early nineteenth centuries:

Coulomb's law (electrostatic force law): $F = \dfrac{q_1 q_2}{4\pi r^2}$ 1770s

The Poisson equation (electrostatic potential): $\nabla^2 V = \dfrac{\rho_e}{\varepsilon_0}$ 1813

Oersted's discovery – the magnetic field created by an electric current. 1820

The Biot-Savart law (magnetic field strength): $dB = \dfrac{\mu_0 I d\underline{l} \wedge \hat{r}}{4\pi r^2}$ 1820

Ampère's theorem (magnetic field and currents): $\oint \underline{H}.d\underline{s} = \sum_{enclosed} i$ 1825

Field lines for electric and magnetic fields (introduced by Faraday). 1830s
Electromagnetic induction (Faraday's experiments)). 1830-1845

Lenz's and Neumann's laws: $E = -\dfrac{d\Phi}{dt}$ 1845

$$\text{curl } \underline{E} = -\frac{\partial \underline{B}}{\partial t}$$

Maxwell's equations: $\text{curl } \underline{H} = \underline{J} + \dfrac{\partial \underline{D}}{\partial t}$ 1865

$$\text{div } \underline{D} = \rho_e$$
$$\text{div } \underline{B} = 0$$

You may well wonder why we are considering a unification which took place in the nineteenth century in a book that is meant to be focused on the twentieth century. This is for the very good reason that electromagnetism is the first example of a successful unified field theory and shows how we might hope to approach unification in other areas. It is also because the success of electromagnetism made it a serious contender for a theory of everything at the turn of the century and many physicists were busy trying to construct field-based models of matter in which mass emerges as a kind of electromagnetic drag factor in the all-pervading ether and particles are some kind of spacetime knot. These ideas influenced Einstein and relativity was born from the conflict between Maxwell's electromagnetism and that older universal theory, Newtonian mechanics. On another front the continuous nature of electromagnetic waves and the discrete charges that emit and absorb them led to problems in the thermodynamics of radiation from which the *ad hoc* ideas of old quantum theory emerged.

9.1.3 A Unified Field Theory – Einstein's Quest

" ... *before Maxwell people conceived of physical reality – in so far as it is supposed to represent events in nature – as material points, whose charges consist exclusively of motions, which are subject to total differential equations. After Maxwell they conceived physical reality as represented by continuous fields, not mechanically explicable, which are subject to partial differential equations. This change in the conception of reality is the most profound and fruitful one that has come to physics since Newton; but it has at the same time to be admitted that the program has by no means been completely carried out yet. The successful systems of physics which have been evolved since rather represent compromises between these two schemes, which for that very reason bear a provisional, logically incomplete character, although they may have achieved great advances in certain particulars.*"

(Albert Einstein, Maxwell's Influence on the Idea of Physical Reality, On the One Hundredth Anniversary of Maxwell's Birth. Published 1931, in *James Clerk Maxwell: A Commemoration Volume*, CUP)

For Einstein, the universe consisted of matter with mass and charge moving in electromagnetic and gravitational fields. In his special theory of relativity he showed that Newton's assumptions of an absolute space and time was not consistent with the predictions of Maxwell's electromagnetism. For example, if light moves at a constant velocity through absolute space (or the 'ether') then its velocity relative to any moving observer would depend on the observer's own velocity. Experiments carried out by many physicists in the late nineteenth century (especially Michelson) had failed to detect this change in the relative velocity of light. Einstein's solution was to scrap the idea of an absolute reference frame and postulate that the laws of physics are the same for all inertial observers – this is the Principle of Relativity. He was not the first person to propose such an idea, Galileo proposed a similar principle for mechanics in the seventeenth century and Henri Poincaré had put forward an almost identical principle in 1904. Einstein, however, was the first person to realise the deep implications of such and principle and to explore them in detail. His 1905 paper 'On the Electrodynamics of Moving Bodies' is masterful in its clarity and vision. One consequence of the principle is that the velocity of light, which emerges naturally from Maxwell's equations, must be the same for all inertial observers. At a single stroke this removed the need for absolute space or a luminiferous and all-pervasive ether (a hypothetical medium to support electromagnetic waves). It also meant that time and distance were the quantities measured by clocks and not some kind of universal measure accessible only to a being perched outside our universe. Every observer measured space and time according to his own instruments that might disagree with those of other observers and yet always combined to confirm the laws of physics. And what were these laws? In 1905 relativity included electromagnetism and a modified form of mechanics, but not gravitation. Einstein's greatest achievement was to extend the ideas of special relativity to include gravitation in the general theory published in 1915. This did not amount to a unified theory of electromagnetism and gravitation. The former was still described by Maxwell's equations, and the latter was a distortion of the geometry of the space-time continuum. Einstein hoped that electromagnetism might be explained geometrically too, producing a unified field theory from which all observable aspects of physical reality might be derived.

We have already said that he failed to complete this task, but there was one key development that seemed to suggest he was on the right track. In 1919, the same year Eddington verified that light is deflected by gravity (confirming a prediction of the general theory of relativity), Einstein received a letter from Theodr Kaluza in which this relatively unknown mathematician suggested extending general relativity to five dimensions by adding an extra spatial dimension. By doing this the original equations survived and Maxwell's equations joined them. In essence gravity was now a distortion in four-dimensional space-time and electromagnetism was a distortion in a higher fourth spatial dimension. Of course, this is all very well, but where is the extra dimension? We experience three spatial dimensions and time, but there are no obvious experiences associated with a fifth dimension. Kaluza had a neat answer. It has collapsed down to a tiny circle and our experience effectively averages over it. The great mathematician Oscar Klein improved the theory and suggested that the circumference of the circle would be equal to the Planck length (about 10^{-35} m) and so far too small to generate observable effects even in high-energy experiments (this was one of the main problems with the theory – the energies needed to test it were so high it had to be taken on trust or rejected, most physicists took the latter option). He even went so far as to suggest that rotation once around the fifth dimensional circle introduced a periodicity to the electromagnetic field equivalent to a quantum condition, thus explaining the atomicity of charge. The Kaluza-Klein theory inspired Einstein to pursue five-dimensional unified field theories for more than a decade. Later on, frustrated by not being able to tie up the loose ends or link it to the developing mainstream of quantum theory and particle physics he turned his attention to generalisations of 4D space-time, but with no obvious success. The advent of quantum theory, nuclear interactions and discoveries in particle physics seemed completely at odds with the aims of Kaluza-Klein theory and a geometrical unified field theory so very few others followed Einstein on the lonely path he had chosen. He died in 1955. The day before he had returned to his notes on the unified field theory, the unified field theory was (and remains) unfinished business, but 40 years later the quest to find it has been taken up once again.

9.1.4 Symmetry

Hermann Minkowski was a Russian-German mathematician who became interested in mathematical physics when he worked with David Hilbert at Gottingen between 1902 and 1909. In 1907 Minkowski published a book called *Space and Time* in which he showed that the special theory of relativity can be derived from geometry if time is incorporated in a special way as a fourth dimension. This emphasised the symmetry between space and time at the heart of Einstein's theory and was the foundation for the development from the special to the general theories. It was also the spur that turned Einstein's attention toward the formal beauty of the underlying mathematics in physical theories. In his early work Einstein pursued his physical intuitions, later he came to rely more and more heavily on the internal beauty of the mathematics. This may have been a mistake as far as Einstein's attempts at a unified field theory are concerned, but mathematical beauty and simplicity continue to guide and inspire theoretical physicists and the formal link between symmetry and conserved quantities is one of the most powerful ideas used to construct new theories.

In 1956 Sir Arthur Eddington said:

"We need a super-mathematics in which the operations are as unknown as the quantities they operate on, and a super-mathematician who does not know what he is doing when he performs these operations. Such a super-mathematics is the Theory of Groups."

(Sir Arthur Eddington, 1956, *World of Mathematics*, J.R. Newman ed., Simon and Schuster, NY)

The theory of groups was invented by the young French mathematician, Evariste Galois in about 1830. Unfortunately Galois became involved in a duel over a young lady in 1832. Realising both the danger he faced and the importance of his work he wrote a summary of his discoveries the night before the duel and these were published within the year. He died in the duel. He was twenty years old. It was some time before the ideas were fully absorbed and disseminated, but it was gradually realised that here was a 'super-mathematics' that could deal with fundamental symmetry operations regardless of the system to which they refer. The theory was taken further by Cauchy and Lie but was essentially complete by the start of the twentieth century. However, few physicists at this time were aware of it or mathematically equipped to deal with symmetry groups, so the introduction

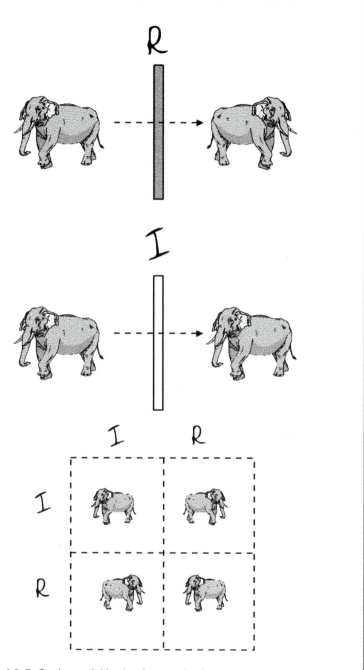

Figure 9.3 Reflection and identity form a simple symmetry group.

of these ideas into theoretical physics was very slow at first.

The idea of a symmetry group can be illustrated quite simply. Imagine reflecting objects in a plane mirror. In some cases the reflections are identical to the object and in other cases they are not. For example, a circle reflects to a circle and a square to a square, but a reflected '2' is different to an unreflected '2'. Objects whose reflections are identical to themselves possess bilateral symmetry or reflection symmetry. Call the operation of reflection R and the operation of 'doing nothing' I (this is called the identity operator). For an object possessing reflection symmetry both R and I leave the object unchanged. For an object with no reflection symmetry the I leaves it unchanged and R changes it. We can also consider combinations of operations, e.g. reflection followed by identity or two reflections one after the other. The interesting point is that any combination of these two operations in any order is still equivalent to one of the original operations. For example $R.R = I$, $R.I = R$ and $I.R.I.R = I$. The symmetry operations R and I form a 'closed group'. The way these operations combine can be summarised in a simple table (entries at intersections represent the combined effect of the two operations:

	I	R
I	I	R
R	R	I

The same pattern of outcomes can be represented in two distinct ways if we assign numerical values to I and R and interpret the sequence of operations, $I.R$ as the product of their numerical representations.

Representation 1: $I = 1$ and $R = 1$ 'symmetric'
Representation 2: $I = 1$ and $R = -1$ 'antisymmetric'

The power of group theory is that the symmetry group can be detached from any particular physical process and applied to all processes that share the same underlying symmetry. For example, if a physical system consists of two identical particles A and B then its wavefunction can be written down as a linear superpostion of the state with A in position 1 and B in position 2 and of the state with A in position 2 and B in position 1 (i.e. swapped over). This is a kind of reflection symmetry, the exchange of particles under this symmetry can be represented in either of two ways, a symmetric or asymmetric way. Nature uses both. The wavefunction for pairs of particles with a half-integer spin (the fermions) is asymmetric, that for particles with integer spin is symmetric (the bosons). This makes an enormous difference to the way the particles behave when there is more than one present. Fermions cannot go into identical quantum states – so electrons in an atom have to fill a sequence of energy levels rather than clustering in the lowest level. Without this fermionic statistics we would have no Periodic Table, no chemistry, no life. Bosons, on the other hand can condense into a common lowest available energy state (a bose condensate), producing the kind of coherent group behaviour seen in lasers, superconductivity and superfluidity. It is clear that symmetry is quite literally vital.

The link between laws of physics and symmetry is very strong. In 1918 the mathematician, Emmy Noether proved that the existence of a symmetry in the mathematical description of the universe implies the existence of a conservation law related to that symmetry. For example, it seems reasonable to assume that the behaviour of a particle in empty space is unchanged if we move the particle to a new position. This translational symmetry, i.e. the observation that translation leaves the underlying laws unchanged, leads directly to the conservation of linear momentum in the direction of translation. Extend the idea to three dimensions and you have conservation of vector momentum (since each component is independently conserved). We would also expect the behaviour of the particle to be unchanged if we observe it at a different time. This leads to conservation of energy. In 4D space-time translational symmetry in 4D results in the conservation of the 4-vector momentum (4-momentum) whose invariant magnitude generates probably the most useful relation in particle physics:

$$E_0^2 = E^2 - p^2 c^2$$

The power and significance of group theory rests on two facts:

- If some group of transformations operating on a physical object or system preserve the inherent symmetry of the system then the possible quantum states of the system correspond directly to the possible representation of the underlying group.
- Mathematicians have worked out and classified all well-behaved groups and their possible representations.

Imagine an atom in empty space. All directions are alike, so the system has complete rotational symmetry in three dimensions. This group of symmetry operations belongs to a group labelled O_3. One of the representations of O_3 is a triplet state. Physically this would correspond to three allowed quantum states of equal energy and is found to represent the allowed states for atoms with spin-1. However, if the atom is placed in a magnetic field rotational symmetry is destroyed because of the preferred axis defined by the direction of the magnetic field. This alters the allowed energy states and the spectrum of atoms in magnetic fields shows a triplet of spectral lines, indicating that the spin-1 states are now at slightly different energies.

Turn this approach around. If we were to find a triplet state existing in nature it might be reasonable to assume it corresponds to a triplet representation of the O_3' symmetry group – even if the symmetries are not related to spatial rotations (remember it is the internal structure of the group of transformations that matters, not the system that has them). If so, the relations and transformations between members of the physical system should map onto those characteristics of the group O_3'. This happened with the discovery of the force-carrying mesons in the nucleus, pions. These come in three kinds, positive, negative and neutral all with very similar masses and nuclear interactions. On this basis the properties of the pions can be predicted. *They were* (even before the first pion was discovered!) and experiments confirmed the predictions. Pions *do* behave like a physical representation of the O_3' symmetry group. This group is called the 'isotopic-spin group' related to a conserved property in certain nuclear interactions.

The most startling and compelling success for abstract group theory applied to particle physics is in the classification of the strongly interacting particles (or hadrons). We have mentioned isotopic spin. This is a symmetry that connects the pions. It also connects other strongly interacting particles and they can be grouped together by isotopic spin in the same way that we might group particles with the same electric charge. As more and more hadrons were discovered another new symmetry was discovered and called hypercharge. If the hadrons were grouped according to isotopic spin, electric charge and hypercharge intriguing and suggestive patterns began to emerge. As with the pion triplet it was thought that these particle patterns might correspond to representation of some important abstract symmetry group. The 'eightfold way' was proposed by Murray Gell-Mann and Yuval Ne'eman to classify hadrons according to the mathematical symmetry group labelled SU(3). There are other representations of SU(3) involving 10 and more members. Gell-Mann pointed out that 9 of the known baryons would fit the symmetry of the 10-member set and worked out the properties that the missing tenth member must have. The discovery of the omega-minus particle at Brookhaven in 1964 confirmed the theory. The differences between the masses of these particles then arise from a process of spontaneous symmetry breaking. This is linked to the idea of Higgs field which endows all particles with mass.

Despite the successes of group theory its very generality may turn out to be its chief weakness. As Freeman Dyson said:

"The trouble with group theory is that it leaves so much unexplained that one would like to explain. It isolates in a beautiful way those aspects of nature that can be understood in terms of abstract symmetry alone. It does not offer much hope of explaining the messier facts of life, the numerical values of particle lifetimes and interaction strengths – the great bulk of quantitative experimental data that is now waiting for explanation. The process of abstraction seems to have been too drastic, so that many essential and concrete features of the real world have been left out of consideration. Altogether group theory succeeds just because its aims are modest. It does not try to explain everything, and it does not seem likely that it will grow into a complete or comprehensive theory of the physical world."

(Freeman J. Dyson, September, 1964, in Scientific American)

Of course, the eightfold way and SU(3) was not the end of the story of the hadrons.

Murray Gell-Mann and George Zweig both realised that the eightfold way patterns could arise if the baryons were each composed of three smaller fundamental particles which were at various times referred to as partons, aces or quarks. This in no way detracts from the group theoretical approach to the baryons and symmetry principles have played a central role in understanding the nature of quark interactions through the exchange of gluons (in QCD, quantum chromodynamics).

9.1.5 The Electroweak Unification

We have already mentioned Maxwell's unification of electricity, magnetism and optics. The development of the quantum theory of atoms and radiation led to a quantum field theory of electromagnetism called quantum electrodynamics (QED), the most useful and accurate physical theory so far developed. In QED the electromagnetic force is transmitted by an exchange of massless spin-1 force-carrying particles – photons. Electromagnetism and hence QED have a powerful internal symmetry called a gauge symmetry, a term meaning 'measure' first used by Hermann Weyl in his attempts to find a unified field theory during the 1920s. The idea is abstract but it can be illustrated by a simple example. Imagine carrying out an experiment to generate electricity by dropping a mass and using it to turn a dynamo. If you did the experiment at sea level and then repeated it in a laboratory on the third floor of a building you would be surprised if the results differed. All that matters is the height through which the mass falls (ignore the variation of g with altitude). And in lifting the mass to its starting position it makes no difference to the energy generated if you move it directly into position or if you wave it about a bit first. The energy output depends only on the difference in positions in the field between the start of the drop and the end. This is reminiscent of a spatial symmetry which leaves the physical system unchanged; in this case we can add any amount of potential energy on to our starting value and the results of the experiment are unaffected – we can re-gauge the energy and retain the symmetry of the situation. Symmetries like this are called gauge symmetries. In gravitation theory it allows us to introduce a gravitational potential Φ from which the field strength at any point can be derived:

$$g = -\frac{d\Phi}{dt}$$

If the potential function is regauged to $\Phi' = \Phi + K$ where K is a constant extra potential then it is clear that:

$$g' = -\nabla\Phi' = -\nabla(\Phi + K) = -\nabla\Phi = g$$

A similar gauge symmetry is associated with electrostatics allowing us to define and use the electrostatic potential. This gauge symmetry can be extended to include static and moving charges and if we reverse the process all the properties

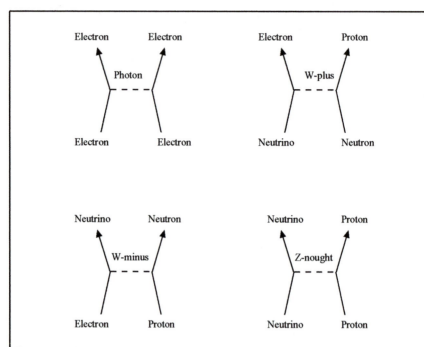

Figure 9.4 Electroweak unification relates the massless photon to three other force-carrying particles, the W-plus and W-minus and the Z-nought. The final interaction, involving the Z-nought, allows neutrinos to scatter off protons without changing the nature of any of the particles or transfering any charge. This process is called a weak neutral interaction (or weak neutral current) and was first seen in the Gargamelle bubble chamber at CERN in 1972. This implied the existence of the Z-nought which was finally discovered in 1983 (also at CERN).

of electromagnetism can be derived from the extended gauge symmetry. Among other things this results in the existence of a massless force-carrying particle, the photon.

The unification of electromagnetism with the weak nuclear force was achieved by extending the gauge symmetry still further, generalising the exchange of force-carrying particles to include particles which themselves carry mass and charge. This resulted in the prediction of new force-carrying particles called the W^+, W^- and Z^0 vector bosons which were discovered in 1983 at CERN. The symmetries which underly QED are effectively a sub-group of those involved in the electroweak theory. Furthermore electroweak theory involves a process known as 'spontaneous symmetry breaking' which results in the W and Z particles gaining mass. This had to be the case because interactions mediated by massless bosons, such as electromagnetism and gravitation, are infinite in range whereas the weak nuclear force was known to be a very short range and so must be carried by heavy particles. One of the first experimental confirmations of electroweak theory was the detection of 'weak neutral currents', the scattering of weakly interacting particles by the exchange of a Z^0.

The general mathematical structure of exchange forces was studied by C.N. Yang and his student, R.L. Mills in the early 1950s. This is effectively a theoretical generalisation of electromagnetism, and is known as Yang-Mills field theory. The W and Z particles are quanta of the Yang-Mills field in the same way that photons are quanta of the electromagnetic field. The Yang-Mills field has many more components than the electromagnetic field and this leads to far greater complexity and formidable mathematics. The colour force between quarks is also described by a Yang-Mills field theory and involves eight new force-carriers, the gluons. These are also quanta of a Yang-Mills field.

9.2 BEYOND THE STANDARD MODEL

9.2.1 Higher Symmetries

The idea of an abstract symmetry related to physical objects is important when we consider the intrinsic spin of a particle. Electrons are spin-$\frac{1}{2}$ particles. This means they possess an intrinsic angular momentum of $h/2\pi$ Js. However, quantum mechanical spin is not just a microscopic analogue of macroscopic spin (e.g. the rotation of the Earth). If a large-scale object such as the Earth rotates once it returns to its starting condition. However, the symmetry associated with spin-$\frac{1}{2}$ particles (fermions) requires a rotation through $720°$ before the original state is again reached. This seems to imply that fermions possess symmetry in a different kind of geometrical space than that of our normal human experience. The spin states of fermions turn out to be representations of the SU(2) symmetry group. The same symmetry group is associated with protons and neutrons and is linked to isotopic spin. This, as we have seen, was one of the new particle properties that led physicists toward the eightfold way and ultimately the quark model. SU(2) is also the symmetry associated with the weak force which governs interactions involving electrons and neutrinos, muons and muon neutrinos and taus and tau neutrinos. This doublet structure of the leptons is directly associated with a representation of SU(2). Electromagnetism itself is associated with a smaller symmetry group, U(1) and the electroweak unification has a symmetry created by combining these two symmetries to form U(1)×SU(2). The idea behind Grand Unified Theories (GUTs) is to combine these individual symmetries into a larger all-inclusive symmetry.

Attempts to incorporate electroweak unification and the strong interactions in a single theory have achieved some success, but many problems remain. The simplest attempts simply multiply up the individual symmetries to produce a composite: SU(3)×SU(2)×U(1). Unfortunately this is not really unification because the representations arising from each individual symmetry do not fit into a single representation of some larger symmetry group. This leads to the Standard Model in which about 19 separate parameters such as particle masses and strengths of interactions must be put in 'by hand'. An alternative approach is to postulate a larger symmetry to start with and try to derive the patterns of interactions and transformations discovered in nature from representations of this symmetry group. One candidate is SU(5). The consequence of a larger symmetry group is that particles fall into larger 'families' (or multiplets) which are linked to each other by symmetry transformations. In electromagnetism there is a single type of charge (from the '1' in U(1)). In the weak interaction there are electron-neutrino doublets and quark doublets (the '2' in SU(2)). Particles in each doublet can be transformed by the exchange of force-carrying bosons, the Ws and Zs. These transformations are a manifestation of the symmetry operations and are analagous to rotations in the rotation group:

e.g.   electron to electron-neutrino $e^- \leftrightarrow v_e + W^-$
          down quark to up quark $d \leftrightarrow u + W^-$

In QCD there are quark triplets (the '3' in SU(3)). These can also transform into other members of the triplet by exchanging gluons. In SU(5) there are quintuplets which combine the quark triplets with lepton doublets. The implication is clear, this sort of GUT allows the transformation of quarks into leptons and vice versa by the exchange of new force-carrying particles (SU(5) predicts 12 new force carriers labelled X and Y particles). If this is true protons must decay. In SU(3) they cannot decay because they are the lightest baryon. In SU(3) they could decay to pions by converting a quark to a lepton by the exchange of an X particle:

proton decay (?): $p \leftrightarrow \pi^0 + e^+$
underlying decay: $d \leftrightarrow e^+ + X^{-4/3}$

The eightfold way was confirmed by the discovery of the omega minus and a GUT based on SU(3) would be strongly supported if proton decays were ever discovered. So far, despite a number of serious attempts this has not happened. Neither have we found any evidence for the existence of X or Y particles. However, this is not surprising. Their mass, if they exist, is expected to be around 10^{16} GeV/c^2, compared to the 10^4 GeV that will be reached by the LHC early in the twenty-first century. These truly exotic particles would have been around in the extreme high-energy conditions that existed about 10^{-37} s after the Big Bang and would have been responsible for the continual transformation of particles within the SU(5) quintuplet. In fact it is thought that the differences between electromagnetic, weak and strong forces observed in our relatively cool low energy era would not have been present in the early universe. Forces governed by SU(2) and SU(3) weaken as energy increases whereas those governed by U(1) grow stronger. In the early universe all three forces would have been of equal strength. This is another example of a broken symmetry. The apparent diversity in the forces surrounding us derives from a unity at much the higher temperatures that existed in the very early universe.

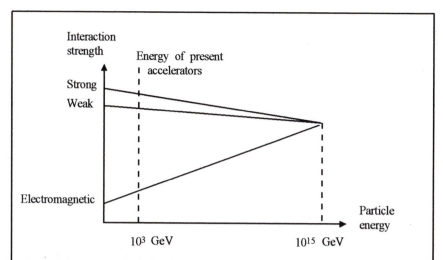

Figure 9.5 Grand unified theories predict that electromagnetic forces get stronger at higher energies whilst the strong and weak interactions become weaker. This leads to all three interactions having comparable strength at energies of about a thousand billion times the energy of our most powerful present-day accelerators. This energy is also in the region where quantum gravity would be important.

9.2.2 Supersymmetry

You may have noticed that the multiplets associated with all the symmetry groups discussed so far contain families of spin-$\frac{1}{2}$ particles, fermions. What about the integer spin bosons? SU(5) introduced transformations from quarks to leptons but both are fermions. Is it possible to transform fermions into bosons? In the 1970s a new symmetry, called supersymmetry, was discovered. This linked each fermion with a boson and vice versa, offering the possibility of classifying all particles into a single scheme. Of course it also doubled the number of particles in nature by introducing supersymmetric partners to all existing particles ('sparticles'). However, the inclusion of bosons in the particle multiplet gave some hope that gravity might be brought in alongside the other forces of nature. The transformation of fermions into bosons and back was analogous to the kinds of geometrical transformations needed for general relativity. In 1976 supergravity was proposed and attempts were made to combine it with supersymmetric versions of the Standard Model to stand as a Theory of Everything. In order to do this the theory needed an 11-dimensional space and was a kind of super-Kaluza-Klein theory designed for the late twentieth century. Unfortunately its aesthetic appeal was not backed up by experimental confirmation and so far not one of the predicted supersymmetric particles has turned up (definitely) in any high-energy experiment. On top of this no consistent quantum theory of supergravity could be formed, as it ran into the same problems with infinities that plagued the early attempts to formulate QED. Gradually interest waned, but the enthusiasm for unification continued to grow.

Matter Particles and their Supersymmetric Partners			
Particle	**Spin**	**Superpartner**	**Spin**
Electron	$\frac{1}{2}$	Selectron	0
Muon	$\frac{1}{2}$	Smuon	0
Tau	$\frac{1}{2}$	Stau	0
Neutrino	$\frac{1}{2}$	Sneutrino	0
Quark	$\frac{1}{2}$	Squark	0
Force-carriers and their Super Partners			
Graviton	2	Gravitino	$\frac{3}{2}$
Photon	1	Photino	$\frac{1}{2}$
Gluon	1	Gluino	$\frac{1}{2}$
$W^{\pm}$	1	Wino$^{\pm}$	$\frac{1}{2}$
Z^{0}	1	Zino	$\frac{1}{2}$
Higgs	0	Higgsino	$\frac{1}{2}$

9.2.3 Strings

The story of quantum field theories and unification run into the same problem time and time again. It is a recurring nightmare for the theoretical physicist, and it is called 'infinity'. The earliest pre-quantum attempt to build a model of matter out of fields, the Abraham–Lorentz theory of the electron, blew up when the field close to the electron was included. It is easy to see why this is. Electric fields store energy (their energy density is proportional to the square of the field strength) but the field strength varies as an inverse square, so it increases without limit as we approach a point-like particle. On the other hand, extended particles fall foul of relativity. There seems no way out. When quantum theory came along the problem shifted to the quantum field theories and was only dealt with by a rather *ad hoc* procedure called renormalisation, in which an infinite correction is made to the infinities that arise so that a finite residue, representing physically measurable properties, remains. This was and remains a rather unsatisfactory aspect of fundamental physics.

In the late 1960s the first 'string' theories were investigated as a way of unifying the strong interactions with the weak and electromagnetic interactions (at this time it was not known that the fundamental entities were strings). These theories needed to assume the existence of higher spatial dimensions in order to unify the symmetries of fermions and bosons into a supersymmetry, but they still broke down when they were applied to point-like particles because of the infinities that came out of the calculations. However, it was gradually realised that if the fundamental particles are replaced by one-dimensional extended strings then the theory avoids infinities. The combination of supersymmetry in 10 dimensions and the interpretation of 'particles' as excitations of 2D strings is called superstring theory. Not only does it become renormalisable but, at low energies, it reduces to the familiar field equations of general relativity! This is all very interesting, but if 'particles' are actually strings and space-time consists of nine spatial dimensions plus time (one of the few combinations that leads to a sensible string theory) then why do particles behave like points in high-energy collisions and why does the universe appear to contain just four dimensions? The answer to both questions lies in the scale of the strings and the compactification of the extra dimensions. In the original Kaluza-Klein theory it was suggested that the fifth dimension had somehow curled up into a circle on the scale of the Planck length. If this has happened for each of the additional six dimensions then it is hardly surprising we do not notice them. In a similar way, the strings themselves must be extremely small to account for the observed weakness of gravitation and are thought to have an extension comparable to about 100 times the Planck length (about 10^{-33} m) so modern high-energy experiments, which resolve detail down to about 10^{-18} m, will not be able to distinguish strings from particles.

Explaining why we cannot detect the stringiness of strings is all very well, but what does string theory predict that might be used to test the theory? It certainly leads to general relativity at low energies and to the Yang-Mills gauge theories which explain the electroweak and strong interactions. But these were already part of the Standard Model. Supersymmetry, on the other hand, is new and, since none of the particles we have already discovered can be a supersymmetric particle of any of the others, the theory predicts a supersymmetric partner for all known particles. It is thought that spontaneous supersymmetry breaking has endowed these 'sparticles' with higher mass than their partners, probably in the range 100 GeV/c^2 to 1000 GeV/c^2 (100 to 1000 times more massive than a proton). This explains why we have not yet detected them, but, unlike the structure of the strings, these are just within reach of current and planned accelerators. It will be a surprise (and a disappointment) if they do not turn up in the LHC. If they do exist they may account for a some of the 'dark matter' astronomers are searching for.

Superstring theory unifies the two major strands of twentieth century physics, general relativity and quantum theory. For a long time it was thought that this unification would require a change to quantum theory, which has always seemed to lack the formal beauty of Einstein's geometrical general relativity. However, if superstring theory is correct, quantum theory will survive pretty well unchanged but general relativity will be modified. This opens up another interesting possibility. In general relativity the topology of space-time is built into the universe and unchangeable. However, at high-energies and small lengths quantum theory should lead to wild fluctuations in local geometry and topological changes. The smooth space-time of general relativity turns out to be an average over tiny quantum fluctuations in superstring theory so that low-energy approximations give the fixed smooth space-time of Einstein's theory whilst at higher energies and on smaller scales geometry and topology are far more fluid.

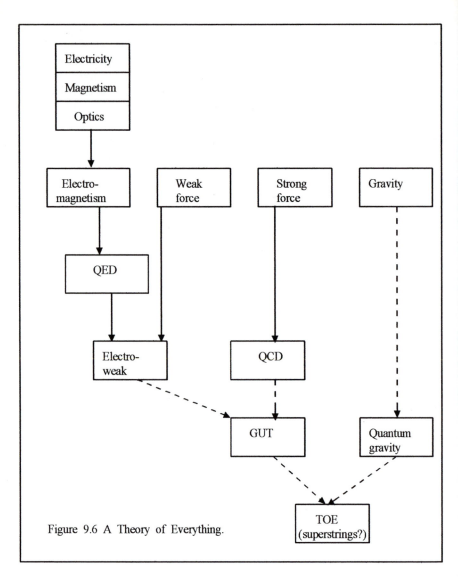

Figure 9.6 A Theory of Everything.

Where does this leave us? Superstring theory has not yet gained general acceptance as a theory of everything, but it has made a very strong claim to that title. It is a geometrical theory which incorporates gravitation, electromagnetism, Yang-Mills field theories and matter (together with a host of extra super symmetric particles we are yet to discover). Unlike GUTs based on the Standard Model it is renormalisable. It retains both general relativity (at low energies) and quantum theory whilst remaining well-behaved and self-consistent. So what are the problems? There are two:

- The higher dimensions and structures of the strings themselves are and will remain beyond the dreams of experimental physics for the forseeable future, perhaps for all time.
- The mathematical field theory of strings is formidable, and whilst its complexity leads to a vast number of possible solutions, we do not know the appropriate criteria for selecting those that apply to our universe.

If the second problem can be overcome superstring theory holds out the prospect that the fundamental constants of nature may be calculated from first principles.

GUIDE TO CHAPTERS 10 TO 12
SPACE AND TIME

10 The Speed of Light
11 Special Relativity
12 General Relativity

10

Maxwell showed that the speed of light derives from the laws of electromagnetism.
Michelson refined measurements of the speed of light and searched for a medium to support it.
The Michelson-Morley experiment failed to detect this 'luminiferous ether'.
The speed of light is a universal speed limit.

11

Newtonian mechanics and Maxwell's electromagnetism are not consistent in their treatment of space time and motion.
Einstein proposed that the laws of physics should be the same for all observers regardless of their motion.
The principle of relativity leads to many counter-intuitive results: time dilation, length contraction; mass-energy equivalence etc ...
Minkowski reinterprets relativity as a theory of 4D space-time geometry.

12

How can gravity and acceleration be included in relativity?
Einstein realised that free-falling observers do not feel gravity.
The equivalence principle linked gravity and acceleration and hinted at new phenomena: deflection of light by gravity, gravitational time dilation and the advance of perihelion.
Gravity can be interpreted as a distortion of Minkowski's 4D space-time: 'matter tells space how to bend and space tells matter how to move'.
The theory predicts gravitational waves and black holes.

10
THE SPEED OF LIGHT

10.1 MEASURING THE SPEED OF LIGHT

10.1.1 Measuring the Speed of Light

"... light is always propagated in empty space with a definite velocity c which is independent of the state of motion of the emitting body."

(Albert Einstein, 'On the Electrodynamics of Moving Bodies' Annalen der Physik, 17, 1905)

It may seem odd to devote a chapter to light in a book on twentieth century physics, but the speed of light is central to our modern ideas about space and time. It is now believed that this speed is a limit that can never be reached or exceeded by particles with non-zero rest mass and it is the greatest speed at which information can pass from one place to another. The speed of light provides a fundamental link between measures of space and time that was first recognised by Einstein in his theory of relativity (1905).

The importance of this speed and the accuracy with which we can measure it was recognised in 1983. In that year the General Conference on Weights and Measures met in Paris and adopted a defined value for the speed of light. This value was set at exactly 299 792 458 ms^{-1} and will not be changed as a result of any future measurements. The second is also a defined quantity, equal to exactly 9 192 631 770 periods of the radiation emitted during the transition between two hyperfine levels of the ground state of the cesium-133 atom. The metre is derived from these two as the distance travelled by light in 1/(299 792 458) of a second, so there is a very real sense in which light is now used to measure distance. The metre is now realised using lasers stabilised by certain specified molecular transitions.

The switch from an experimentally defined metre to a standard speed of light came about because of the increased accuracy with which the speed of light could be measured using lasers with high spectral purity. These were available from the 1970s on and their frequencies could be compared with cesium 'clocks' by harmonic mixing and they improved the accuracy of measurements by a factor of 100 in just over a decade.

10.1.2 From Galileo to Michelson

"The rapid motion of sound assures us that that of light must be very swift indeed, and the experiment that occurred to me was this. I would have two men each take one light inside a dark lantern or other covering, which each could conceal and reveal by interposing his hand, directing this toward the vision of the other. Facing each other at a distance of a few [metres], they could practice revealing and concealing the light from each other's view, so that when either man saw a light from the other, he would at once uncover his own. After some mutual exchanges, this signalling would become so adjusted that without any sensible variation either would immediately reply to the other's signal, so that when one man uncovered his light, he would instantly see the other man's light.

This practice having been perfected at a short distance, the same two companions could place themselves with similar lights at a distance of two or three miles and resume the experiment at night, observing carefully whether the replies to their showings and hidings followed in the same manner as near at hand. If so, they could surely conclude that the expansion of light is instantaneous, for if light required any time at a distance of three miles, which amounts to six miles for the going of one light and the coming of the other, the interval ought to be quite noticeable. And if it were desired to make such observations at yet greater distances, of eight or ten miles, we could make use of the telescope, focusing one for each observer at the places where the lights were to be put into use at night. Lights easy to cover and uncover are not very large, and hence are hardly visible to the naked eye at such distance, but by the aid of telescopes previously fixed and focused they could be comfortably seen."

(Galileo Galilei, in *Dialogue Concerning Two New Sciences*, 1638)

Galileo's experimental attempt to measure the speed of light failed because it is much greater than he anticipated. However, the method outlined in his wonderful book is in principle the same as that used by the great American experimental physicist, Albert Abraham Michelson, in 1926. Michelson spent much of his scientific career developing ever more precise methods to measure the speed of light and investigating how the Earth's motion might affect its speed relative to terrestrial laboratories. He won the Nobel Prize for Physics in 1907 *"for his optical precision instruments and the spectroscopic and metrological investigations carried out with their aid"*. He was the first American to win a Nobel Prize.

The earliest successful methods to measure the speed of light were astronomical. The reason for this is obvious. The distances over which light can travel in a terrestrial experiment are short, a few kilometres at most, so very accurate timings of extremely short intervals are required. This did not become possible until the latter half of the nineteenth century. Astronomy, on the other hand, offered

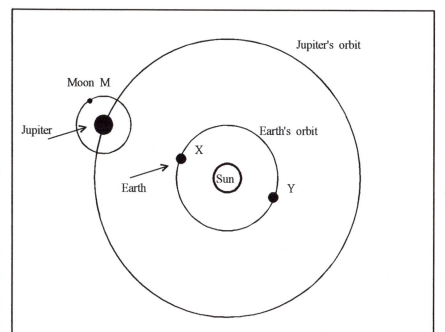

Figure 10.1 Roemer's calculation of the speed of light used a Moon of Jupiter as a 'clock', giving regular time signals when it came out from behind Jupiter itself. If the distance between Earth and Jupiter was constant the time signals would be regularly spaced. However, the distance changes by an amount equal to the diameter of the Earth's orbit (MY-MX). The extra time taken by light to cross the orbit shows up as a periodic variation in the observed reappearance times of the Moon. This variation has an amplitude of about 8 minutes, showing that light takes about this time to travel one astronomical unit.

enormous distances so that time differences could be measured in minutes rather than microseconds. The first person to take advantage of this was the Danish astronomer, Römer. He observed that times of eclipses of the Moons of Jupiter (caused as they pass behind the planet) had a systematic variation in addition to their period of orbit that seemed linked to the relative positions of the Earth and Jupiter in their respective orbits. In 1676 he predicted that a certain eclipse would be ten minutes later than expected (on the assumption of a fixed period) because of a delay in the light reaching the Earth as it crosses the Earth's own orbit. At that time this distance was not accurately known, but still led to a value for the speed of light in excess of 2×10^8 ms^{-1}. The English astronomer James Bradley carried out the second astronomical determination of the speed of light. He observed the annual variation in apparent positions of stars (called stellar aberration) and accounted for it by assuming the direction of incoming light is affected by the motion of the earth in its own orbit. The angular displacement of the star is then directly related to the ratio of the earth's orbital velocity to the speed of light. He came up with a speed of 2.84×10^8 ms^{-1} (the modern interpretation of aberration uses the relativistic transformation of velocities, but leads to the same value).

The first laboratory method for c was carried out by the French physicist, Louis Fizeau, in 1849 using a method that was improved by Cornu in the 1870s. The basic idea was to chop a light beam up using a rapidly rotating toothed wheel, and send it several kilometres to a mirror that reflects it straight back to the toothed wheel. On its return the wheel has rotated through a small angle. At a certain rate of rotation the returning light will hit the tooth next to the gap through which it departed. If the rate of rotation is measured the time of flight of the light can be calculated and the speed deduced from this. Cornu achieved a value of 2.99990×10^8 ms^{-1} in 1874. A related method was suggested by Arago in 1838 and carried out by Foucault in 1862. This involved bouncing light off a rapidly rotating mirror, sending it on return journey to a distant mirror and bouncing it off the original rotating mirror once again to form an image. The faster the mirror rotates the more it will turn during the time of flight of the light and the greater the angular displacement of the observed image. The speed of light could be calculated from the rate of rotation, the length of the path and the angular displacement. Foucault used this method to show that the speed of light in water is less than in air.

Figure 10.2 **James Clerk Maxwell** 1831-1879. The Maxwell equations are a cornerstone of mathematical physics. The Maxwell-Boltzmann distribution is at the heart of classical statistical thermodynamics.   Artwork by Nick Adams

In the mid-nineteenth century James Clerk Maxwell published his equations for electromagnetism and demonstrated that fluctuations in the fields (created by accelerating charges) propagate outwards from their source at a speed given by:

$$c = \frac{1}{\sqrt{\varepsilon_0 \mu_0}}$$

This speed was immediately recognised as that of light and light's electromagnetic nature was thereby confirmed. This raised the possibility that the speed of light could be determined by measurements on non-visible components of the electromagnetic spectrum, an idea that was followed up in the twentieth century.

In the meantime Michelson and Newman improved the rotating mirror method and obtained values of around $2.9985 \pm 0.0060 \times 10^8$ ms^{-1} in the mid-1880s.

Michelson also showed that the ratio of the velocity of light in air to the velocity in water is 1.33 (the refractive index of water). However, Michelson's crucial

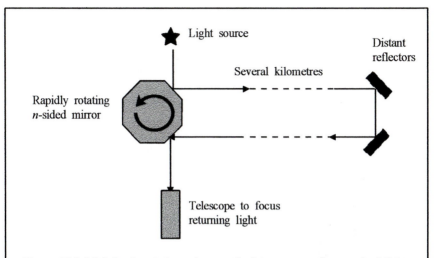

Figure 10.3 Michelson's rotating mirror method to measure the speed of light. An observer will only see a clearly focused image when the returning light bounces off a mirror face aligned in the same way as if the mirror was at rest. The lowest frequency at which this occurs is when the n-sided mirror turns through $1/n$ of a revolution during the time taken for the return light trip to the distant mirror.

contribution was the development of a *null method* in which the light reflects off and returns to a multi-sided rotating mirror whose rotation rate is adjusted so that the returning light strikes a face parallel to the original one. This reflects the beam along exactly the same path as if the multi-sided mirror were at rest (hence a 'null method' – experimenters had to find the conditions for zero image displacement).

From 1924 to 1926 Michelson directed a series of measurements of the speed of light in air using a rotating mirror method and sending light from the Mount Wilson Observatory over a return distance of about 70 km. The baseline for these experiments was surveyed by the U.S. Coast and geodetic survey to an accuracy of nearly 1 part in 7 million (about 1 mm in 70 km) and the rate of rotation was measured using stroboscopes. His result was $2.99796 \times 10^8 \pm 0.00004$ ms^{-1}.

Not content with this amazing result he collaborated with G.F. Pease and F. Pearson in his final years to carry out a direct measurement of the speed of light in a vacuum. A 1 mile tunnel was used and evacuated to a pressure of 60 Pa. The time for the beam of light to travel up and down the tunnel several times was measured. This gave a value of 2.99774×10^8 ms^{-1} (1932).

In 1940 Birge reviewed all the experimental values for the speed of light, paying particular attention to the errors involved in each determination. He concluded that the best value for the velocity of light was $c = 299\ 776\ 000 \pm 4000$ ms^{-1}.

The development of new technologies, especially those linked to the invention of radar during the second world war, allowed post-war physicists to try some new approaches. In 1958 Froome used a microwave interferometer to obtain a value of 299 792 500 ± 100 ms^{-1}. By 1975 results collected by several metrology laboratories set the value at 299 792 458 ± 1 ms^{-1}. In 1983 the uncertainty was dropped and this value was adopted as the defined value of the speed of light. This removes the need for any future measurements of the speed of light.

Figure 10.4 **Albert Abraham Michelson** 1852-1931 won the 1907 Nobel prize for his experiments on the speed and nature of light.

Artwork by Nick Adams

10.1.3 The Michelson-Morley Experiment and the Ether

"Whatever difficulties we may have in forming a consistent idea of the constitution of the ether, there can be no doubt that the interplanetary and interstellar spaces are not empty, but are occupied by a material substance or body, which is certainly the largest and probably the most uniform body of which we have any knowledge."

(James Clerk Maxwell, Ether, Encyclopaedia Brittannica, 1890)

"For well nigh a century we have had a wave theory of light; and a wave theory of light is almost certainly true. It is directly demonstrable that light consists of waves of some kind or other, and that these waves travel at a certain well-known velocity - achieving a distance equal to seven times the circumference of the Earth every second; from New York to London and back in the thirtieth part of a second; and taking only eight minutes on the journey from the Sun to the Earth. This propagation in time of an undulatory disturbance necessarily involves a medium. If waves setting out from the Sun exist in space eight minutes before striking our eyes, there must necessarily be in space some medium which conveys them. Waves we cannot have, unless they be waves in something."

(Oliver Lodge, *The Ether of Space*, Harper and Brothers, London and New York, 1909.)

It seemed obvious to most physicists that waves of any kind must be waves in something, the something must act as a medium through which they pass. This was simply an extension of the Newtonian view of mechanical waves in which particles of a material medium vibrate in sequence as a wave passes through. There were no known media that would transmit vibrations at any speed approaching that of light so they invented a new medium – the 'luminiferous ether' or just ether (a more convenient term for those, such as Oliver Lodge, who saw it as the seat of far more than just electromagnetism). However, if the ether really existed than the speed of light was presumably a speed relative to the ether and the speed of light relative to us (on the moving Earth) should be variable. For example, if the Earth is moving through the ether at velocity v in the same direction as a particular beam of light, then the light beam will have a relative velocity $(c - v)$ with respect to measuring apparatus in a terrestrial laboratory. In a similar way the relative velocity should be $(c + v)$ when the directions of the light beam and Earth are opposed. Maxwell was aware of this, but pointed out that the effect would be concealed (to some extent) in terrestrial measurements because these involve timing rays which travel out and back along the same path (the effects do not cancel, but the overall delay compared to no motion through the ether becomes a second order term in v/c). He thought this was too small an effect to be measured experimentally. Michelson decided to prove him wrong.

Michelson was involved with several experiments to detect the presence and effect of the ether. The most famous of these was, without doubt, the one he carried out with Edward Morley in 1887. It was a refined attempt to measure the velocity of the Earth through the ether by detecting the difference in time of flight of light beams travelling along perpendicular paths in an interferometer.

If the Earth really does move through the ether then there should be an 'ether wind' through terrestrial laboratories and this wind should change direction as the Earth itself progresses around its orbit. In the Michelson-Morley experiment a ray of light moving back and forth along a line parallel to the Earth's orbital motion should be delayed more than one moving perpendicular to it. This delay would be detected by its effect on interference fringes when the two returning rays superpose. The 1887 experiment was certainly sensitive enough to detect a shift in fringes due to an ether wind velocity comparable to that of the Earth's orbital velocity. But no shift was observed. The experiment has been repeated many times since, and no significant shift has ever been detected. This is probably the most famous null result in the history of physics and it threw the classical view of light, as an undulation in the luminiferous ether, into chaos. It was almost twenty years before Einstein solved the problem by abolishing the ether, and this (as we shall see) resulted in a radical reinterpretation of space and time.

In the 1930s R.J. Kennedy and E.M. Thorndike carried out a Michelson–Morley type experiment which differed from the original in two very important ways.

- The interferometer arms were of different lengths.
- The apparatus was set up and maintained for several months while they

looked for a shift in the interference fringes.

The first difference was to rule out one of the early explanations of the null result – the Lorentz-Fitzgerald contraction. This was a proposal that the apparatus contracts in its direction of motion through the ether by an amount which is exactly enough to conceal the reduction in average light speed for the return journey in that direction (it only cancels exactly if the arms are of equal length). The second difference was to try to observe the fringe shift as the Earth's orbital motion rotated the apparatus in the ether and altered the orientation of the arms with respect to the ether wind (changing the time of flight for both beams by a different amount and so changing their final phase difference). The results showed that any shift was due to a change in light speed of less than 2 ms^{-1} on the round trip. This agreed with special relativity to an accuracy of about 1 part in 10^8.

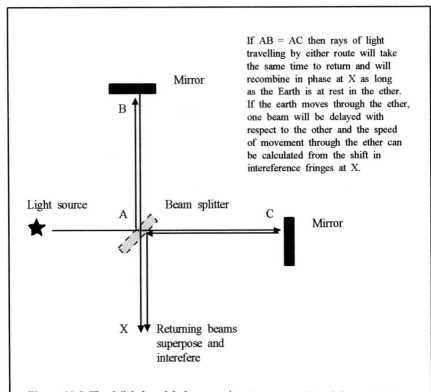

If AB = AC then rays of light travelling by either route will take the same time to return and will recombine in phase at X as long as the Earth is at rest in the ether. If the earth moves through the ether, one beam will be delayed with respect to the other and the speed of movement through the ether can be calculated from the shift in intereference fringes at X.

Mirror

B

Light source

A

Beam splitter

C

Mirror

X   Returning beams superpose and interefere

Figure 10.5 The Michelson-Morley experiment was an attempt to measure the effect of the Earth's motion through the ether on the apparent velocity of light. It revealed no such effect and ultimately the ether hypothesis was abandoned.

10.2 FASTER THAN LIGHT?

10.2.1 A Trick of the Light

According to the special theory of relativity no massive body can be accelerated up to the speed of light, no matter how much energy is supplied or how long the period of acceleration lasts. This limitation is intimately linked to causality. If faster than light motion is possible then the order of cause and effect in one reference frame could be reversed when observed from another reference frame moving faster than the speed of light with respect to the first one. However, there are several physical situations that *do* result in super-luminal velocities, some of which are discussed below and none of which actually violate causality.

Figure 10.6 Although all the mechanical parts of a pair of scissors are limited to sub-light speeds, the point where the blades cross could (in principle) move faster than the speed of light.

Before discussing apparent super-luminal speeds in actual physical objects it is worth considering a couple of well-known thought experiments. The first of these involves a very large pair of scissors! As these are closed the point at which they cross moves rapidly along from the pivot toward the blade tips. If the blades are extremely long the velocity of the crossing point can increase without limit since the closing blades become almost parallel and a small motion at the tips results in a large movement of the crossing point. In principle the crossing point can move faster than the speed of light. A similar argument applies to a line of Christmas tree lights that flash in sequence, if the intervals between flashes are short enough the pulse of light moves along the chain faster than the speed of light. There are many other equivalent examples including the tip of a searchlight as it sweeps across the clouds and the motion of a planetary shadow when it is projected onto the surface of another planet by the Sun. The electron beam in a TV scans across the screen and could in principle 'move' faster than *c* (some oscilloscope traces already approach this speed). The reason why none of these motions is a problem for relativity is that the *motion* is in each case an illusion. We are making a connection between successive positions of *different objects* rather than the actual sequence of positions of one particular object. For example, in a TV the electrons responsible for the beam at the left of the screen are not the same electrons responsible for the beam at the right of the screen – no electron moves across the screen at the scan speed. In other words, there is no causal connection between the points taken to connect the motion.

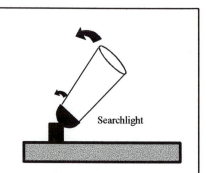

Figure 10.7 Photons obviously travel at the speed of light, but it is still possible (in principle) for the end of the beam to scan across the clouds at a speed greater than this.

Quasars are galaxies whose nuclei emit enormous amounts of energy and so can be seen at very great distances. They are among the most distant and energetic objects studied by astronomers. Some are powerful radio sources. These radio waves are emitted by synchrotron radiation as energetic electrons spiral around the strong magnetic fields generated by jets of charged material emerging from the galactic nucleus. In the 1960s astronomers measured the velocities of quasar jets and found that some seemed to be moving at many times the velocity of light. These velocities were calculated using the apparent displacements of the jet in a sequence of radio images of the quasar (the distance of the quasar is calculated from its red-shift and the images are formed using long baseline interferometry).

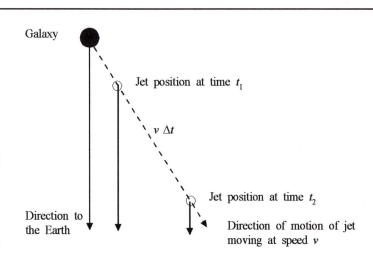

Figure 10.8 Some galactic jets appear to be moving away from their emitting galaxy at speed well in excess of c. This is an illusion caused by their component of motion toward us. Light from later positions has a shorter distance to travel to reach us, so it arrives soon after light from much greater distances, giving the impression that the jet is moving away from the galaxy much faster than is actually the case.

These apparent super-luminal speeds can be explained if the jet is actually moving toward us (see Fig. 10.5 and the Maths Box below).

Maths Box: Super-luminal Jets

Radio images at times t_1 and t_2 will not be a true representation of the relative motion of the jet relative to the galaxy. It will have approached us during this time, so that the time of flight of light from the jet is less than from the galaxy. This implies that we are comparing the galaxy's position at a particular time with the position of the jet some (considerable) time later. This means the apparent velocity of the jet perpendicular to our line of sight v_{app} is greater than its actual velocity v. The relation between v_{app} and v is simple to derive.

Consider two positions of the jet separated by a time interval Δt on Earth. During this time the jet moves a distance $v\Delta t$. Assume this velocity is at angle θ to our line of sight. The jet approaches us by an amount $v \cos \theta \, \Delta t$. It also moves perpendicular to our line of sight by $v \sin \theta \, \Delta t$.

The time of flight of light from the jet to us is reduced by $\dfrac{v \cos \theta \Delta t}{c}$.

Radio images for these positions will be separated by an Earth time:

$$\Delta t' = \Delta t - \frac{v \cos \theta \Delta t}{c} = \Delta t \left(1 - \frac{v \cos \theta}{c} \right)$$

The apparent velocity is given by:
$$v_{app} = \frac{v \sin \theta \Delta t}{\Delta t'} = \frac{v \sin \theta}{\left(1 - \dfrac{v \cos \theta}{c} \right)}$$

This is equal to v if $\theta = 90°$. For smaller angles and large v the apparent velocity can exceed the speed of light. For example, if $v = 0.9c$ and $\theta = 10°$ the apparent velocity is $v_{app} = 11.4c$.

There are also super-luminal velocities associated with supernovae. One of these involves the intense burst of neutrinos emitted from the exploding star. Intergalactic space is not empty and the light burst from the same explosion is delayed slightly because of its interaction with this tenuous medium (its velocity through space is less than its velocity through a true vacuum, as it is in any transparent medium). When Supernova 1987A exploded neutrinos reached the Earth on 22 February 1987 and the light arrived some 20 hours later. Neutrinos only interact through the weak force so were able to travel through space faster than light (but not faster than c). This is similar to the Cerenkov effect in which charged particles travel faster than the speed of light through a transparent medium. They do interact electromagnetically and emit a characteristic electromagnetic shock wave called Cerenkov radiation.

In the years following 1987 astronomers observed two concentric expanding rings centred on the supernova. The rings are formed by the reflection of light from interstellar gas clouds (this is an example of a 'light echo'). The apparent radial velocity of these rings was greater than c. This too is a consequence of their movement toward us and a proper geometrical analysis of the motion results in a sub-luminal velocity for the expanding gas cloud. Think about a point on the ring, if the ring is moving toward us the point moves along a path comparable to that of the jet in the previous example, with the same consequences for apparent and actual velocities.

10.2.2 Quantum Theory and the Speed of Light

Einstein realised very early on that the Copenhagen Interpretation of quantum theory implied super-luminal effects. For example, imagine a pair of correlated particles emitted from a single event. The total angular momentum of the pair of particles is fixed, so the determination of the spin-state of either one of them determines the spin-state of the other. In the Copenhagen Interpretation the particle pair is represented by a spreading wavefunction. Prior to any experimental measurement neither particle 'possesses' a definite angular momentum, the wavefunction represents a superposition of all possibilities. It is only when the measurement is made that the particles adopt definite angular momentum states (e.g. spin-up or spin-down). Let us assume that particle A travels off to the left while particle B travels off to the right. We carry out a measurement on A and a short time afterward we measure the spin-state of B. If the time is so short that a signal could not pass between the two particles at sub-light speed then there are two possibilities:

- The spin state of B might be completely random, i.e. the correlation with A has been lost.
- The spin state of B might correlate with the results of measurements on A.

Both possibilities pose serious questions. If the first is correct then conservation of angular momentum is violated. If the second is correct then we have to ask how B 'knew' the result of A's measurement (what Einstein referred to as a 'spooky action-at-a-distance'). Classically this would not be a problem because both states are determined at the moment the particles are ejected from the source. However, this cannot hold in quantum theory because the states only become actualised at the moment of observation – as borne out by interference effects in which the superposed states are essential to predict the final outcome. Experiments confirm that the spin-states *are* correlated. If we interpret this as the result of a signal passing from A to B after the measurement then this signal would have to travel instantaneously (or at least very much faster than c). Of course, the wavefunction itself is not an observable object, so connections within it that are maintained faster than the speed of light do not necessarily violate causality. In fact the correlation between A and B does not require the transmission of information faster than c. If the experiment is repeated many times the sequences of results at A and at B are both random, so neither contains any information. If experiments at A were to be carried out in a sequential way so that information is present in the experimental sequence (e.g. by encoding a message in the sequence of measurements of different components of A's spin) then measurements at B could not decode this information unless B also had access to A's results – and this would have to be sent by conventional means (e.g. radio) at or below the speed of light. The information is locked into the correlation between the results at A and at B and can only be accessed if both strings of data are compared.

One further quantum mechanical effect which may lead to super-luminal velocities is quantum tunnelling. When a classical particle strikes a barrier it bounces off and moves away from the barrier. In a similar way classical waves bounce off a reflective barrier. However, when electrons or light hit barriers their behaviour is governed by quantum theory, not classical theory (by 'barrier' we mean a potential energy barrier). One way to think about this is to represent the incident object by a wave packet whose waves are partly reflected and partly transmitted at the barrier. If the barrier has finite thickness and height the amplitude of the quantum wave decays exponentially into the barrier and has finite amplitude on the far side. The statistical interpretation of the wavefunction says that the intensity of the wave is proportional to the probability that the particle is present in that region of space. Partial reflection and partial transmission therefore implies that the incident particle has a probability of passing through the barrier, even though a classical particle could not do so. This is described as quantum mechanical tunnelling because the kinetic energy of the associated particle is less than the potential energy it requires to pass 'over' the energy barrier – it 'tunnels' through. If the wavefunction collapses everywhere at the same time then the time between the particles being present on one side and then the other side of the barrier could be very short indeed, implying that the particle shifts through the barrier at a speed in excess of c. The time taken to pass through a barrier like this is called the 'interaction time' or 'tunnelling time'. There is no consensus about whether the interaction time is zero, constant or proportional to barrier width, but many physicists assume it will allow transmission at faster then the speed of light. In the 1990s a number of experiments were carried out to measure this interaction time and calculate the propagation speeds of the particles involved. Their results seem to suggest:

- That the interaction time reaches a maximum value as the barrier width increases
- Transmission can take place at super-light speeds.

In March 1995 Gunter Nimtz sent a frequency modulated microwave signal 12 cm at a speed 4.7 c. The modulated waves were carrying Mozart's 40th symphony. This caused a great deal of interest and the jury is still out on what it actually says about faster than light communications. However, few physicists believe that this

11
SPECIAL RELATIVITY

11.1 THE PHYSICS OF SPACE AND TIME

11.1.1 Mechanics and Electromagnetism

Galileo and Newton were both aware that the laws of mechanics take the same form in all uniformly moving reference frames. This idea is called 'Galilean Relativity'. It implies that mechanics does not distinguish between inertial reference frames. Another way of putting this is to say that no experiments carried out in one inertial reference frame can establish its absolute state of motion – rest and uniform motion are indistinguishable. If all phenomena are reducible to mechanics (as most physicists before Einstein assumed) then the idea of an absolute rest frame would be meaningless, or at least unnecessary.

The development of electromagnetism in the mid-nineteenth century appeared to change all this. Maxwell's equations predict that electromagnetic waves travel at a speed c and physicists assumed that this speed must be measured relative to some supporting medium that they called the ether. If this assumption were correct then the speed of electromagnetic waves relative to an observer's measuring equipment would depend on the speed of the observer relative to the ether. This follows from the Galilean transformation for position. Imagine a light pulse moving along the x-axis at speed c in the ether (i.e. in a reference frame at rest in the ether). The position of the pulse relative to its stationary source is given by $x = ct$. Now, if an observer moves in the same direction at speed v and passes the light source at the instant it emits the pulse, then the position of the pulse with respect to this observer is $x' = ct - vt$ and the relative velocity of light seen by the moving observer is:

$$c' = \frac{ct - vt}{t} = (c - v)$$

This result is based on two assumptions about space and time:

- The ether itself defines an absolute spatial reference frame.
- There is an absolute time common to all observers whatever their motions.

Figure 11.1 **Albert Einstein** 1879-1955 won the 1921 Nobel Prize for Physics for his work on the photoelectric effect in which he introduced the idea of light quanta or photons. He made many crucial contributions to quantum theory despite deep misgivings about its meaning and interpretation. However his best known work concerns the nature of space, time and matter developed in the special theory of relativity (1905) and the general theory of relativity (1916). His famous equation, $E=mc^2$ explains the energy released in radioactive decay and other nuclear transformations such as fission and fusion. These ideas were used to explain how stars shine, to develop nuclear weapons and nuclear power stations and are responsible for the creation of new particles in accelerators. General relativity explained the expansion of the universe in terms of geometry and led to the prediction of black holes and gravitational waves. Einstein's later work concerned the quest for a unified field theory and a critique of the foundations of quantum theory. He was an active pacifist and in 1952 he was invited to become president of Israel on the death of Chaim Weizmann. He declined.   Artwork by Nick Adams

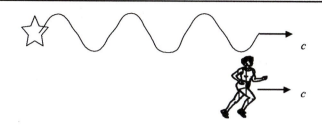

Figure 11.2 Einstein wondered what a light wave would look like if you could move alongside it at the speed of light. He was sure this must be impossible, but nothing in Newtonian mechanics forbids it.

We have already seen that experiments to measure the variation of the speed of light caused by the Earth's motion through the ether produced a series of null results, the most famous of these being the Michelson-Morley experiment. It is not clear to what extent the young Einstein was influenced by these results or by the theoretical attempts to explain them by Fresnel, Lorentz, Poincaré and others, but he was certainly concerned by the problem. When he was sixteen years old he asked himself what a light wave would look like if he chased after it at the speed of light. According to the assumptions above he should observe a stationary wave crest (since $c - v = 0$). However, the Maxwell equations have no such solution, so either:

- Maxwell's equations do not apply for a rapidly moving observer.
- **OR**: It is impossible for an observer to move at the speed of a light beam.

Either way, something had to give to make mechanics and electromagnetism mutually consistent.

Before we discuss Einstein's solution to this problem it is worth mentioning the work of Lorentz and Poincaré. They anticipated Einstein's theory and produced much of the same mathematical formalism, but neither of them took the final and crucial step which transformed our views of space and time. Lorentz explained the null results of experiments designed to detect motion through the ether by introducing a series of physical effects resulting from the motion of apparatus and measuring equipment through the ether. Lorentz's basic assumption was that all material bodies are held together by electromagnetic forces which are themselves supported by the ether. The motion of an extended body through the ether affects these interactions and introduces spatial distortions, the most famous of which is known as the Lorentz contraction. He also introduced a local time co-ordinate for moving reference frames so that the laws of electromagnetism retained their original form for the moving body. However, his retention of the ether and his insistence that local times were just a mathematical device made the whole approach rather *ad hoc*. He worked on it for over a decade and wrote down a set of transformation equations for space and time co-ordinates in moving reference frames. These Lorentz transformations are central to special relativity, although Einstein arrived at them from more general assumptions. Lorentz referred to his theory as one of corresponding states and published his most complete version of the theory in 1904 in a paper entitled Electromagnetic Phenomena in a System Moving with any Velocity less than that of Light. Lorentz's work included many of the mathematical formulae that reappeared in Einstein's 1905 paper, but lacked the authority invested in Einstein's work by his adoption of relativity as a universal principle.

Meanwhile, Poincaré was concerned with the philosophical foundation of the subject. By 1904 he had become convinced that it was impossible to detect our motion relative to the ether by any experimental method whatsoever. He summarised this by stating a Principle of Relativity that would not be out of place in Einstein's own paper: " ... the laws of physical phenomena must be the same for a 'fixed' observer or for an observer who has uniform motion of translation relative to him: so that we have not, and cannot possibly have, any means of discerning whether we are, or are not, carried along in such a motion. "

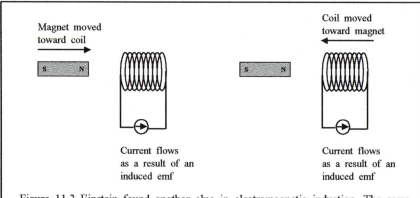

Figure 11.3 Einstein found another clue in electromagnetic induction. The same induced emf is induced when a magnet approaches a coil as when the coil approaches the magnet at the same rate. This suggests that the underlying laws are the same for an observer moving with the magnet as for one moving with the coil.

This insight, coupled with his familiarity with the work of Lorentz (it was Poincaré who showed that the Lorentz transformations form a group) has led many historians of science to wonder why he did not take the next step himself and produce special relativity. But it is easy to be wise after the event and the fact that such a brilliant and intuitive physicist did not take this step underlines Einstein's genius in creating such a radical theory. Poincaré, like Lorentz, believed that the impossibility of experimentally detecting the ether was a result of physical effects due to motion through the ether rather than a fundamental principle in its own right. He retained his belief in the ether and in an absolute or true time, which is measured on a clock at rest in the ether. For Poincaré the principle of relativity had to be *explained* in more fundamental terms.

11.1.2 The Principle of Relativity

Einstein developed his ideas on relativity while he worked as a Patent Officer in Berne, isolated from the mainstream of scientific thought and with restricted access to books and journals. Despite this he published a number of scientific papers and each one of his three contributions to Annalen der Physik, 17, 1905, would have been enough to establish his reputation. The first paper dealt with the statistics of Brownian motion and provided an analytic way to confirm the existence of atoms and molecules (which were still regarded by some physicists as mathematical artefacts). The second (for which he won the Nobel Prize) introduced light quanta (photons) to explain the photoelectric effect. The third paper, On the Electrodynamics of Moving Bodies, set out the ideas of special relativity. Einstein's introduction to this remarkable paper is quoted below:

"It is known that Maxwell's electrodynamics – as usually understood at the present time – when applied to moving bodies, leads to asymmetries which do not appear to be inherent in the phenomena. Take, for example, the reciprocal electrodynamic action of a magnet and a conductor. The observable phenomenon here depends only on the relative motion of the conductor and the magnet, whereas the customary view draws a sharp distinction between the two cases in which either the one or the other of these bodies is in motion. For if the magnet is in motion and the conductor at rest, there arises in the neighbourhood of the magnet an electric field with a certain definite energy, producing a current at the places where parts of the conductor are situated. But if the magnet is stationary and the conductor in motion, no electric field arises in the neighbourhood of the magnet. In the conductor, however, we find an electromotive force, to which in itself there is no corresponding energy, but which gives rise – assuming equality of relative motion in the two cases discussed – to electric currents of the same path and intensity as those produced by the electric forces in the former case.

Examples of this sort, together with the unsuccessful attempts to discover any motion of the earth relatively to the 'light medium', suggest that the phenomena of electrodynamics as well as of mechanics possess no properties corresponding to the idea of absolute rest. They suggest rather that, as has already been shown to the first order of small quantities, the same laws of electrodynamics and optics will be valid for all frames of reference for which the equations of mechanics hold good. We will raise this conjecture (the purport of which will hereafter be called the 'Principle of Relativity') to the status of a postulate, and also introduce another postulate, which is only apparently irreconcilable with the former, namely, that light is always propagated in empty space with a definite velocity c which is independent of the state of motion of the emitting body. These two postulates suffice for the attainment of a simple and consistent theory of the electrodynamics of moving bodies based on Maxwell's theory for stationary bodies. The introduction of a 'luminiferous ether' will prove to be superfluous inasmuch as the view here to be developed will not require an 'absolutely stationary space' provided with special properties, nor assign a velocity-vector to a point of the empty space in which electromagnetic processes take place.

The theory to be developed is based – like all electrodynamics – on the kinematics of the rigid body, since the assertions of any such theory have to do with the relationships between rigid bodies (systems of co-ordinates), clocks, and electromagnetic processes. Insufficient consideration of this circumstance lies at the root of the difficulties which the electrodynamics of moving bodies at present encounters."

(Albert Einstein, On the Electrodynamics of Moving Bodies, Annalen der Physik, 17, 1905)

It is immediately clear where this diverges from the work of Lorentz and Poincaré. Einstein has no need for absolute space, a 'luminiferous ether' or absolute time he is proposing a new fundamental principle that can be paraphrased:

- The laws of physics are the same in all inertial reference frames.

His second postulate, that the speed of light is constant regardless of the motion of the light source can be considered as a consequence of the first if Maxwell's electromagnetism is taken as a 'law of physics'. The consequence for relative measurements of space and time co-ordinates remains to be worked out by a detailed critique of the experimental methods used to obtain them. It is these methods to which Einstein refers in the final paragraph of his introduction.

Einstein's paper is divided into two parts. The first is kinematical, dealing with simultaneity, the relativity and transformation of co-ordinates and times, velocity addition and the interpretation of the equations. This last is particularly important since Einstein is the first to make it clear that co-ordinate transformations relate to actual readings on moving measuring instruments so that differently moving rods and clocks give different results. The second part deals with electrodynamics and shows how to transform Maxwell's equations, accounts for relativistic Doppler effects and explains stellar aberration. It also considers the dynamics of a slowly accelerated electron, showing that its mass and energy content increase with speed. Later in 1905, also in Annalen der Physik, Einstein published a paper entitled, Does the Inertia of a Body depend upon its Energy-Content? in which he demonstrated that the mass of a body must change when it emits light energy. This leads to Einstein's most famous result – mass energy equivalence: $E = mc^2$.

Later work by the mathematician Hermann Minkowski showed that the Lorentz transformation equations of three spatial and one time co-ordinate are equivalent to pseudo-rotations in a four-dimensional space-time geometry. This powerful insight was soon adopted by Einstein and set him on the path to a geometrical interpretation of gravitation central to the general theory of relativity (which he published in 1915).

In the following sections we shall look at some consequences of the principle of relativity, Minkowski's space-time interpretation and the application of relativistic ideas in particle physics.

11.1.3 The Relativity of Simultaneity

Einstein struggled with the ideas of relativity through the decade leading up to his triumphs of 1905, but it was his realisation that the real problem was rooted in the philosophy of time that led to his major breakthrough:

"My solution was really for the very concept of time, that is, that time is not absolutely defined but there is an inseparable connection between time and the signal velocity. Five weeks after my recognition of this, the present theory of special relativity was completed."

(Albert Einstein, Kyoto Address, 1922)

Most people are familiar with estimating the distance of a lightning strike by counting the seconds between the arrival of the flash (which travels at the speed of light) and the bang (which travels at the speed of sound). The distances involved are only a few kilometres so the light arrives pretty well instantaneously. The distance is therefore approximately $x = vt$ where v is the speed of sound (about 330 ms^{-1} making the distance roughly $3t$ kilometres). However the flash and the bang originate in the same electrical discharge, and the fact that the two signals arrive at different times does not detract from the fact that they were emitted *simultaneously*. We can always correct for the time of flight of the signals.

Now think about two lightning flashes that strike at different places, how can we decide whether they are simultaneous or not? One way would be for an observer to record when he sees each lightning flash (this would require accurate electronic equipment), work out his distance from each strike, and use this information with the speed of light to calculate the actual times (on his clock) of the strikes. If the two times work out to be the same he can conclude that the strikes were simultaneous:

Data:	t_1 = time at which first strike is heard
	t_2 = time at which second strike is heard
	d_1 = distance from first strike
	d_2 = distance from second strike
	c = speed of light
Calculation:	Time of first strike $\qquad T_1 = t_1 - d_1/c$
	Time of second strike $\quad T_2 = t_2 - d_2/c$
	If $T_1 = T_2$ then the lightning strikes were simultaneous.

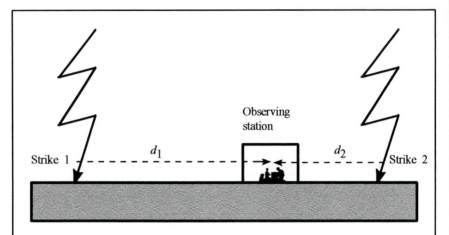

Figure 11.4 Two lightning bolts strike at different points along an east-west line. Observers on the same line record the time of arrival of the light from each flash. They then correct for the difference in distances to each strike before they conclude the strikes were simultaneous. However, being simultaneous in one frame of reference does not imply that the strikes are simultaneous in all reference frames.

If time is absolute this calculation establishes that the two lightning strikes are simultaneous for *all* observers (since they share a common time and there is a unique value of this time at which the strikes occurred). But this would be an additional assumption (and one that would lead to a contradiction). Einstein's approach is to consider how a moving observer receives the signals and corrects them for their time of flight.

Imagine an observer moving along the line from strike 1 toward strike 2 at high speed. If she passes the point mid-way between the two strikes just as light arrives there she can do a similar calculation to work out when each strike must have occurred. She could argue as follows:

- The speed of light from both strikes is c.
- I am now half-way between the two charred patches of ground, so the original flashes must have occurred when I was closer to strike 1 than strike 2.
- I am receiving the light from both flashes at the same time, but light from flash 2 has travelled further than light from flash 1.
- Strike 2 occurred *before* strike 1.

It is important to remember that the moving observer sees the event of a lightning strike at a fixed position in her own reference frame, not at the position the original observer 'carries with him'. In other words, the lightning might strike the ground and create a burnt patch which is the position of the strike for all time in

Two lightning strikes occur and are judged to be simultaneous in a particular reference frame. An observer in a moving reference frame passes at velocity v.

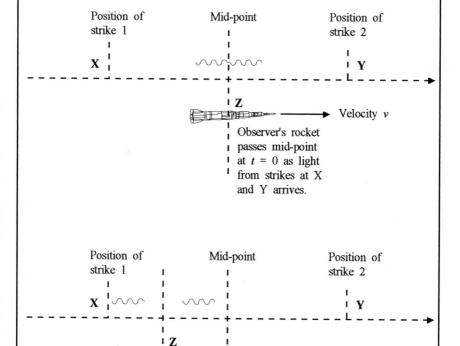

From the moving rocket both light pulses must converge on the observation point Z. This means that, prior to detection (i.e. $t<0$) the two pulses were equidistant from Z (since they travel at speed c relative to the rocket). But prior to detection the rocket was closer to strike 1. This implies that (for this observer) the light travels farther from strike 2 than strike 1 before it is detected. Strike 2 must have occurred before strike 1. The strikes are not simultaneous (in this reference frame).

Figure 11.5 The relativity of simultaneity. Two events that are simultaneous for a particular observer in a particular inertial reference frame are not simultaneous for observers moving with respect to this reference frame.

the original reference frame, but the position of the strike (i.e. the source of the flash) only coincides with the burnt patch at one instant in the moving reference frame the instant of the strike).

The conclusion is clear, *simultaneity is relative*. This has deep significance. It means that two observers in relative motion will not agree on the times at which events occur. This means that two clocks separated by some distance and synchronised in a particular reference frame will not be synchronised for observers moving relative to that frame. This is not simply a matter of correcting for time of flight of light. They cannot be synchronised in more than one reference frame.

This can be seen very clearly if we consider the operations involved in synchronising a pair of distant clocks (see Fig. 11.5). Imagine an observer A who wishes to synchronise a pair of clocks. He chooses two clocks that can be started automatically when they receive a light pulse. He also takes a measuring tape and a flashbulb. First he zeroes the two clocks, separates them by a measured distance to positions X and Y, and places the flashbulb half-way between them. When the flashbulb goes off light travels an equal distance to each clock and both start *at the same time*. For observer A they are synchronised. If at any time thereafter (assuming they are good clocks) a pair of events occur next to them so that the recorded time of each event is the same, then these events are simultaneous for A.

Now consider B, who is racing past the flashbulb travelling parallel to XY. At the instant the flash occurs B is beside it, but according to B clock X is moving back, away from the flash, at speed v and clock Y is moving toward the flash at speed v. The light (which travels at the constant speed c relative to B) will reach Y first and then X. For observer B, Y will always be ahead of X. What does this mean for the pair of events which A judges to be simultaneous? B will agree that the two clocks read the same time for the pair of events, but will not see the events as simultaneous because B sees a time difference between A's clocks. For B they are not synchronised. The event that occurs at Y when the clock there reads T occurs *before* an event at X when X reads T, because (for B) Y started before X.

Furthermore, a third observer, moving rapidly along the line YX would see the time order of these two events *reversed*, concluding that the event at X occurs before the event at Y. The size of these disagreements increases with the spatial separation of the clocks and the speed of the observer. However, whilst it is possible for moving observers to see separate events in different orders, e.g. X then Y, Y then X or X and Y together, it is not possible for the disagreement to be large enough to reverse the causal order of events. That is, if X' caused Y' then no observer could travel fast enough to see Y' before X' (after correcting for time of flight of light, etc.). The reason for this will become clear later and is linked to the velocity of light as a limit.

To summarise: an inertial observer is justified in claiming that two events are simultaneous in his reference frame if they occur at the same local time on reliable clocks sychronised in his reference frame (using a procedure equivalent to the one above). He is not justified in claiming that these events are synchronised for observers in other inertial reference frames.

Synchronization of clocks, A's point of view:

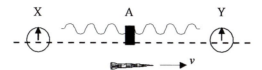

A synchronises two clocks by sending a light pulse out from a point mid-way between them and using the pulse to activate both clocks.

Synchronization of clocks, B's point of view:

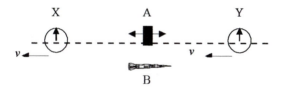

B passes the mid point of XY just as the light pulse is emitted. B sees A's system travelling to the left at velocity v, but light travels at velocity c away from its point of origin. This means that B will see the light travelling to the right move toward a clock moving toward it at velocity v, and light emitted toward the left approach a clock moving away from it at speed v.

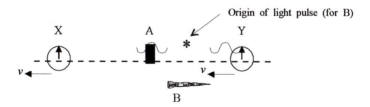

B sees the light reach clock Y before it reaches clock X, so Y starts before X and the two clocks are not synchronized.

Figure 11.6 A very reasonable procedure for synchronizing clocks in A's reference frame fails miserably in B's.

11.1.4 Time Dilation

People often wonder why the speed of light plays such an important role in relativity. Why not the speed of sound or the speed of neutrinos or some other speed. What is it about light? The speed *c is* the speed of light, but its significance seems to be deeper than this. Light travels at this speed because the speed itself is fundamental rather than it being a fundamental speed because of light or electromagnetism. We believe that all particles with zero rest mass travel at this speed (including gravitons). The speed itself gives the link between space and time and is the fastest rate at which a causal influence can be propagated. The fact that we first came across it in the context of light is not particularly important.

   This deep significance of *c* makes it the ideal invariant quantity (i.e. same for all inertial observers) to compare how differently moving observers measure time and space. Here we are going to use light to show why moving clocks run slow. A clock is a device with some intrinsic periodic change that can be counted to measure time (e.g. a swinging pendulum or a vibrating quartz crystal). We are going to use a light clock in which a pulse of light bounces up and down between parallel mirrors. This is particularly convenient because all inertial observers will agree on the speed with which the light pulses move (although not necessarily the path these pulses take).

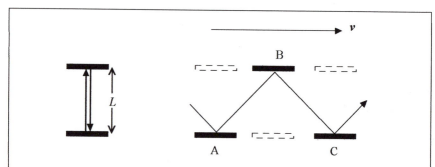

Figure 11.7 A light clock (left) measures time as light bounces back and forth between two parallel mirrors. If the clock moves (right) the light path is longer between ticks, so it runs slow with respect to a stationary clock. However, time for an observer moving with the clock seems normal, so the time in the moving reference frame runs slow when viewed by the stationary observer.

   If you are standing beside one of these light clocks its period *T* will be simply $2l/c$ where *l* is the vertical separation of the two parallel mirrors. However, if an identical light clock is fitted inside a rocket that passes you at high speed you will see the light path in the moving clock stretched out in its direction of travel. The light has farther to travel to complete one period, *but it still travels at the same speed c relative to you*. It must take longer, so the period between 'ticks' on the moving clock is greater than on your own 'stationary' clock. The moving clock runs slow. Now, this is all very well, but what effect does this have on the passage of time for a moving observer in the rocket with the light clock? As far as she is concerned she is in an inertial reference frame and so everything is normal. Her clock does not seem to run slow, it keeps up with all other measures of time inside the rocket reference frame, including her own metabolism and rate of aging. After all, if the moving observer could measure a slowing of their light clock relative to any other clock within her system she would have a way to calculate her speed relative to some absolute reference frame (e.g. the ether) and this would imply that experiments such as the Michelson-Morley experiment should not produce a null result, and would violate the principle of relativity.

So, the moving observer feels fine, but how does she appear *to you*? You have already noticed that the moving clock runs slow and yet she is keeping time with her own clock. You must see her metabolism running slow. The moving observer will age more slowly than you. Time slows down in the moving reference frame and if the velocity is very high this effect becomes extreme. It is called *time dilation*. The ratio of clock rates in each reference frame is given by the gamma-factor:

$$\gamma = \frac{1}{\sqrt{1 - \dfrac{v^2}{c^2}}}$$

so the time T' that passes on your clock during one period of the moving clock is:

$$T' = \gamma T$$

For low velocities ($v \ll c$) $\gamma \approx 1$. For higher velocities $\gamma > 1$ and as $v \to c$ $\gamma \to \infty$.

The point about special relativity is that there are no preferred inertial reference frames, so we should now ask how the moving observer sees time passing in *your* reference frame. The situation is completely symmetric. She will see your clock run slow for the same reasons that you saw hers run slow. She will also conclude that you are aging more slowly than she is. This seems strange, how can her clock run slower than yours and yet yours run slower than hers? There is no problem unless we try to bring them together to compare them, but to do that we will need to change the motion of at least one of the clocks. This is at the heart of one of the most famous 'paradoxes' in relativity, the 'twin paradox'.

11.1.5 The Twin Paradox

Twin brothers celebrate their twenty-first birthday together on Earth and then one of the twins embarks upon a high-speed return rocket journey to a distant star during which time the gamma-factor is 2. The twins are reunited after twenty Earth years have passed. The Earth twin calculates that the travelling twin has aged at half his own rate, so should only have aged by ten years. The Earth twin, who is 41 at the reunion, expects to meet his brother of 31 (a difference that should be clearly noticeable). This is not the 'paradox'. The travelling twin also does some calculations, assuming that the Earth twin has aged at half his rate. If twenty Earth years have passed then forty spaceship years have gone by and he would expect to be 61 when he meets his brother of 41 (an even more noticeable difference, but in the other direction). This is the paradox – we get different outcomes depending on whose point of view we adopt. What actually happens?

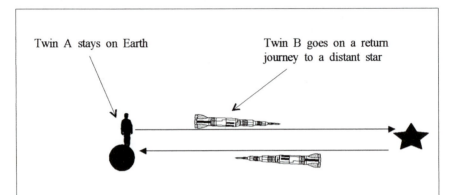

Twin A stays on Earth

Twin B goes on a return journey to a distant star

Figure 11.8 The Twin Paradox. According to A, B is moving and so B's time runs slow. A concludes that B should be younger than A when they reunite. B argues that it is A who is moving and who should age less. Who is right? A. Why? Because A remains in the same reference frame throughout the process whereas B changes motion (accelerates) several times. B ages less.

An experiment like this has actually been carried out. However, in place of twin brothers identical atomic clocks were sent on different journeys in jet aircraft. In 1971 two American physicists, Hafele and Keating synchronised the times on four identical caesium clocks with standard clocks at the US naval Observatory in Washington DC. They flew two of the clocks eastwards around the world and two westwards and compared their times with the standard clocks on their return. When additional effects due to gravitational time dilation had been removed the special relativistic effects were in agreement with Einstein's predictions and confirmed that the travelling clocks lose relative to the stationary ones. In other words, the travelling twin would be younger, as calculated by the earth twin. But where does this asymmetry come from? Why is it not valid to assess the situation from the point of view of the travelling twin?

Put yourself in the position of the twin who stays on Earth. If we simplify things by ignoring the Earth's own motion this twin remains in the same inertial reference frame throughout their separation. This is therefore a valid reference frame from which to assess the situation. The travelling twin, on the other hand, begins and ends in this reference frame but accelerates to high speed as he travels away, decelerates at the star and then accelerates back to Earth, where he once again slows down. There are three different inertial reference frames involved in this sequence of motions. There is no unique inertial reference frame from which the travelling twin can assess things. This calls the predictions of the travelling twin into question and shows quite clearly that the two views are not symmetric.

To move from one inertial reference frame to another involves acceleration and these accelerations result in inertial forces that can be felt inside the rocket (e.g. being 'pressed back into his seat' on take-off). To analyse the time shift from the point of view of the rocket twin we would need to include the effects of acceleration on clock rates. This can be done (one way is to integrate over the momentary inertial frames through which the accelerating rocket must pass, another is to treat the accelerated rocket as if it is in a gravitational field and apply the gravitational time dilation formula from general relativity) and the result brings the predictions of the travelling twin into line with those of the Earth twin. When they reunite the Earth twin really will be 41 and the rocket twin 31.

This brief discussion of simultaneity, time dilation and the twin paradox illustrates the radical results that follow from Einstein's interpretation of the relativity principle:

- There is no absolute time.
- Each clock records its own proper time – moving clocks run slow.
- Observers in relative motion may observe events in a different order.
- Time travel into the future is possible (in fact, it is commonplace).

The last of these needs amplification. In one sense we are all travelling in time since our experience is of a sequence of moments stretching from the distant past to the far future. But what we usually mean by time travel is a difference in the amount of time we experience compared to others. In this respect high-speed motion gives a sure-fire way of travelling into the future. Recall the rocket twin in the example above. He has experienced ten years while his brother (and everyone else on Earth) has lived through twenty years. The rocket twin has travelled ten years into the Earth's future. In principle extremely high velocities would allow people to travel very far into the future. For example, if the gamma-factor was 1000 and a rocket passenger travelled at this speed for 10 years he would return to earth 10 000 years in the future. Everyone he had known would be long dead and the society he left would almost certainly have vanished.

11.1.6 Length Contraction

When a rocket makes a high-speed journey to a distant star less time elapses for the rocket passengers than for their colleagues who remained behind on Earth. For example, if the rocket travels for most of its journey at $0.99c$ the gamma-factor is about 7 so 1/7 as much time passes for the passengers as passes on Earth. Imagine a journey at this speed from Earth to Canopus, about 22 light years. For observers on Earth just over 22 years would pass during the outward journey. A little over three years would pass for the travellers. But this poses a problem. If the rocket travellers calculate their speed for the journey using the time of 3 years and distance of 22 light years they arrive at a velocity of about $7c$, violating the principle of relativity. There is a way round this. If we assume that they too conclude they travelled at close to the speed of light then they can use this speed and their journey time to calculate the journey distance – about 3 light years. This is an example of another relativistic effect called length contraction. The distance from earth to Canopus has contracted by the gamma-factor as a result of the rocket's motion. This has an interesting consequence. We might think that the finite speed of light and our finite lifetimes would put a practical limit on how far we can explore in a single generation – less than about 70 light years. This is not the case. The faster the rocket travels the greater the length contraction and the greater its range within one human lifetime.

Length contraction arises naturally in special relativity rather than as an *ad hoc* add-on in the Lorentz theory (the Lorentz-Fitzgerald contraction) although its mathematical expression is identical. The length of a moving rod is given by:

$$l' = \frac{l}{\gamma}$$

where l is the length the rod would have when measured by instruments in its own rest frame and l' is its length measured on the same instruments when it flies past them at speed v.

Lorentz explained the null result of the Michelson-Morley experiment by introducing this contraction factor due to motion of the apparatus through the ether. Its reappearance in special relativity also explains why the null result is obtained (namely that the length of the interferometer arm parallel to the ether wind contracts by just enough to offset the expected delay). However, relativity gives us a much simpler and more convincing explanation – the speed of light is the same in all directions in all inertial reference frames, so we should expect light to take exactly the same time along equal length interferometer arms in any inertial laboratory whatever their orientation. There is no ether wind.

11.1.7 Velocity Addition

Common sense suggests that if we walk along a moving train our speed relative to the track (w) will be the sum of our speed relative to the carriage (u) and the speed of the carriage relative to the track (v): $w = v + u$. It is simple to see that his classical velocity addition formula cannot hold in special relativity. In place of the person walking, imagine a torch held by a stationary passenger and shone inside the carriage parallel to the direction of motion. If the classical velocity addition formula is valid for this example then the speed of light relative to the track outside would be: $w = v + c$. This would violate the principle of relativity. The classical velocity addition formula would also result in two sub-light velocities combining to give a super light velocity. For example, if a projectile was fired at speed $0.75c$ inside a rocket travelling at $0.75c$ relative to the earth then the classical formula would predict a projectile velocity of $1.5c$ relative to the Earth.

If we look carefully at the two velocities we are trying to combine we will see why the classical approach is likely to give the wrong answer. Think back to the first example with a person walking along the train. We are adding two velocities:

- v – the velocity of the train relative to the track, which is distance travelled along the track divided by time taken as measured by instruments at rest relative to the track.
- u – the velocity of the walking passenger relative to the train, which is the distance they move along the carriage divided by the time they take as measured by instruments inside the carriage, i.e. at rest with respect to the train.

We have already seen that moving clocks run slow, moving objects contract along their direction of motion and that the synchronisation of sets of clocks in different inertial frames is impossible. It is hardly surprising that the naive classical approach fails to combine u and v correctly – they have been measured by instruments in two different reference frames.

So how do we combine velocities in relativity? One way would be to correct the velocity u for the effects of motion on instruments in the train. This would give a value u' which is the relative velocity of the person to the train *as it would be measured by instruments at rest relative to the track*. Then the classical formula could be used to combine v and u'. The usual way is to derive the velocity addition formula for v and u from the Lorentz transformation equations (summarised below). This results in the relativistic velocity addition formula:

$$w = \frac{v+u}{1+\dfrac{uv}{c^2}}$$

If u and v are small compared to c, as they are in our train example, then the denominator is effectively 1 and the classical formula can be used without serious error. However large u or v become (up to c) their sum remains less than or equal to c. For example, if we combine two velocities of $0.75c$ each we get:

$$w = \frac{0.75c+0.75c}{1+\dfrac{0.75^2c^2}{c^2}} = 0.96c$$

In 1913 William de Sitter argued that the images of binary star systems give strong support for the velocity addition formula, at least when light is involved. Some binary stars have orbits orientated so that one star is sometimes moving directly toward us and sometimes directly away. If the velocity of light is affected by the velocity of its source then light from the approaching star should travel toward us more rapidly than light from the same star when it is moving away. This could result in light from the approaching star on a later orbit overtaking light from earlier orbits when it is receding. If this were the case 'ghost images' would be seen when the binary system is observed from the Earth. They have never been observed. There is a counter argument to this – light is continually absorbed and emitted on its journey through space. The re-emission of light (from intergalactic gas, for example) leaves a source that does not share the velocity of the original source, so the 'memory' of the binary system's internal motions may be lost before it can result in ghosts.

A more direct and more accurate test of the velocity addition formula has been carried out using the light emitted by high-energy pions created in particle collisions. The sources are travelling at over $0.99c$ and yet the photons they emit are still travelling at c relative to apparatus at rest in the laboratory (the measurement was made directly by time of flight over a distance of about 30 m).

11.1.8 Mass Energy and Momentum

The idea that inertial mass might depend on velocity predates Einstein. In the Abraham-Lorentz theory of the electron the particle acquires an electromagnetic mass as a result of a self-interaction, a coupling to its own electromagnetic field. A colleague of Abraham, Walter Kaufmann, confirmed that the mass of the electron does indeed increase with velocity, but the theory had serious intrinsic problems. Most of these were related to the geometrical shape and size of the electron. Abraham treated the electrons as small rigid spheres and fixed the radius to make the electromagnetic mass agree with the experimentally measured value. Rigid bodies conflict with relativity and yet a point electron would have an infinite self-energy and mass. This brought the Abraham-Lorentz theory into conflict with Einstein's, that also predicts an increase of mass with velocity. However, the form of the variation of mass with velocity is different in the two theories, so they could be distinguished experimentally. Kaufmann and others carried out careful measurements of the specific charge (e/m) of electrons as they increased the accelerating voltage. The results showed that electron mass increases in agreement with Einstein's predictions.

The velocity addition formula gives a clear indication that inertial mass must increase with velocity. The effect of acceleration is to increase velocity. This can be considered as a step-by-step process in which increments Δv are added to the existing velocity v in the momentary inertial reference frame associated with the projectile while it moves at velocity v. The result of such an increment is:

$$v' = \frac{v + \Delta v}{1 + \dfrac{v\Delta v}{c^2}}$$

which is clearly less than $v + \Delta v$ and certainly less than c. The closer v becomes to c the smaller the effect of such an increment. This implies that a graph of velocity versus time for a projectile must curve over to approach the horizontal line $v = c$ as an asymptote as t approaches infinity. Inertial mass is a measure of the reluctance of a body to change its state of motion, so the inertia of such a body is increasing and approaches infinity as the velocity approaches c.

Einstein derived his result by assuming that mass depends on velocity and that mass and momentum are conserved in all inertial reference frames. Einstein's equation for mass and velocity is:

$$m = \gamma m_0$$

where m is the mass at velocity v and m_0 is the rest mass (i.e. the mass of the object measured by apparatus at rest with respect to it).

The increase of mass with velocity is also an increase of mass with kinetic energy. Einstein showed that an increase in mass Δm is linked to the increase in energy ΔE by the equation:

$$\Delta E = c^2 \Delta m$$

He also showed that when something radiates energy ΔE it must lose mass Δm where these quantities are linked together by the same equation.

It is easy to see why radiation transfers energy and mass. Electromagnetic waves are emitted by accelerated charges. The polarisation of the waves is parallel to the oscillation direction and the magnetic field vector of the electromagnetic wave is therefore perpendicular to this motion. When charges move perpendicular to a magnetic field they experience a force given by the left-hand rule. In this case an emitting particle experiences recoil force or 'radiation reaction'. When an electromagnetic wave is absorbed it makes charges vibrate parallel to its electric field vector. Once again there is a magnetic field perpendicular to the moving charges and this time it exerts a force on them parallel to the direction in which the wave was travelling. What are the dynamical consequences of these forces? Imagine a wave pulse of duration Δt that exerts a force F on the emitter and absorber. Taking the direction of energy transfer as positive the net effect of the impulses is to reduce the momentum of the emitter and increase the momentum of the absorber by an amount $F\Delta t$. In other words the wave has transferred a momentum $F\Delta t$ from the emitter to the absorber. Momentum transfer implies mass transfer, so the mass of the emitter has fallen and the mass of the absorber has gone up by the same amount. How much? The speed of transfer was c, and the momentum associated with the light pulse was $\Delta p = F\Delta t$. This makes the mass transfer $\Delta m = F\Delta t/c$. Using a simple mechanical argument we could say that the work done to create the light pulse acts on a wave train of length $c\Delta t$, so the work done $\Delta E = Fc\Delta t$.

$$\Delta m = \frac{F\Delta t}{c} = \frac{\Delta E}{c\Delta t} \times \frac{\Delta t}{c} = \frac{\Delta E}{c^2}$$

Leading to: $\Delta E = c^2 \Delta m$

Pulling all this together Einstein came to the conclusion that all forms of energy have mass and that all mass is equivalent to a certain amount of energy. The famous equation is:

$$E = mc^2$$

This equivalence means that we no longer regard conservation of mass and conservation of energy as separate laws, they are equivalent. It is also worth pointing out that this relation is universal. It is true that c^2 is enormous, so that mass changes associated with common energy transfers are utterly negligible. Nonetheless they do occur, and the chemical energy released by burning fuels is associated with a change of mass in exactly the same way that the energy released in a nuclear reactor is.

11.1.9 The Lorentz Transformation

So far we have dealt with relativistic effects in a rather *ad hoc* way, relating them all back to fundamental principles, but not linking them in any consistent mathematical structure. The Lorentz transformation equations convert the co-ordinates of an event measured in one inertial reference frame into the co-ordinates of the same event as they would be measured in another inertial reference frame.

Imagine two inertial observers A and B passing one another with a relative velocity v parallel to their x and x' axes. Both observers start their clocks at the instant their origins coincide. Later they each record the co-ordinates of the same event (e.g. a supernova explosion). How are their results related?

According to observer A the event occurs at (x,y,z,t)
According to observer B the event occurs at (x',y',z',t')

The Lorentz transformation equations are then (see appendix for a derivation):

$$x' = x - vt \qquad x = \gamma\left(x' + vt'\right)$$
$$y' = y \qquad\qquad y = y'$$
$$z' = z \qquad\qquad z = z'$$
$$t' = \gamma\left(t - \frac{xv}{c^2}\right) \qquad t = \gamma\left(t' + \frac{x'v}{c^2}\right)$$

All the results discussed in previous sections can be derived from these equations. Here we shall give just one example, the time dilation equation.

Maths Box: Deriving the Time Dilation Formula

Identical clocks are placed at the origins of A and B. How do the clock rates compare when viewed from either reference frame? To find out we use the Lorentz transformation to convert time co-ordinates (that is, clock readings) from one frame into time co-ordinates for the other frame. We know that $x = x' = 0$ *when* $t = t' = 0$ and that all motion is along the positive x- and x'-axes.

ACCORDING TO A:
$$t = \gamma\left(t' + \frac{vx'}{c^2}\right)$$

This relates the time t' measured on A's clock to the time t' that is indicated on B's clock at that instant in A's frame. But B's clock is at $x' = 0$ (the origin in B's frame) so:

$$t = \gamma t'$$

Since $\gamma > 1$ the reading on A's clock must be greater than on B's. B's clock, which is moving with respect to A, runs slow. This is the time dilation formula we derived previously.

ACCORDING TO B:
$$t' = \gamma\left(t - \frac{vx}{c^2}\right)$$

but A's clock is at $x = 0$ so
$$t' = \gamma t$$

According to B, B's clock always shows a greater time than A's. So B sees A's clock run slow. The Lorentz transformation leads to the same symmetrical results we discussed previously.

11.2 SPACE-TIME

11.2.1 The Fourth Dimension

"The views of space and time which I wish to lay before you have sprung from the soil of experimental physics, and herein lies their strength. They are radical. Henceforth space by itself, and time by itself, are doomed to fade away into mere shadows, and only a kind of union of the two will preserve an independent reality."

(Hermann Minkowski, address to the 80th assembly of German Natural Scientists and Physicians, 1908).

Minkowski had been one of Einstein's mathematics teachers in Zurich, but by 1907 he was at the University of Goettingen. In November of that year he gave a colloquium in which he pointed out the formal similarity between the mathematics of rotation in three dimensions and the Lorentz transformation regarded as a pseudo-rotation in four dimensions. He developed this idea into a new formal representation of special relativity as the mathematics of 'events' in a four dimensional *space-time continuum* and showed that electromagnetic fields must be represented by second rank tensors. Einstein was initially dismissive of the space-time approach, regarding it as unnecessary abstraction, but he soon realised that it gave a powerful meaning to the equations and adopted this geometrical approach in his development of the general theory. (Readers who wish to avoid the mathematics should skip ahead to section 11.2.2).

An event in space-time is something that happens at a particular place and time, e.g. an explosion or a flash of light, and it is represented by four co-ordinates:

Co - ordinates of event as seen in reference frame A : (x, y, z, t)

Co - ordinates of event as seen in reference frame B : (x', y', z', t')

In general these co-ordinates will be different even though they describe the same event. However, they are related by the Lorentz transformation and Minkowski showed that this is equivalent to a pseudo-rotation in Euclidean space or a rotation in pseudo-Euclidean space. The difference between these two interpretations depends on how the time co-ordinate is defined.

Euclidean space : (x, y, z, ct)

pseudo - Euclidean space : (x, y, z, ict)

The multiplying factor c converts time in seconds to time in metres so it is consistent with the measure of the three spatial dimensions. The i is the square root of -1 and its inclusion makes the analogy between rotations in 3D and in 4D space-time perfect. We shall adopt this pseudo-Euclidean approach. In Minkowski space-time the Lorentz transformation can then be written down as a matrix. This is shown below for both a 2D transformation and a 4D transformation. The former is convenient in situations where all relative motion occurs along the x and x' axes so that y and z co-ordinates of all events are the

Figure 11.9 **Hermann Minkowski** 1864-1909 introduced the geometrical interpretation of special relativity.

Artwork by Nick Adams

same in both reference frames. This is the significance of the 1s in the matrix.

$$\begin{pmatrix} x' \\ ict' \end{pmatrix} = \begin{pmatrix} \gamma & i\beta\gamma \\ -i\beta\gamma & \gamma \end{pmatrix}\begin{pmatrix} x \\ ict \end{pmatrix}$$

$$\begin{pmatrix} x_1' \\ x_2' \\ x_3' \\ x_4' \end{pmatrix} = \begin{pmatrix} \gamma & 0 & 0 & i\beta\gamma \\ 0 & 1 & 0 & 0 \\ 0 & 0 & 1 & 0 \\ -i\beta\gamma & 0 & 0 & \gamma \end{pmatrix}\begin{pmatrix} x_1 \\ x_2 \\ x_3 \\ x_4 \end{pmatrix}$$

x_1 to x_4 are x, y, z, and ict.

Comparison of the 2D matrix with a simple rotation of x-y axes in 2D Euclidean space is interesting:

$$\begin{pmatrix} x' \\ y' \end{pmatrix} = \begin{pmatrix} \cos\theta & \sin\theta \\ -\sin\theta & \cos\theta \end{pmatrix} \begin{pmatrix} x \\ y \end{pmatrix}$$

The Lorentz transformation would have the same form as a 2D rotation if:

$$\sin\theta = i\beta\gamma$$
$$\cos\theta = \gamma$$
$$\tan\theta = i\beta$$

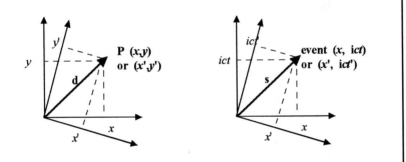

Figure 11.10 Rotation of spatial axes changes x and y co-ordinates but does not affect the distance **d** (black arrow) of a point from the origin. In a similar way, motion (rotation of space-time axes) changes the times and position of events but leaves 4-vectors (such as intervals **s**, black arrow) unchanged.

This link to rotations leads to one of the most powerful ideas in relativity, the realisation that whilst co-ordinates of events are usually different for different observers (i.e. they measure them to occur at different times and in different places) certain relationships between the co-ordinates are invariant (i.e. the same for all inertial observers). This can be understood using the analogy with 2D rotations, or even by using a very simple model.

Take a piece of graph paper and mark y and x-axes on it. Now take a coin or other flat circular disc that fits entirely on the graph paper and mark two points A and B on the circumference of the disc. Place the disc on the graph paper and read off the co-ordinates of the points. Now rotate the graph paper through some angle and read off the new co-ordinates for the same two points.

Co-ordinates of points A and B on original axes $\quad (x_A, y_A) \quad (x_B, y_B)$

Co-ordinates of point A and B on rotated axes $\quad (x'_A, y'_A) \quad (x'_B, y'_B)$

These two sets of co-ordinates are related by the rotation matrix above. However, it is obvious that the distance across the coin is unchanged by rotating the axes, even though the co-ordinates of the points themselves have changed. Distance AB is an invariant quantity with respect to 2D rotations. But distance can be calculated using Pythagoras's theorem.

Original axes : $\quad d^2 = \left(x_B{}^2 - x_A{}^2\right) + \left(y_B{}^2 - y_A{}^2\right) = \Delta x^2 + \Delta y^2$

Rotated axes : $\quad d'^2 = \left(x'_B{}^2 - x'_A{}^2\right) + \left(y'_B{}^2 - y'_A{}^2\right) = \Delta x'^2 + \Delta y'^2$

Distances are invariant under rotation so:

$$d^2 = \left(x_B{}^2 - x_A{}^2\right) + \left(y_B{}^2 - y_A{}^2\right) = \left(x'_B{}^2 - x'_A{}^2\right) + \left(y'_B{}^2 - y'_A{}^2\right) = d'^2$$

A similar expression can be constructed in 4D space-time. It is Pythagoras's theorem applied to pseudo-Euclidean geometry:

$$s_{AB}{}^2 = \Delta x_1{}^2 + \Delta x_2{}^2 + \Delta x_3{}^2 + \Delta x_4{}^2 = \Delta x_1'{}^2 + \Delta x_2'{}^2 + \Delta x_3'{}^2 + \Delta x_4'{}^2 = s'_{AB}{}^2$$

where $\Delta x_1 = x_B - x_A$, etc. and $\Delta x_4 = ic\Delta t = ic\left(t_B - t_A\right)$

The quantity s_{AB} is effectively the 4D 'distance' between events in the space-time 'world'. It is called the interval and has the same value for all inertial observers. The four-dimensional displacement from event A to event B in space-time has components of the form $\Delta x_1 = x_1(B) - x_2(A)$ and its magnitude is given by s_{AB}. This is an example of a '4-vector' in space-time. The magnitudes of all 4-vectors are calculated from their components in the same way (i.e. using Pythagoras), and the magnitude of all 4-vectors are invariants.

The transformation laws for energy and momentum can be derived from the Lorentz transformations by assuming they hold good in all inertial reference frames. This leads to a result that illustrates the immense power of the four-dimensional approach. If momentum and total energy are represented by:

$$\text{Momentum} = mv = \gamma m_0 v$$

$$\text{Total energy} = mc^2 = \gamma m_0 c^2$$

then they transform like components of a 4-vector.

$$
\begin{pmatrix} p_1' \\ p_2' \\ p_3' \\ p_4' \end{pmatrix} = \begin{pmatrix} \gamma & 0 & 0 & i\beta\gamma \\ 0 & 1 & 0 & 0 \\ 0 & 0 & 1 & 0 \\ -i\beta\gamma & 0 & 0 & \gamma \end{pmatrix} \begin{pmatrix} p_1 \\ p_2 \\ p_3 \\ p_4 \end{pmatrix}
$$

where $\quad p_1 = mv_1 \quad p_2 = mv_2 \quad p_3 = mv_3 \quad p_4 = \dfrac{iE}{c}.$

The magnitude of this 4-vector is an invariant, taking the same value in all inertial reference frames:

$$
p_1{}^2 + p_2{}^2 + p_3{}^2 - \frac{E^2}{c^2}
$$

This can be rearranged as follows:

$$
p_1{}^2 + p_2{}^2 + p_3{}^2 = m^2\left(v_1{}^2 + v_2{}^2 + v_3{}^2\right) = m^2 v^2 = p^2
$$

where p is the magnitude of the 3-momentum, so we can write:

$$
p^2 - \frac{E^2}{c^2} = \text{invariant}
$$

If this is applied to a particle in its own rest frame then for this frame energy is rest energy E_0 and 3-momentum is zero. This leads to one of the most useful equations in particle physics:

$$
p^2 - \frac{E^2}{c^2} = 0^2 - \frac{E_0{}^2}{c^2}{}^2
$$

leading to: $E^2 - p^2 c^2 = E_0$

This is used to relate the energy and momentum of a moving particle to its rest energy (or rest mass).

In space-time the components of the energy-momentum four-vector are separately conserved in all inertial reference frames. In ordinary mechanics and electromagnetism the laws of conservation of energy and momentum are distinct principles but in relativity they are aspects of a deeper unity, the conservation of the momentum-energy 4-vector.

Minkowski's geometrical vision is startling. He described the space-time continuum as a 'world' through which events associated with any particular particle in space-time forms a continuous 'world-line'. This world-line is the history of that particle and the universe as a whole can be regarded as a tapestry of interweaving world-lines for all space and all time. This view is not a snapshot of the now, but a geometrical block model of past, present and future as if they exist independently of our own moment-by-moment perception of them. Different observers will record positions and times of events differently, but they can all calculate the invariant connections between them in a completely consistent way. Relativity seemed to reveal an ideal world beyond the ephemeral one presented to our senses, close to the Platonic view of a world of perfect forms.

We can illustrate the power of the geometrical approach by using it to derive the time dilation formula (see Fig. 11.11). Consider two events which occur at the origin of B's laboratory moving at velocity v in A's positive x-direction. (As usual we assume that the origins coincide at $t - t' - 0$ and that event 1 occurs at the origin as the origins cross.)

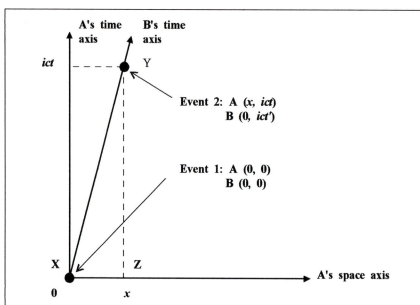

Figure 11.11 Space-time geometry. Triangle XYZ can be used to derive the time dilation formula.

For B: co-ordinates of event 1 are $(0, 0)$ and of event 2 are $(0, ict')$
For A: co-ordinates of event 1 are $(0, 0)$ and of event 2 are (x, ict)

Invariant interval s between events is: $s^2 = 0 - c^2 t'^2 = x^2 - c^2 t^2$

But $x = vt$ so $-c^2 t'^2 = v^2 t^2 - c^2 t^2$

leading to
$$t = \frac{t'}{\sqrt{1-\dfrac{v^2}{c^2}}} = \gamma t'$$

This is the time dilation formula! The two events at B's origin could be ticks of a lightclock. The period t' would be the proper time between ticks and t would be the dilated time observed from A's reference frame. The moving clock takes longer to tick than a similar stationary clock (i.e. more than one period of A's clock elapses in A's frame during one period of B's clock). Although the result is familiar the method used to derive it is now geometric. If we cast relativity into a space-time form then time dilation is a consequence of the invariance of the interval.

We can further emphasize the role of space-time geometry by deriving the same formula pictorially using Pythagoras's theorem.

Co-ordinates of tick 1:   (0, 0)   for A   and   (0,0)   for B
Co-ordinates of tick 2:   (vt, ict) for A   and   (0, ict') for B

By Pythagoras's theorem in triangle XYZ $(vt)^2 + (ict)^2 = (ict')^2$

Leading to the time dilation formula $t = \dfrac{t'}{\sqrt{1-\dfrac{v^2}{c^2}}}$ as before.

Minkowski died from appendicitis in 1909. He was 44 years old. His work in Goettingen influenced the great mathematician David Hilbert and interested him in the fundmental problems of theoretical physics, but his major conribution was to recognise the geometrical nature of special relativity and alert Einstein to its power. In fact Einstein came to regard geometry as a branch of experimental science (an idea entertained by Gauss) and used it as the foundation of his masterwork, the general theory of relativity.

"geometry ... is evidently a natural science; we may in fact regard it as the most ancient branch of physics. Its affirmations rest essentially on induction from experience, but not on logical inferences only ... The question whether the practical geometry of the universe is Euclidean or not has a clear meaning and its answer can only be furnished by experience ... I attach special importance to the view of geometry which I have just set forth, because without it I should have been unable to formulate the theory of relativity."

(Lecture before the Prussian Academy of Science, January 27 1921.)

Here Einstein was referring primarily to his general theory, but it was the geometrisation of special relativity by Minkowski that first set him on this road.

11.2.2 Relativity in Action

Special relativity is not just an abstract theory that applies under such extreme conditions that we do not ever need to worry about it. In modern high-energy physics the effects of relativity must be allowed for and are demonstrated everyday. People often assume that the point of a particle accelerator is to make the accelerated particles go at ever higher velocities. There is some truth in this, but it is better to think of them as machines designed to concentrate an enormous amount of energy (and therefore mass) into a sub-atomic particle. To understand this shift of emphasis consider the following. A 25 GeV proton accelerator at CERN in 1959 accelerated protons to $0.9993c$. A 1000 GeV proton accelerator at Fermilab in 1990 squeezed them up to $0.9999995c$, an increase in velocity of less than 0.1 % for a 40-fold increase in energy and an enormous capital investment. This shows that the velocity of light really does act as a limiting velocity and the host of massive particles created in particle collisions relies on the conversion of kinetic energy to rest energy in accordance with Einstein's equations. Colliders such as LEP and the proposed LHC at CERN are ideal because particles with similar energy make head-on collisions so there is no momentum in the centre of mass frame and all the kinetic energy can be converted to rest energy of new particles (unlike a fixed target machine in which some of the input kinetic energy must remain in that form in order to satisfy the conservation of linear momentum).

Time dilation effects are important in particle detectors. Many of the particles created in collisions have extremely short lifetimes but they are created with very high velocities. This means their 'internal' time appears to run slow when viewed from the laboratory reference frame. This means they live many times longer in the laboratory than they do in their own rest frames, often long enough to leave a discernible track in a vertex detector close to the point of collision. A similar effect has been measured in muons created in the upper atmosphere by cosmic ray bombardment. These have a lifetime of about 2 microseconds so would travel only about 600 m in their own reference frame before decaying. If this were the case in the Earth's reference frame then effectively none of them would reach the surface. In practice about one in 8 travel the 50 or so kilometres to the surface. This is because of time dilation. Two microseconds on the 'muon clock' lasts much longer than this in the Earth's reference frame. Turning this argument around, a muon that passes through the atmosphere and hits the surface must 'see' the atmosphere severely contracted (to a few hundred metres) in its own reference frame. This is, of course, the Lorentz contraction.

Dirac's attempts to make the Schrodinger equation consistent with special relativity led to the Dirac equation and the prediction of antimatter. This was confirmed in 1932 when Anderson identified positrons in cosmic ray tracks. The first anti-atoms were created in the laboratory at CERN in 1997. The symmetry between matter and antimatter is only seen from the perspective of relativity.

The idea of mass-energy equivalence has had a greater impact on the political history of the twentieth century than just about any other scientific discovery.

When Einstein first proposed the idea he thought it might explain the enormous reserves of potential energy in radioactive sources. However, these sources release this energy over a very long period of time so the power output is usually low (but can be large enough to maintain radioactive batteries in heart pacemakers and spacecraft). The discovery of nuclear fission in the late 1930s revealed another way to release large amounts of energy from individual nuclei, but Rutherford was skeptical, suggesting that the idea of extracting large amounts of energy from the nucleus was 'moonshine'. He was wrong. Induced nuclear fission and a chain reaction provided a way to generate a huge amount of power continuously, as in a nuclear reactor, or explosively, as in a nuclear bomb. The fission weapons detonated at Hiroshima and Nagasaki are a chilling testament to the awesome power of even relatively low yield nuclear explosions. Fusion weapons, so far only detonated in tests, but having yields up to a thousand times those of their fission relatives, are truly terrifying. The early nuclear reactors were designed as much to produce weapons grade plutonium (a by-product of fission when fast neutrons are absorbed by uranium-238) as to generate electricity. Nonetheless, and despite regular fears over nuclear waste, links to weapons programmes and operational safety, many nations now depend on nuclear power as a major source of electricity (about 25% of the UK's electricity is generated in nuclear reactors). At present all commercial reactors use nuclear fission, but a great deal of research has been carried out into nuclear fusion, although the prospect of a commercial reactor still seems far off. There are two main approaches, electromagnetic confinement of a heated plasma in a Tokamak (toroidal magnetic field machine designed by Andrei Sakharov) and inertial confinement, the implosion of fuel pellets under intense laser bombardment.

12

GENERAL RELATIVITY

12.1 EQUIVALENCE

12.1.1 The 'Happiest Thought'

"Then there occurred to me the ... happiest thought of my life, in the following form. The gravitational field has only a relative existence in a way similar to the electric field generated by magnetoelectric induction. Because for an observer falling freely from the roof of a house there exists – at least in his immediate surroundings – no gravitational field. Indeed, if the observer drops some bodies then these remain relative to him in a state of rest or of uniform motion, independent of their particular chemical or physical nature (in this consideration the air resistance is, of course, ignored). The observer therefore has the right to interpret his state as 'at rest'.

Because of this idea, the uncommonly peculiar experimental law that in the gravitational field all bodies fall with the same acceleration attained at once a deep physical meaning. Namely, if there were to exist just one physical object that falls in a gravitational field in a way different from all others, then with its help the observer could realise that he is in a gravitational field and is falling in it. If such an object does not exist, however – as experience has shown with great accuracy – then the observer lacks any objective means of perceiving himself as falling in a gravitational field. Rather he has the right to consider his state as one of rest and his environment as field-free relative to gravitation.

The experimentally known matter independence of the acceleration of free fall is therefore a powerful argument for the fact that the relativity postulate has to be extended to co-ordinate systems which, relative to each other, are in non-uniform motion."

(Albert Einstein, The Morgan manuscript, unpublished article for Nature, 1921)

By 1907 Einstein realised that special relativity could explain all mechanical and electromagnetic phenomena but one: gravity. Relativity had shown there was an intimate connection between mass and energy, but gave no explanation for the equality of inertial and gravitational mass, a curious experimental fact that had been noted by Galileo and commented upon by Newton. To understand why this needs explaining consider an experiment attributed to Galileo. He is said to have taken two spheres of similar radius but different mass to the top of the tower of Pisa and dropped them over the side at the same instant. They hit the ground at more or less the same time, with any slight difference being due to the different effect of air resistance. A similar experiment was carried out on the Moon when a hammer and feather were dropped through the vacuum above the Moon's surface and hit the ground at the same instant. Mass plays two distinct roles in this experiment:

- Gravitational mass m_g determines the weight of the sphere in the gravitational field of strength g, and hence the resultant force on it by $F = m_g g$.
- Inertial mass m_i determines the acceleration of the sphere under the influence of the resultant force by Newton's second law, $a = F/m_i$.

Put these together and the free-fall acceleration is given by:

$$a = \frac{F}{m_i} = \frac{m_g g}{m_i} = \left(\frac{m_g}{m_i}\right) g$$

If the free-fall acceleration is independent of the nature of the falling body and is determined only by the field strength g then the ratio (m_g/m_i) must be a constant for all bodies. In other words the inertial and gravitational masses are directly proportional to one another and, if suitable units are chosen, are equal. Nothing in Newtonian mechanics or special relativity leads to this conclusion. It hints at a deeper physics that can unify the inertial forces associated with acceleration and gravitation.

In 1907 while Einstein continued to work in the Patent Office he was struck by the idea that if a person falls freely he will not feel his own weight. Think of astronauts in an orbiting spacecraft. They float about as if there is no gravitational field and yet, seen from the Earth, they and their spacecraft are both moving in an orbit determined by gravitational attraction (for example, at the altitude of the Hubble space telescope the gravitational field strength is about 8 Nkg^{-1} compared with its surface value of about 10 Nkg^{-1}). Orbiting spacecraft and their occupants are projectiles in free fall, but within that freely falling reference frame there are no gravitational effects. The importance of this observation is startling. It implies that a freely falling reference frame is just like the inertial frames of special relativity. Einstein had found a way to link gravitation and acceleration to uniform motion.

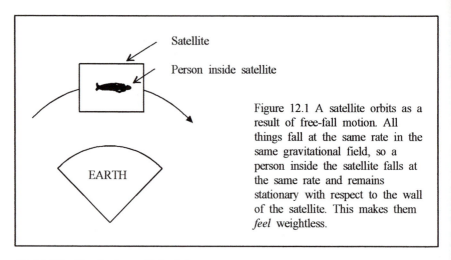

Figure 12.1 A satellite orbits as a result of free-fall motion. All things fall at the same rate in the same gravitational field, so a person inside the satellite falls at the same rate and remains stationary with respect to the wall of the satellite. This makes them *feel* weightless.

12.1.2 The Equivalence Principle

In his own popular explanation of relativity *(Relativity, the Special and the General Theory*, Methuen, 1920) Einstein describes a thought experiment. Imagine a region of empty deep space far away from all stars and galaxies. In this region there is a laboratory with no windows and inside the laboratory is a man. Unknown to the man, an alien has attached a strong tow-rope to the roof of the laboratory and is towing it upward at a rate of 9.8 ms^{-2}. How does the man inside the laboratory interpret his experiences? If he releases a small object he will see it accelerate toward the floor of his laboratory. If he jumps upwards he will fall back down. If he stands on weighing scales they will read his weight. All his mechanical experiments and experiences are exactly as they would be if his laboratory were resting on the surface of a planet with the same gravitational field strength as the Earth. Of course, we know this is not the case. We know he is being accelerated by external forces. But all his experiences are consistent with the idea that he is positioned in a uniform gravitational field. He feels the floor pressing up underneath his feet because of his own 'weight' pulling him 'down'. If he hangs something from a spring balance the spring will stretch.

"Ought we to smile at the man and say he errs in his conclusion? I do not believe we ought to if we wish to remain consistent; rather we must admit that his mode of grasping the situation violates neither reason nor known mechanical laws. Even though it is being accelerated with respect to the 'Galilean space' first considered, we can nevertheless regard the chest as being at rest. We have thus good grounds for extending the principle of relativity to include bodies of reference which are accelerated with respect to each other, and as a result we have gained a powerful argument for a generalised postulate of relativity."

(Albert Einstein, 1920, *Relativity, the Special and the General Theory*, Methuen)

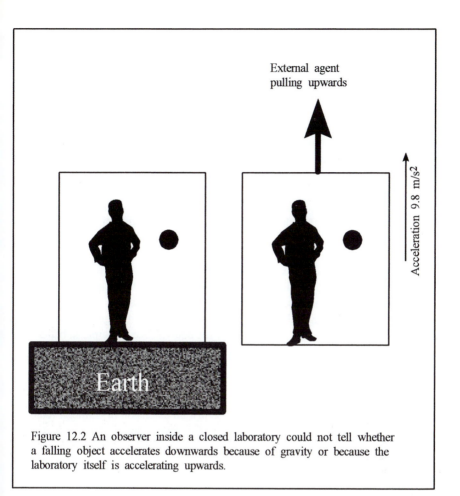

Figure 12.2 An observer inside a closed laboratory could not tell whether a falling object accelerates downwards because of gravity or because the laboratory itself is accelerating upwards.

This is all very reminiscent of the arguments that led to special relativity. There he extended the mechanical equivalence of all inertial reference frames and postulated that the laws of physics are the same for all inertial observers. Here he is pointing to the mechanical equivalence of uniform gravitational fields and constant acceleration. The next step follows logically, it is called the *equivalence principle*:

- The laws of physics in a uniform gravitational field are the same as in a reference frame undergoing uniform acceleration.

A careful and consistent application of the principle of relativity led to new physics with effects such as time dilation, length contraction, etc. In a similar way Einstein was able to predict new physical effects by applying the equivalence principle.

12.1.3 Testing Equivalence

The simplest way to test the equivalence principle is to take a number of different objects and see if they fall at the same rate. A more accurate test was first carried out by the Hungarian physicist Baron Roland von Eötvos in the late nineteenth century. He used a torsion balance set up in his laboratory in Budapest to balance the combined inertial forces due to the Earth's rotation and gravitational forces on a pair of masses against one another. He set up the balance in one orientation and then rotated it 180° trying to detect a change in the couple (turning effect) on the beam. If gravitational and inertial forces are proportional then there will be no change in turning effect when the whole apparatus is rotated. If they are independent of one another then there will be a turning effect in both orientations but one way it will be clockwise, the other anticlockwise. Eötvos used platinum on one side of the balance and a variety of other materials on the other side. He found no change in turning effect when the apparatus was rotated. He concluded that gravitational mass and inertial mass are the same thing to an accuracy of better than 1 part in 10^8. Later experiments (1960s and 70s) based on a similar idea but using inertial forces due to the Earth's orbital motion and the gravitational forces from the Sun, confirmed the predictions of the equivalence principle to better than 1 part in 10^{11}.

12.1.4 Gravity and Light

The special theory of relativity is in a sense just an extension of Galilean relativity to include the effects of electromagnetism. The equivalence principle is also an extension of Galileo's ideas, and leads to some very interesting results when applied to light. In the thought experiment above the occupant of the accelerated laboratory was unable to distinguish uniform acceleration from a uniform gravitational field by any mechanical experiment. If the equivalence principle is correct then optical experiments will not be able to distinguish between them either.

Imagine a light beam moving in a direction parallel to the floor of the laboratory and entering through a small hole in its left hand wall. Watched from outside the light beam is travelling horizontally and the laboratory is accelerating upwards. The laboratory will rise as the light beam crosses it so that the beam hits the right hand wall at a lower point in the laboratory than the hole through which it entered. From our inertial frame of reference the beam travels in a straight line, for the observer inside the laboratory the light follows a parabolic path, just like a material projectile. But the observer inside the laboratory can interpret his own situation in two equivalent ways:

- He could be at rest in uniform downward gravitational field of strength g.
- He could be accelerating upward at rate g.

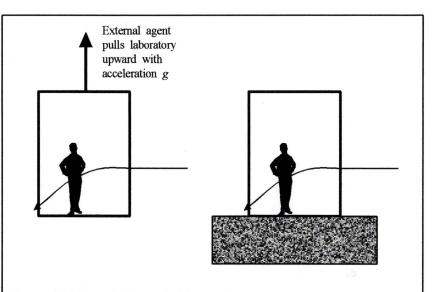

Figure 12.3 The equivalence principle. An observer accelerating at rate g past a beam of light will see it follow a parabolic path. An observer in a gravitational field of strength g should see light follow a similar path. This implies that light is deflected by gravity, a prediction supported by many observations.

In the first case the bending of the light beam is interpreted as a dynamical effect. In the second it must be a gravitational effect. Light is deflected by a gravitational field. Light *falls*, just like anything else, and at the same rate. This is a clear-cut physical prediction of the equivalence principle – gravity should deflect light.

Einstein suggested that the deflection of light by gravity could be tested by observing stars close to the occulted disc of the Sun during a total eclipse. Light travelling from such a star to an observer on Earth must graze the surface of the Sun and fall toward it a bit. This means the apparent position of the star would shift away from the Sun. By comparing this with the expected position if the Sun were not present (based on observations of the star's position relative to the background stars at night) its deflection could be measured and compared with Einstein's prediction.

In 1911 Einstein used the equivalence principle to calculate that light at grazing incidence to the Sun would be deflected by 0.88 seconds of arc and that the deflection decreases with the reciprocal of angular distance from the Sun. By 1919, when the astronomer Sir Arthur Eddington set out to measure the effect, Einstein had doubled the prediction to 1.75 seconds. The doubling is because of an additional deflection due to the warping of space-time around the Sun. It is interesting to note that the original deflection follows simply from the fact that light energy has mass and mass falls in a gravitational field (it had even been predicted on the basis of classical mechanics by the German astronomer, Johann

Georg von Soldner back in 1801!). The additional deflection is a purely relativistic effect.

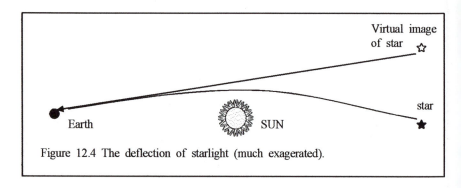

Figure 12.4 The deflection of starlight (much exagerated).

Eddington's expedition was lucky to get any measurements at all, the weather on the day of totality was terrible. But it did clear, briefly, around the time of the eclipse, and some photographs were taken of which two were useable. When analysed back at Oxford they showed a deflection of 1.60 ± 0.31 seconds, in reasonable agreement with Einstein's revised prediction. Similar measurements were repeated by other groups during other eclipses, but the method gives poor accuracy and there has been only a modest improvement on Eddington's original measurements. However, radio astronomy has provided another way to test the prediction and this has the great advantage that radio measurements do not require a total eclipse. All that is needed is a suitable radio source that is periodically occulted by the Sun. Radio measurements have confirmed Einstein's predictions to an accuracy of better than 1%.

The deflection of light by gravity has many similarities to the refraction of light when it moves through a medium of variable density, and leads to similar consequences. Lenses use this effect to form images and it is possible for the masses of galaxies and stars to deflect light in such a way that we can see double or multiple images of very distant objects. This remained a theoretical idea suggested by Sir Oliver Lodge in the 1920s and investigated in detail by Einstein and Fritz Zwicky in the 1930s, but the extremely long focal lengths required for such 'gravitational lenses' suggested they would never be observed. However, the discovery of extremely distant extremely powerful quasars provided suitable objects. In 1979 a 'double-quasar' Q0957+561 was identified as a double image. many more have been identified since then and the Hubble Space Telescope has provided brilliant photographs of multiply imaged galaxies. On a smaller scale some gravitational microlenses have also been discovered. These produce interesting optical effects when a small massive body moves across our line of sight to a more distant object.

12.1.5 Gravity and Time

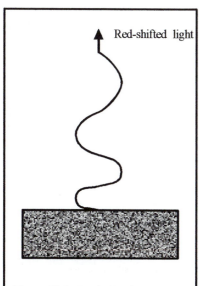

Figure 12.5 Gravitational red-shift. Photons climbing out of a gravitational field lose energy. This reduces their frequency and increases their wavelength.

Imagine shining a torch from the floor of a room and receiving the light on the ceiling. The energy transferred has mass so work must be done to lift it from the floor to the ceiling. This means the light arriving at the ceiling delivers less energy than was emitted by the source on the ground. How can this be? The easiest way to understand what is happening is to think of individual photons. Each photon has an energy $E = hf$. If the photons arriving at the ceiling have less energy they must have lower frequency. This is a shift toward the longer wavelength end of the spectrum – a gravitational red shift.

Now imagine the 'torch' at gound level is actually emitting monochromatic laser light that is used to calibrate an accurate light clock so that the light clock at ground level completes one period in N periods of the emitted light. The frequency of this light when it arrives at the higher level is reduced so an observer there would receive less than N waves during one period of her own light clock (calibrated against a similar light source at the same altitude). The high altitude observer concludes that the low level clock is running slow. On the other hand, if she sends light down to the ground it will get a gravitational blue shift (since it gains energy as it 'falls') and more than N waves will arrive at ground level during one period of the ground level clock. The person on the ground concludes that the higher level clock is running fast. As in special relativity, both observers are equally justified in assuming that observations in their own reference frame are valid. Both observers agree that the higher clock (which at a greater gravitational potential) runs fast, the lower clock (lower gravitational potential) runs slow.

- Time runs faster at higher gravitational potentials.

In the strong gravitational field close to the surface of neutron stars or black holes this becomes a significant effect, but it is measurable for atomic clocks which spend a significant time at different altitudes (e.g. if two clocks are synchronised

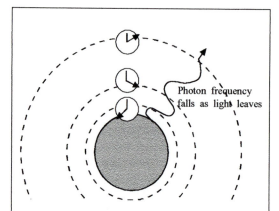

Photon frequency
falls as light leaves

Figure 12.6 Clocks in a strong gravitational
field run slow. This is a consequence of
gravitational red shift. It means a star or planet
is effectively surrounded by layers of 'slow
time'. When light passes through this region its
velocity, relative to a distant inertial observer,
changes and it 'refracts'. This explains the
deflection of light by gravity.

and then one is taken on a high altitude trip in an aircraft). This experiment has been carried out and the gravitational part of the time delay is in agreement with Einstein's predictions. Incidentally, if you wanted to travel into the future one way to do it would be to travel to a black hole and orbit just outside the event horizon (so you can get away again). In the strong field (and low potential) your clock and time would run slow relative to the outside universe (e.g. your family back on Earth). When you return to Earth much more time will have passed in your absence than you experienced in your own reference frame. You have travelled into the Earth's future (see Maths Box on p.301).

The gravitational red shift has provided another terrestrial test of general relativity. This involves a procedure much like the thought experiment in which a torch beam is shone from the floor to the ceiling of a laboratory. The idea is simple. Excited atoms emit photons with well-defined frequencies determined by the energy level structure within the emitting atom. A similar but unexcited atom in the ground state is able to absorb these photons as electrons make quantum jumps to higher levels. Now place some excited atoms at the bottom of a tower and some unexcited ones at the top. The photons travelling up the tower will be red shifted and the atoms at the top will have a reduced chance of absorbing them. However, if the atoms at the top are moved slowly downwards the Doppler effect will shift the apparent frequency of the photons to a slightly higher value. At a particular speed the shift due to the Doppler effect will exactly cancel the gravitational red shift and the photons will be strongly absorbed. This speed therefore gives a measure of the gravitational red shift for photons passing up the tower. A very similar experiment can be carried out for photons passing down the tower, but in this case the photons are blue shifted and the platform supporting the unexcited atoms at the bottom must now be moved downwards to compensate.

There is a problem with this experiment. The gravitational shift in a terrestrial tower is tiny and the spread of frequencies due to thermal motion of the atoms and recoil on emission was many orders of magnitude larger. In 1958 Rudolph Mössbauer made a discovery that won him the Nobel Prize (1961) and made the

---

Maths Box: Gravitational Time Dilation

Imagine sending a photon from a point in a gravitational field at potential ϕ (a negative value) to a receiver in an inertial reference frame at infinity (potential 0). The photon must do an amount of work $W = (0 - m\phi)$ to reach the receiver.

For a photon:

$$m = \frac{E}{c^2} = \frac{hf}{c^2}$$

The photon energy on arrival will be:

$$E' = E - W = E + \frac{hf\phi}{c^2}$$

This can be used to relate frequencies:

$$hf' = hf + \frac{hf}{c^2} = hf\left(1 + \frac{\phi}{c^2}\right)$$

If the receiver is at a distance

R from a star of mass M then:

$$\phi = -\frac{GM}{R^2}$$

and

$$hf' = hf\left(1 - \frac{GM}{Rc^2}\right)$$

The frequency at the receiver (at infinity) is lower than the frequency at the emitter. But this light frequency is a measure of clock rates in the gravitational field. The observer at infinity would see clocks closer to the surface of the star run slow. It is clear that clocks run slow in a gravitational field so that, when a time t at infinity a shorter time t' passes in the field:

$$t' = t\left(1 - \frac{GM}{Rc^2}\right)$$

For two points in a gravitational field separated by a gravitational potential difference $\Delta\phi$ the clock at higher potential will run faster than the clock at lower potential and times on the two clocks will be related by:

$$t' = t\left(1 - \frac{\Delta\phi}{c^2}\right)$$

These effects are called gravitational time dilation.

---

terrestrial test of gravitational red shifts possible. At low temperatures the nucleus of an atom can become fixed in the surrounding lattice so that the linear momentum of the emitted gamma ray is taken up by the whole lattice rather than just one nucleus. The effective mass of the recoiling lattice is enormous compared to the individual nucleus, so the wavelength and frequency of the emitted photon is defined to very high precision (about 1 part in 10^{12}). A corresponding effect for an unexcited atom means it will absorb an incoming photon only if it is extremely close to the same frequency. In other words the emitter and absorber act as an extremely sharply tuned system. This process is called recoilless nuclear resonance absorption, but is usually known as the Mössbauer Effect. Rudolph Mössbauer shared the 1961 Nobel Prize for Physics *"for his researches concerning the resonance absorption of gamma radiation and his discovery in this connection of the effect that bears his name"*.

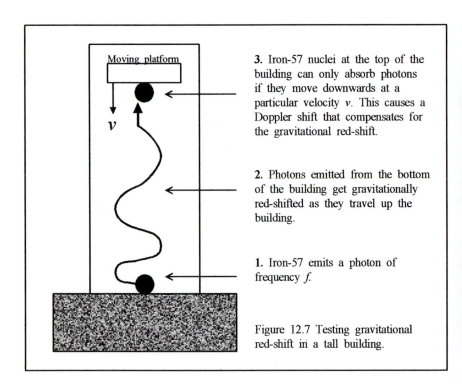

3. Iron-57 nuclei at the top of the building can only absorb photons if they move downwards at a particular velocity v. This causes a Doppler shift that compensates for the gravitational red-shift.

2. Photons emitted from the bottom of the building get gravitationally red-shifted as they travel up the building.

1. Iron-57 emits a photon of frequency f.

Figure 12.7 Testing gravitational red-shift in a tall building.

Between 1960 and 1965 Pound, Rebka and Snider carried out experiments in which they measured the gravitational red and blue shifts for gamma-ray photons emitted by iron-57 nuclei. The gamma rays travelled up or down the Jefferson Physical Laboratory Tower at Harvard. Their results agreed with general relativity to about 1%. R. Vesot and M. Levine carried out a much more accurate experiment in 1976. They used microwaves transmitted to Earth from an altitude of 10 000 m from a source inside a Scout D rocket. Their results agreed with general relativity to about 0.1%.

It is worth pointing out that the effect of gravity on clocks has to be allowed for in modern satellite navigation systems such as the Global Positioning System (GPS). The clock rates on Earth and at the altitude of a geostationary satellite differ by a few parts in 10^{10}. The positions of these satellites must be known to an accuracy of a few metres and the time for signals to travel to and from them must be accurate to a few nanoseconds, so general relativistic corrections must be made.

12.2 GRAVITY AND GEOMETRY

12.2.1 Distorting Euclid

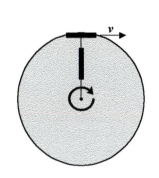

Figure 12.8 A rod on the circumference of a rotating reference frame is Lorentz contracted relative to an inertial observer. One along a radius is not. Observers within the rotating reference frame must see a different geometry to those in the inertial frame.

The link between gravity and geometry is shown by another of Albert Einstein's thought experiments. Imagine a rotating circular roundabout and assume you are looking at it from an inertial reference frame positioned directly above the axis of rotation. The circumference is related to the radius by the familiar Euclidean equation $c = 2\pi r$. But think what would happen if someone *on the moving roundabout* tried to measure the radius and circumference. Any measuring rod they placed along a radius would agree with your own rods because their rod is moving perpendicular to its length and so does not suffer a Lorentz contraction. However, a rod placed on the circumference flies past you and contracts along its length. It will require more of the moving rods to stretch around the circumference than your own inertial rods. The observer in the rotating reference frame will conclude that $c > 2\pi r$ and the roundabout does not obey the rules of Euclidean geometry.

This rather surprising conclusion must be related to the rotation of the reference frame. The moving observer is in a state of acceleration (centripetal acceleration) as she is continually deflected from a straight-line path by forces exerted on her feet by the floor of the space station. This implies that:

- Space-time in an accelerating reference frame is non-Euclidean.

If we apply the equivalence principle to this result we are led to conclude that space-time must also be non-Euclidean in a gravitational field. Einstein took this one step further.

- Gravity *is* a distortion of space-time geometry.

Maths Box: Gravitational length contraction

Let the roundabout in the example described above have an angular velocity ω. The measuring rods lined up at the circumference at radius r will all be moving with speed $v = r\omega$ so they will be contracted relative to an inertial observer so that their length l' is related to the length of a similar rod in the inertial frame by:

$$l' = l\sqrt{1 - \frac{v^2}{c^2}}$$

The moving rods have a centripetal acceleration a relative to the inertial reference frame: $a = r\omega^2$. This create an 'artificial' gravitational field of strength $g = r\omega^2$. To link the rotating frame to an extended gravitational field it is convenient to introduce a potential function so that potential differences in the rotating frame can be related to potential differences in other gravitational fields. This can be done if we consider moving a test mass m from the central axis (taken as potential zero) to a position at radius r on the rotating roundabout.

$$\text{Work done} = \int_{r=0}^{r=R} -ma\,dr = \int_{r=0}^{r=R} -mr\omega^2\,dr = \left[-\frac{mr^2\omega^2}{2} \right]_{r=0}^{r=R} = -\frac{mR^2\omega^2}{2}$$

This is a change of potential of : $\Delta\phi = -\frac{R^2\omega^2}{2} = -\frac{v^2}{2}$

This can be substituted into the length contraction formula to give :

$$l' = l\sqrt{1 - \frac{v^2}{c^2}} = l\sqrt{1 + \frac{\Delta\phi}{c^2}} \quad \text{(where } l' \text{ is at the lower potential)}$$

This result can be applied to the field of a spherically symmetric star of mass M.

Potential at radius R from the star: $\qquad \phi = -\frac{GM}{R}$

For a distant observer ($\phi = 0$) rods near the star contract to l':

$$l' = l\sqrt{1 + \frac{\phi}{c^2}} = l\sqrt{1 - \frac{GM}{Rc^2}}$$

For weak fields this can be expanded using the binomial theorem:

$$l' \approx l\left(1 - \frac{GM}{2Rc^2}\right)$$

In Newton's theory there is an absolute background of Euclidean space and gravity is an additional force field superimposed on top of it. In Einstein's theory the geometry of space-time is non-Euclidean and its curvature at each point determines the gravitational field. This can be understood by analogy. Imagine a flat billiard table with a dent in its surface. Balls will roll across the surface in perfect straight lines unless they pass over the dent. Then they will follow the curvature of the surface and deflect around it. A suitably projected ball might even orbit for a while (until friction takes its toll). The effects of matter are similar to the dent on the surface of the table. Matter distorts space-time so that the curvature is greatest close to the mass and gets smaller further away. Free bodies move along the straightest paths available in space-time; these are called geodesics and are similar to the great circle routes around the globe in that they represent the shortest paths between two points on the surface. Hence John Wheeler's statement:

- Matter tells space how to curve.
- Space tells matter how to move.

The orbit of the Earth, which is near circular when viewed from the perspective of three-dimensional space plus time, is actually a geodesic path in the four-dimensional non-Euclidean space-time near the Sun. The Earth is simply moving along the shortest route through a distorted geometrical continuum. So are all the other planets.

Einstein's 1911 calculation of the deflection of starlight was half his 1915 prediction. The earlier result was on the basis of the equivalence principle alone acting in the flat Euclidean space-time of special relativity. The extra deflection comes about from the distortion of space-time close to the Sun, so the experimental results are a test of the full prediction of general relativity. However, the same deflection can also be derived using a weak field approximation to general relativity. To do this the effects of gravity on rods and clocks affects light like a change of density (e.g. when light passes from air to glass) and causes the velocity of light relative to the Earth to change as it passes the Sun. It is as if there is a variable refractive index in the space-time close to the Sun.

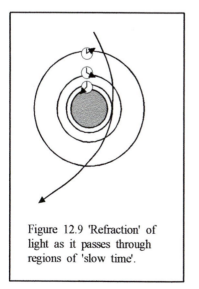

Figure 12.9 'Refraction' of light as it passes through regions of 'slow time'.

Maths Box: The 'Refractive Index' of Space-time

When we look at light passing the Sun the significant deflection takes place close to the Sun in a weak field region of lower potential than our own. The approximate potential difference between us and the surface of the Sun is just:

$$\Delta\phi = -\frac{GM}{R}$$

where R is the radius of the Sun. (The residual effect of the Sun's field at the Earth's orbit and the Earth's own field can be neglected). This means that lengths contract and time clocks run slow close to the Sun (relative to our own measuring instruments) by an amount:

$$l' = l\left(1 - \frac{GM}{2Rc^2}\right)$$

$$t' = t\left(1 - \frac{GM}{2Rc^2}\right)$$

However, for a freely falling observer near the surface of the Sun there will be no gravitational field and the speed of light will have its usual value. For this to be the case we must see light slow down because the clocks and rods used to measure it do not agree with our own. Imagine the procedure carried out by a freely falling observer as they measure the speed of a light beam. They must time it to move through a measured distance. However, their measuring rods are shorter than ours so 3×10^8 m for them is less than that for us. Also their clocks run slow, so 1 second for them is *more* than that for us. Both effects lead to their measurement for c being an *overestimate* compared to ours. But their result must be 3×10^8 ms^{-1}, so we must obtain a lower value c':

$$c' = c\left(1 - \frac{GM}{2Rc^2}\right)^2 \approx c\left(1 - \frac{GM}{4Rc^2}\right)$$

The 'refractive index' of space-time is therefore: $n = \dfrac{c'}{c} = 1 - \dfrac{GM}{4Rc^2}$

If this value is used to calculate the deflection of a ray of light passing at glancing incidence to the Sun it gives a result:

$$\Delta\theta = \frac{4GM}{Rc^2}\text{ radians, about } 1.74 \text{ arcseconds}$$

12.2.2 Mercury

In 1858 Le Verrier made a careful calculation of the orbits of the planets using Newton's theory of gravitation and allowing for perturbations to the orbits due to gravitational interactions between the planets themselves. To his surprise Mercury seemed to be drifting from the Newtonian predictions as perihelion (closest approach to the Sun) in its ellitical orbit advanced by about 43 seconds per century. One possibility (that Newton himself had considered) was that the inverse square law of gravitation is only an approximation and higher-order force laws have an increasing effect closer to the source. As early as 1859 the British astronomer, James Challis suggested adding an inverse fourth power, and this can in fact solve the problem. However, it is an *ad hoc* solution and so rather unsatisfactory. Einstein's theory also leads to a weak field correction to Newton's law that is approximately an inverse fourth power, and predicts the same advance of perihelion. Although the effect was known before Einstein published his theory the agreement between general relativity and observation was one of the most important early successes for the new theory.

12.2.3 Shapiro Time Delay

Einstein himself suggested the three classic tests of general relativity:

- advance of perihelion;
- deflection of starlight;
- gravitational red shift.

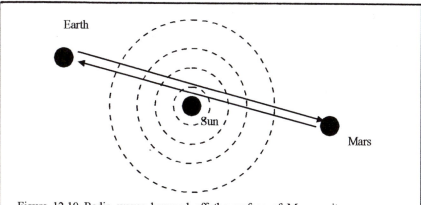

Figure 12.10 Radio waves bounced off the surface of Mars as it passes behind the Sun travel through regions of 'slow time' close to the Sun. This introduces a delay (the Shapiro time delay) above the expected time of flight and can be used to check the predictions of general relativity.

One other test has been added to these. It involves a delay in the round-trip journey time for light passing through a gravitational field and is called the Shapiro time delay (after Irwin Shapiro who was the first person to study the effect). The idea is simple. If radio waves are bounced off the surface of Mars as it passes behind the Sun, the reflected waves will pass close to the Sun through regions of 'high refractive index' and will take longer than if the Sun was not there. The expected time for the round trip without delay was calculated using Newton's theory, which is accurate in the weak field region where Mars orbits. General relativity predicts a delay on a return journey to Mars of about 250 μs. The first results were obtained using the powerful haystack radio antenna in Massachusetts and agreed with general relativistic predictions to within 20%. Later more accurate experiments were carried out using spacecraft orbiting Mars and on the Martian surface. This gave a dramatic increase in precision because the point of reflection is sharply defined and led to an agreement with general relativity to within 0.1%.

12.2.4 Gravity – Fact or Fiction?

Einstein's realisation that freely falling observers feel no gravitational effects puts gravitational forces on a par with inertial forces (the apparent forces you feel inside an accelerating vehicle). The gravitational field at any point in space disappears for an observer freely falling at that point, so the whole of space-time can be imagined as a lattice of inertial reference frames in different relative states of acceleration. The general theory tells us how to relate observations made by measuring instruments in any one of these inertial reference frames to observations made in any of the others. Each freely falling observer defines an inertial reference frame in which there is no gravitational field or inertial forces and in which the laws of physics take the same simple form as in special relativity.

This idea that gravity is an artefact of motion and can be transformed away by choosing a freely falling reference frame seems to imply that gravity is no more than an illusion. However, it is not possible to choose a single freely falling reference frame in which the gravitational field over an extended region is completely transformed away. This is because real fields are not uniform, they have varying strength and direction. Many, like the Earth's field, are more or less spherically symmetric. If we imagine a laboratory falling toward the Earth it will extend over a region in which the field lines are converging, so the field at the sides has an inward component and the field near the floor is slightly greater than the field near the ceiling. Any dust particles falling through the room will tend to collect in piles near the centre of the floor and ceiling. This squeezing and stretching results from tidal forces and is a genuine gravitational effect that cannot be eradicated by any particular choice of reference frame. It has real physical consequences – the lunar and solar tides are example on Earth. Close to a black hole the tidal effect can become so great that an unfortunate astronaut falling into the black hole would be stretched out like spaghetti.

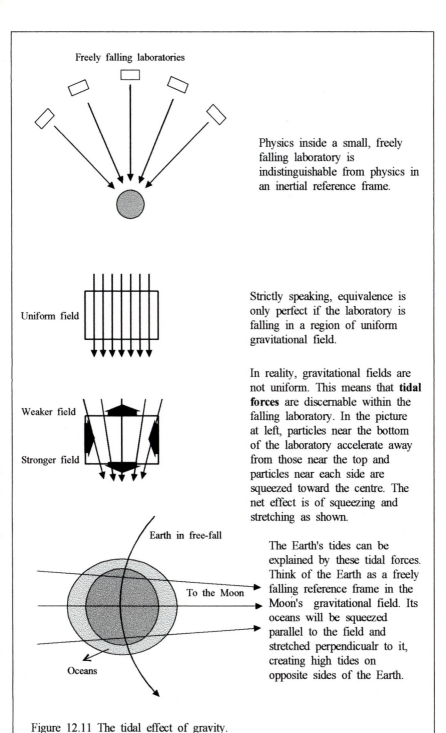

Freely falling laboratories

Physics inside a small, freely falling laboratory is indistinguishable from physics in an inertial reference frame.

Uniform field

Strictly speaking, equivalence is only perfect if the laboratory is falling in a region of uniform gravitational field.

Weaker field

Stronger field

In reality, gravitational fields are not uniform. This means that **tidal forces** are discernable within the falling laboratory. In the picture at left, particles near the bottom of the laboratory accelerate away from those near the top and particles near each side are squeezed toward the centre. The net effect is of squeezing and stretching as shown.

Earth in free-fall

To the Moon

The Earth's tides can be explained by these tidal forces. Think of the Earth as a freely falling reference frame in the Moon's gravitational field. Its oceans will be squeezed parallel to the field and stretched perpendicualr to it, creating high tides on opposite sides of the Earth.

Oceans

Figure 12.11 The tidal effect of gravity.

It is true that the gravitational field at any point in space can be transformed away by transforming to a freely falling reference frame, so g is in a sense an illusion. The tidal effect of gravity however, cannot be transformed away – it is real.

12.2.5 Gravitational Waves

When charges are accelerated they radiate electromagnetic waves – periodic disturbances of the electromagnetic field. In a similar way accelerated masses disturb the 'gravitational field', but this corresponds to disturbances in the geometry of space-time. Anything in the path of these gravitational waves will be squeezed and stretched periodically by tidal forces as the waves pass through

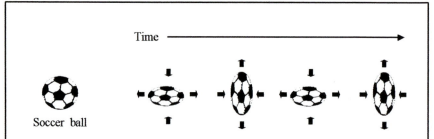

Figure 12.12 The effect of a gravitational wave on a soccer ball. In practice the distortions are likely to be weaker by a factor of about 10^{20}.

(electromagnetic waves are dipole radiation, the tidal effects of gravity are a result of quadrupole radiation, gravitational dipole radiation is undetectable). This squeezing and stretching transfers energy from the source of the gravitational waves to the bodies that absorb them. However, the calculated strength of gravitational waves is tiny, even from violent events such as the collision of stars in our own galaxy. For example, the collapse of a rotating star into a black hole in our own galaxy would produce terrestrial vibrations with an amplitude of about 1 part in 10^{17} of the separation of the particles involved – for two atoms one metre apart this is only 1% of the diameter of a nucleus. Nonetheless, a number of experiments have been set up to try to detect gravitational waves and there are two main techniques:

• Weber bars – large solid cylinders set into resonant vibration by the tidal forces. The size shape and density of the bars effectively tune them to a range of gravitational wave frequencies. Sensitive piezoelectric transducers are attached to the sides of the bars to detect the tiny vibrations induced by gravitational waves.
• Interferometers – these are designed to send laser beams along two similar perpendicular paths and recombine them when they return. Gravitational

waves change the relative lengths of the interferometer arms so that the phase difference between the two returning beams changes periodically. If this can be detected then the amplitude and frequency of the incident waves can be determined. Interferometers are far more sensitive than Weber bars.

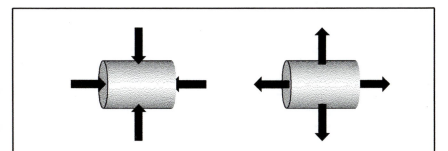

Figure 12.13 Detecting gravitational waves. Weber bars (above) are forced to oscillate by the fluctuating tidal forces set up by a passing gravitational wave. These tiny resonant vibrations will be detected by piezo-electric detectors fixed to the surface of the bar. Another, more sensitive method, is searching for a shift in interference fringes when a long baseline interferometer is distorted by the same tidal forces.

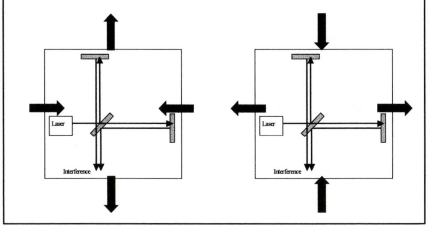

To date gravitational waves have not been detected. If the next generation of gravitational wave detectors is successful then gravity waves will provide astronomers with a new way to explore the universe. They are important because they bring information about the most violent events of all, in particular events involving one of the most mysterious objects in space, the black hole. The most ambitious project planned so far is LISA (Laser Interferometer Space Antenna), an array of six spacecraft arranged in pairs at the corners of an equilateral triangle of side 5 million kilometres following the Earth's orbit but lagging by about $20°$. LISA will be sensitive to vibrations to better than 1 part in 10^{20} and will respond to lower frequencies (a few kHz) than smaller terrestrial detectors. These low frequency gravitational waves are expected to be emitted by particularly powerful sources.

Although we have so far failed to make a direct observation of gravitational waves there is excellent indirect evidence for their existence. In 1974 Russell Hulse and Joseph Taylor discovered the first pulsar that is part of a binary system, PSRB1913+16. The two stars are separated by little more than the radius of the Sun and have orbital speeds of about $0.001c$. This increases the importance of relativistic effects and makes them an ideal subject on which to test general relativity. The pulses of radiation emitted by the pulsar are extremely regular so these can be used to make very accurate measurements on the dynamics of the system. Careful measurement of the advance of periastron (closest approach) led to an incredibly accurate measurement of the total mass of the system, about 2.8275 solar masses. A slight variation in the arrival time of pulses is caused by a combination of the relativistic Doppler effect and gravitational time dilation due to the field of the pulsar's companion. From this the individual masses were found to be 1.42 and 1.40 solar masses. Now comes the most important part – a binary system should radiate gravitational waves, so it should be losing energy. The accurate measurements of essential parameters within the system allowed Taylor and Hulse to predict the rate of loss of energy from the system and the consequent rate at which the orbital period should decrease as the two stars fall closer together. In 1974 the period was 0.059 029 995 271 s and fell by about 1.5 ms in the next two decades. This is in close agreement with the rate calculated from general relativity, which is about 75 μs per year. Taylor and Hulse shared the 1993 Nobel Prize for Physics *"for the discovery of a new type of pulsar, a discovery that has opened up new possibilities for the study of gravitation"*.

12.2.6 Black Holes

Karl Schwarzschild published the first exact solutions of Einstein's equations of general relativity in 1916. The Schwarzschild solution gives equations for the space-time geometry outside a spherical non-rotating mass M. These equations describe how the curvature of space-time changes from point to point. The greater the curvature the more significant non-Euclidean effects (such as gravitational time dilation, light deflection, etc.) become.

One effect is particularly interesting. Think about light rays leaving the surface of a massive star. They will be red shifted as they leave and will be deflected if they travel in any other direction than normal to the surface. Seen from the outside it is as if space-time drains into the star and the light moves on top of space-time. Since light has a constant velocity relative to any free falling reference frame in the gravitational field of the star, it seems to move slowly relative to an external observer. If the star is very massive and compact the effect could be so severe that light cannot escape, space-time 'drains' so quickly that the emitted light stands still (relative to the outside observer) or even falls back. When this condition is reached the core of the star loses contact with the outside world. It has become a black hole. Another way to think about this is to say that the escape velocity from the surface of the star has exceeded the speed of light. When a star collapses it will become a black hole if all its mass is contained inside a critical radius called the Schwarzschild radius (see Maths Box).

Maths Box: The Schwarzschild Radius

The condition for a black hole to form can be calculated using Newtonian equations and the result is the same as that obtained using general relativity.

To escape from the surface a projectile must have positive total energy:

$$TE = GPE + KE \geq 0$$

$$-\frac{GMm}{R} + \frac{1}{2}mv^2 \geq 0$$

$$v \geq \sqrt{\frac{2GM}{R}}$$

For a black hole $v \geq c$

$$\sqrt{\frac{2GM}{R}} \geq c$$

This leads to a limiting radius R_S, the Schwarzschild radius. If all of mass M is inside R_S then the object is a black hole. The average density ρ of matter inside the Schwarzschild radius can also be derived.

$$R_S = \sqrt{\frac{2GM}{c^2}}$$

$$\rho = \frac{M}{V} = \frac{3M}{4\pi R_S^3} = \frac{3c^6}{32\pi G^3 M^2}$$

The Earth, mass about 6×10^{24} kg, has a Schwarzschild radius of about 1 cm and a critical density of about 10^{30} kgm^{-3}. However, the density at which a black hole forms depends on the inverse square of its mass, so more massive black holes can form from matter at ordinary or even very low densities. For example, a galaxy of mass 10^{42} kg and radius of about 10^{21} m will have a Schwarzschild radius of about 10^{15} m and a critical density of about 10^{-4} kgm^{-3}, about one ten thousandth of the density of the atmosphere at sea level.

At the Schwarzschild radius itself light emitted radially will stay frozen at that radius like someone walking up an escalator at exactly the same rate as the escalator moves down. The surface defined by this layer of trapped light is called the event horizon. If the collapsing core left behind by a supernova has a mass greater than about 2.5 solar masses it is expected to continue collapsing and form a black hole. Inside its event horizon the collapse would continue to form a singularity (a point of infinite density) but the outside world would lose contact with the collapsing star as it reached the event horizon. The effects of time dilation would mean that, to an outside observer, the collapse toward the event horizon proceeds ever more slowly so that it would take an infinite time (before any part of the star crosses this boundary).

Until the 1990s there was little evidence to confirm the existence of black holes, although there were a few likely candidates, in particular the strong X-ray source, Cygnus X1. The intense X-rays from Cygnus X1 vary on a timescale less than 1 second. This implies that they come from an object less than 1 light-second in diameter, an order of magnitude smaller than a star like our Sun. The large power output and small size of the source are what would be expected if it was a black hole. The most likely scenario seems to be a binary system in which a star orbits a black hole and the black hole is gradually dragging matter off the star. The acceleration of this material as it falls into the black hole results in intense radiation. More recently the Hubble Space telescope has provided dramatic images that virtually prove the existence of black holes. In 1994 it revealed a black hole in the heart of galaxy M87, and in 1995 it found a huge disc of gas whirling around something with all the characteristics of a black hole in NGC4261. Two giant plumes of gas streaming away from the black hole are propelling it from the centre of the disc.

Part 4
ASTROPHYSICS AND COSMOLOGY

13 Observational Astronomy
14 Stars and Distances
15 Cosmology

13

The Earth's atmosphere and diffraction effects at telescope apertures place restrictions on what we can see from the Earth's surface and how sharply we can form images.
One way to get around this is to put telescopes into space.
Astronomy began with visible light but has been extended to all parts of the electromagnetic spectrum.

14

To measure distances to stars and galaxies we need a range of techniques which depend on a detailed theory of how stars shine and how they live and die.
Stellar spectra contain clues to the nature of stars, the history of the universe and the origin of the elements.
The HR diagram summarises the types and characteristics of stars. Exotic types include red giants, white dwarfs, neutron stars and black holes.
Supernovae seed the universe with heavy elements and are convenient standard candles for measuring large distances.

15

The Milky Way is just one of an enormous number of galaxies separated by very large distances.
The red shifts of galaxies are proportional to their distances – the Hubble Law. The universe is expanding.
The expansion of the universe implies an origin in a hot Big Bang. This is supported by evidence from the microwave background radiation and the abundances of light nuclei.
The universe is about 15 billion years old.
The very early universe may have been in a state of exponential inflation driven by the collapse of an unstable vacuum state.
The fate of the universe depends on its density. At present we can only detect a fraction of the mass that is needed to explain the motion of stars and galaxies.

13
OBSERVATIONAL ASTRONOMY

13.1 LOOKING UP

13.1.1 Seeing Through Air

For millennia humans have looked up at the sky and found clues to the origin of the universe. Until the twentieth century these clues were based on images formed out of visible light by the human eye, aided in the last few centuries by the telescope. Two major developments in the nineteenth century prepared the way for the explosion of discoveries that have marked the last hundred years. The first was the interpretation of optical spectra and its link to atomic and molecular structure. The second was the extension of optics to the vast electromagnetic spectrum and the development of telescopes tuned to different regions of this spectrum. We still turn our attention to the sky and find clues to our ultimate origin and fate, but now these are based on testable scientific theories rather than astrological patterns or metaphysical speculation. And yet the sky has not lost any of its awesome power or mysterious beauty – far from it, the longer we look, the more detail we see, the more deeply it rewards us.

Anyone who has tried to look up at the stars from a city street will appreciate the main obstacle to optical astronomy – the atmosphere itself. Any radiation detected at the surface of the Earth has to travel through many kilometres of a dirty, turbulent, shifting non-uniform medium that absorbs and scatters it in an unpredictable way. In practice the ability of ground-based optical observatories is limited to a much greater extent by the atmosphere than by technology. Cities are probably the worst places from which to look at the sky and it is still possible to see the stars on a clear night far from their smog and light pollution. The best place for an optical telescope on Earth is probably on top of a mountain in Antarctica, but that is not the simplest place to build an observatory! The world's largest and most powerful optical telescopes, Keck I and Keck II are on top of Mauna Kea ('white mountain'), a dormant volcano in Hawaii. This site has been chosen for a number of reasons:

- There is less atmosphere between the telescope and space

- The temperature is very low at that altitude, so most of the atmospheric water vapour has been frozen out (water vapour absorbs strongly at the short wavelength end of the visible spectrum and in the infra-red).
- It is above the turbulence caused by lower altitude winds so telescopes are in a region of relatively stable air (this reduces the 'twinkling' caused by the movement of regions of air of varying density across the line of sight)
- Hawaii is far from other large landmasses and so relatively free from the dust and pollution produced by big population centres.

The way the atmosphere scatters light depends on the ratio of the wavelength of the light λ to the size of the scattering particles d. Larger particles ($d > \lambda$) act like tiny mirrors and cause diffuse reflection. This accounts for the way clouds or mist ($d \approx$ 10-100 μm) scatter light. When the particle size and wavelength are comparable ($d \approx \lambda$) the amount of scattering is proportional to $1/\lambda$ (Mie scattering) so that short wavelengths are more strongly scattered than long wavelengths. When the particle size is much smaller than the wavlength the shorter wavelengths are affected much more severely than the longer wavelengths and the amount of scattering goes as $1/\lambda^4$. This explains why the sky is blue – the light scattered downward from the atmosphere will be reduced in long wavelengths and enhanced in short wavelengths. This also explains sunsets when the light from the Sun has travelled a long way through the atmosphere and most of the blue has been scattered out of it. Although a certain amount of ultra-violet radiation penetrates the atmosphere to sea level, the atmosphere is an effective barrier to shorter wavelengths (e.g. X-rays or gamma-rays).

13.1.2 Satellite Telescopes

The ultimate solution to atmospheric absorption and distortion is to put a telescope into orbit. This an expensive and technologically challenging solution, but it has been carried out many times. The early orbiting observatories detected X-rays and ultra-violet (UV), later infra-red (IR) and radio waves. The Hubble Space Telescope (HST) carries UV, visible and IR cameras and has provided some of the most amazing images of space ever seen by human beings. The Next Generation Space Telescope (NGST) will pick up from Hubble, providing even sharper images of fainter objects and attempting to answer some of the most compelling and subtle questions about the origin, evolution and fate of the universe. Some significant satellite telescopes are listed below.

Satellite	Launched	Comments
Orbiting Solar Observatory OSO	1962	UV, X-ray, OSO-3 detected first gamma-rays from the Sun
Uhuru (SAS-A)	1970	X-rays – first true X-ray telescope able to focus X-rays using reflection

		(produced first all-sky catalogue of X-ray sources)
Copernicus	1972	UV, X-rays
Cos-B	1975	Gamma-rays
International Ultraviolet Explorer Iue	1978	UV (first space telescope that could be controlled like a ground-based telescope)
Iras	1983	IR
Exosat	1983	X-rays – second true X-ray telescope
Rosat	1990	Hard UV, X-rays
Iso		IR
Cosmic Background Explorer COBE		Microwave (background radiation)
Hst	1990	UV, visible,IR

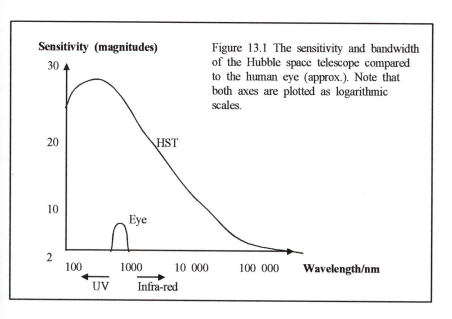

Sensitivity (magnitudes)

Figure 13.1 The sensitivity and bandwidth of the Hubble space telescope compared to the human eye (approx.). Note that both axes are plotted as logarithmic scales.

13.2 OPTICAL ASTRONOMY

When Galileo observed the moons of Jupiter he was using a refracting telescope, one which relies on lenses to form an image of its object. The ability of telescopes to 'see' faint objects and resolve fine detail both depend on the diameter of the aperture. For a refracting telescope this corresponds to the diameter of the objective lens. The amount of light collected is proportional to the area of the lens (and therefore its diameter squared) and the resolving power is proportional to its diameter (the larger the aperture the less significant the spread of light by diffraction). Unfortunately the refractive index for light in glass is frequency dependent, so refraction results in dispersion, i.e. different frequencies in the incident light are deviated through different angles. In practice this results in the blue end of the spectrum being focused closer to the pole of the lens than the red end of the spectrum. There is no single plane in which the entire image is in focus, a problem called chromatic aberration. This can be corrected to some extent by using compound lenses made from transparent materials of different refractive indices, but this also increases the amount of material the light has to pass through, complicating the design and making the whole structure more unwieldy and less sensitive. This is such a serious problem that all large optical telescopes are based on a different technique, proposed by Isaac Newton. Curved mirrors can be used to form images in much the same way as lenses, but they have the advantage that the light does not need to pass through a transparent medium, so there is no chromatic aberration. It is also much simpler to build a large mirror than a large lens and the mass of the resulting structure is reduced so it is less prone to distortion by gravity once in place. All large optical telescopes are reflectors although Schmidt telescopes are a compromise between the two techniques that enables a vast amount of optical information to be recorded in one observation. Here we will look at four large optical telescopes:

- The Anglo-Australian telescope (AAT), a 3.9 m Cassegrain reflector;
- The 1.2 m Schmidt telescope at Siding Spring;
- The 10 m Keck telescope at Mauna Kea;
- The Hubble Space Telescope (HST), a 2.4 m reflector.

13.2.1 The AAT and Schmidt Telescopes

A Newtonian reflector uses a curved mirror as its objective and directs the reflected light back to an eyepiece lens. Most optical telescopes use a different design however, called a Cassegrain reflector, in which the reflected light hits a small mirror above the pole of the objective and is sent back through a small hole in the objective to an eyepiece behind it (often the eyepiece is replaced by a CCD camera for remote viewing and automatic image processing). The Anglo-Australian telescope is a 3.9 m Cassegrain reflector situated on the top of Siding Spring Mountain at an altitude of 1200 m 500 km north of Canberra. It has a sophisticated computer control system and its reflectors are made of Cervit, a vitreous ceramic material which has a very low coefficient of thermal expansion. This is important because temperature changes can distort the mirrors of large telescopes, dramatically limiting their performance. The loading of a large mirror changes as it is turned so the back of the primary mirror is supported 36 continuously adjustable pads and the sides are supported by 24 more. The computer assists observers in tracking objects to an accuracy of about 0.1 seconds of arc in good conditions. The telescope system can be adjusted to form images on

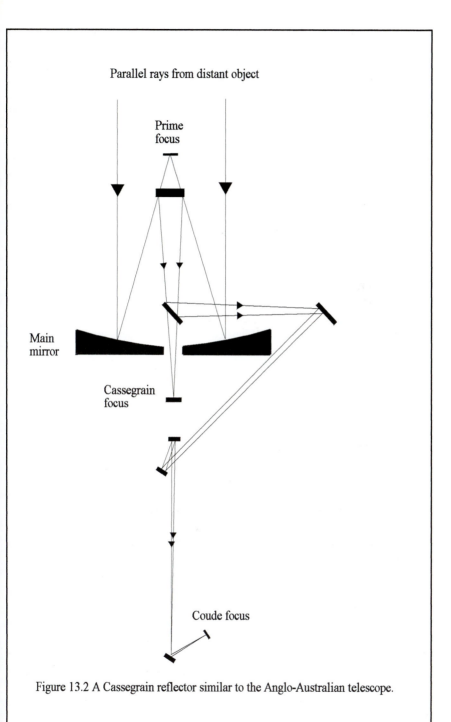

Figure 13.2 A Cassegrain reflector similar to the Anglo-Australian telescope.

a 1024×1024 CCD camera of a region 6.7 arcminutes across or adjusted to collect spectra from 400 objects simultaneously from a region of sky 2° across.

The 1.2 m UK Schmidt Telescope is situated nearby. Schmidt telescopes use a combination of reflection and refraction to provide a much larger angle view of the sky than a conventional Cassegrain reflector (6.6° as compared to just a few arcminutes). This makes Schmidt telescopes ideal for large sky surveys and they are often used to identify objects for more detailed study on the large reflectors. The UK Schmidt spends about one-third of its time connected by 100 fibres to FLAIR, a fibre-fed spectrograph that can record the spectra of up to 100 stars and galaxies simultaneously. It can record images of objects to magnitude 18. Of course, the enormous amount of information recorded by a Schmidt telescope in a single image requires a lot of processing and their use has gone hand-in-hand with the development of advanced computing techniques.

13.2.2 Keck

Keck I and Keck II stand side by side on the summit of Mauna Kea. They are the world's largest optical and infra-red telescopes and each of them has a revolutionary 10 m diameter segmented primary mirror. Each of these primary mirrors is made from 36 individual hexagonal segments. Each segment is 1.8 m in diameter and they fit together to form a single hyperboloid reflecting surface. The curvature of the individual segments is continually adjusted (at 2 Hz) by computer-controlled precision pistons sensitive to changes of just 4 nm. The polishing of the mirrors was so precise that if one hexagonal segment were expanded to the diameter of the Earth the surface bumps would be less than 1 m high! The Keck telescopes use a system of adaptive optics to compensate for distortions of the incoming light caused by atmospheric disturbances. A deformable mirror is positioned in the light path before the detector and its shape is continuously changed to compensate for the shifting atmospheric effects.

Keck's large diameter means it can collect a lot of light, about 17 times as much as the HST and can form images of fainter older and more distant objects. In theory it should also be able to resolve details about 4 times better than the HST, but in practice ground-based telescopes cannot perform to their theoretical diffraction limit – their resolving power is limited by atmospheric distortion. However, the Keck telescope is used by several research teams to look in more detail at objects discovered in HST images. It is being used to investigate new galactic clusters, gravitational lens systems and to probe the core of our own galaxy. At longer wavelengths (in the near infra-red) the telescope can operate near its diffraction limit and captures detailed images of planetary nebulae, protostars, quasars galactic cores and obscured galactic nuclei. Recent observations (1998) of the red-shifts of ultra-distant supernovae seem to imply that the rate of expansion in the universe is accelerating rather than slowing down. If this is confirmed it has major implications for the conventional Big Bang theory and general relativity.

In the 21st century Keck I and Keck II will be connected together to operate as an optical interferometer. This should be able to resolve detail as well as a single optical reflector with an 85 m diameter. One of the main aims of this project is to try to form images of Earth-like planets around other stars.

Keck is only one of a new generation of large optical telescopes that will contribute to astronomy in the twenty-first century. The Gemini project, for example, is building a pair of 8 m reflectors. Gemini North is on Mauna Kea, Gemini South is in Chile; the two together can monitor the northern and southern skies. These differ from Keck in that they both contain single piece mirrors. However, they will have system of actuators and sensors to make continual fine adjustments to the curvature to the mirror surface to compensate for distortions due to wind and loading. In addition a small flexible mirror will be used as part of an adaptive optics (AO) system that should filter out the effects of atmospheric turbulence. This system works by monitoring a local reference star and flexing the mirror to cancel the image changes. The control systems needed for modern optical telescopes were not possible in the past because the powerful computer systems needed to drive them were not available.

The Very Large Telescope (VLT) under construction in Chile consists of four large optical telescopes that can be connected together to form a large optical interferometer. Sometime next century it is hoped a similar telescope could be put into earth orbit and later still several could be connected together to synthesise a huge aperture. One of the aims of such projects would be to form images of faint Earth-like planets around distant stars and perhaps even see surface details!

13.2.3 The Hubble Space Telescope

The Hubble Space Telescope (HST) was deployed from the space shuttle Discovery on 24 April 1990 and was designed to make observations in the visible and ultra-violet parts of the electromagnetic spectrum. It was expected to resolve detail and see much fainter objects than any ground-based instrument. Unfortunately there was an error in the manufacture of the main 2.4 m mirror – it was about 0.002 mm too flat near its outer edge. This dramatically reduced its resolving power. A second problem involved the solar panels that vibrated so much as it moved in and out of the sunlight on its orbit around the Earth that position sensors and correctors on board had problems keeping it pointed at its target object. These vibrations were caused by thermal expansion as the temperature of the panels changed from -90° C in the Earth's shadow to 100° C in sunlight. Some interesting early results were obtained by applying a computed correction to the data collected by the flawed mirror, but the results were still disappointing. In 1993 the space shuttle Endeavour carried out a successful mission to correct the optics on board Hubble (by adding a specially made correction mirror into the light path) and to solve the problem of vibration from the solar panels. This mission was highly successful and the HST has been sending back high quality detailed images of distant (and not so distant) objects ever since.

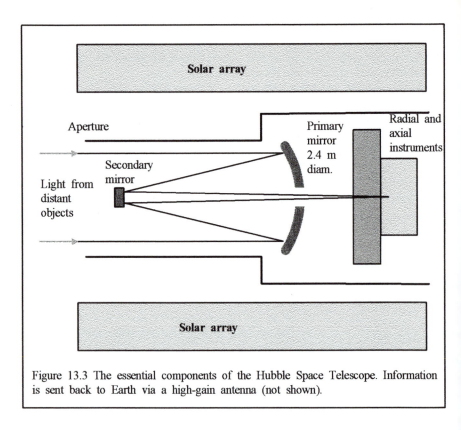

Figure 13.3 The essential components of the Hubble Space Telescope. Information is sent back to Earth via a high-gain antenna (not shown).

Hubble's primary mirror collects light from a region of sky 28 arcminutes across (roughly equivalent to the size of the full Moon in the sky) and the light collected is sent to a number of different instruments. These include:

- Wide-field planetary camera 2 (WFPC2) – to provide digital images over a wide field of view;
- Faint object camera (FOC) – to is a three-stage image intensifier and provides high resolution images of small fields;
- Space Telescope Imaging Spectrograph (STIS) – can be used to collect spectra from the UV to the near IR (115-1100 nm);
- Near Infrared Camera and Multi-Object Spectrometer (NICMOS) – produces IR images on three 256×256 pixel cameras;
- Fine Guidance Sensors (FGS) – these receive light slightly off axis and two of the three are used to correct the spacecraft attitude. HST must be pointed toward its target with very high precision and the FGS are used to 'lock on' to selected guide stars. The third FGS if free to make astronomical measurements and form images.

Hubble has already recorded many important images, but the most famous is the Hubble Deep Field (HDF), an image of faint objects in a region of apparently empty sky near Ursa Major (the Great Bear) about the size of a grain of sand held at arm's length. The HDF was made from 342 separate exposures using the WFPC2 over 10 consecutive days between 18 and 28 December 1995. Separate images were recorded in the UV, blue, red and IR regions of the spectrum and objects down to 30th magnitude (about 10^9 times fainter than the faintest objects visible to the human eye) were captured. Some of these galaxies are over 10 billion years old and the HDF is like an astronomical version of a geological core sample bringing an unprecedented amount of new information about the evolution of galaxies and the universe at large. Many thousands of galaxies are present in the image which is believed to be representative of the sky in any direction.

13.3 RADIO ASTRONOMY

13.3.1 Beginnings

Heinrich Hertz first demonstrated the transmission and detection of radio waves in 1887 and in 1890 Thomas Edison suggested that astronomical objects such as the Sun might be radio sources. The first person to detect radio waves from space was Karl Jansky in 1932. He noticed that radio noise in an antenna he was testing varied during the day and was always most intense from the direction of the galactic centre. By chance 1932 was a year in which there was very little sunspot activity (a quiet Sun) otherwise he probably would have detected radio waves from the Sun too. Solar radio noise was detected in 1942 as intense static on British wartime radar that rose and fell with the rising and setting Sun and disappeared when a large group of sunspots passed to the back of the Sun. These emissions are linked to the acceleration of charged particles in the Sun's atmosphere and their intensity is highly variable, being strongly linked to Sun spots. As the Sun rotates the active regions that emit these waves move across the Sun's disc with a characteristic 27-day period (the rotation period at the Sun's equator).

The first radio maps of the sky were made by a radio amateur, Grote Reber, who built himself a 9 m steerable parabolic reflector in his back garden at Wheaton, Illinois. His maps were published in 1940 and 1942 but the only radio source that could be linked to a definite optical source was the Sun. This emphasises the importance of making astronomical observations in different regions of the electromagnetic spectrum – they provide different information. During the Second World War the development of radar alerted many other scientists and electronic engineers to the possibilities of radio astronomy, so the way was prepared for rapid advances in the post-war years. One scientist working for the Army operational Research group deserves special mention. J.S. Hey made three of the most important early discoveries in radio astronomy:

- that the Sun emits radio waves in the metre wavelength range (from active sunspots);

- that meteors leave ionisation trails in the upper atmosphere (detected by radar echoes);
- the first discrete radio source, the radio galaxy Cygnus A.

We have already discussed the effect of the atmosphere on visible light and the problems this presents to optical astronomy. Life is rather simpler for a radio astronomer. The radio window lies between wavelengths of about 10 mm and 10 m and over this range the atmosphere is pretty well transparent to radio waves. Longer wavelengths are reflected back into space and atmospheric water and oxygen absorb shorter wavelengths. This transparency means that ground-based radio telescopes are actually diffraction limited, unlike their optical counterparts.

When waves pass through a narrow aperture they diffract. This spreads the wavefront beyond the aperture and limits the sharpness of images formed from the waves. A point object observed through a circular aperture forms a set of concentric diffraction rings called Airy's rings (Fig. 13.4). Two point objects close together (i.e. with a small angular separation) may form patterns in which the diffraction rings overlap so that they are not resolved. The first minimum in the diffraction pattern falls at an angle of about $1.22\lambda/D$ so this is taken as the diffraction limit for a telescope with aperture D. In other words two objects which subtend a smaller angle at the aperture will not be resolved. This is known as the Rayleigh criterion and is a useful rule for comparing the resolving powers of different instruments. It is quite clear that an instrument which captures longer wavelengths will need a larger aperture to achieve comparable resolution to one working with shorter wavelength radiation. This explains why radio telescopes (cm and m wavelengths) are so large compared to optical telescopes (around 0.5 μm).

Some examples of diffraction limits are:

Instrument	Aperture	Wavelength	Diffraction limit
Eye	5 mm (pupil)	500 nm	25 arcseconds*
AAT (Aus)	3.9 m	500 nm	0.03 arcseconds*
HST	2.4 m	500 nm	0.05 arcseconds
Jodrell Bank (UK)	76 m	21 cm	12 arcminutes
Effelsberg (Ger)	100 m	21 cm	8.8 arcminutes
Arecibo (Puerto Rico)	300 m	1 m	14 arcminutes
RATAN-600 (Rus)	576 m	1 m	7.3 arcminutes

*The practical limit for these instruments is considerably worse than this.

It is clear from the calculated diffraction limits that radio telescopes need to be very large if they are to approach the resolving power of optical telescopes. For example, the 76 m steerable dish at Jodrell Bank is about 30 times worse than the unaided eye at resolving fine detail. However, making a dish larger also makes it

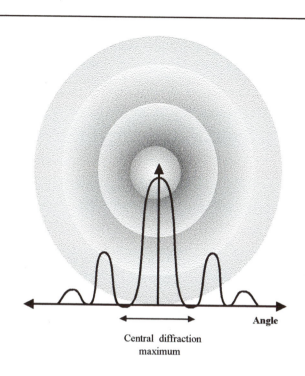

Angle

Central diffraction
maximum

Distant objects which
subtend a small angle at the
aperture to the instrument

Aperture

Screen

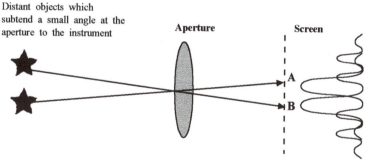

A

B

Without diffraction, points on the objects
would form points on the image.
Diffraction means each point produces a
ring pattern.

Figure 13.4 The diffraction pattern from a circular aperture has a well-defined
intensity distribution of maxima and minima. Images of two objects which are
very close together may overlap because of diffraction. They cannot then be
resolved. The resolving power of an instrument is determined by the ratio of its
aperture size to the wavelength of radiation that it is focusing.

heavier and eventually the distortions in structure brought about by its own weight distort the dish and make it harder to steer. One solution is to excavate a large dish (e.g. Arecibo) but this is not steerable and must rely on the motion of the earth to point it. Another is to make the reflecting surface from a mesh, but his too is limited. Yet another approach is to use an array of radio telescopes and combine their signals to synthesise the resolving power of a much larger telescope. This kind of arrangement is called an interferometer.

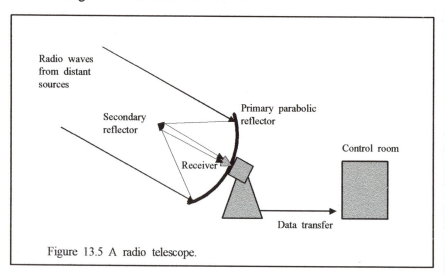

Figure 13.5 A radio telescope.

The simplest kind of interferometer consists of two radio telescopes separated by a fixed distance along a measured baseline. The principle of a long baseline interferometer is quite simple. As the source tracks across the sky radio waves are received from it at both of the radio telescopes. However, there is a path difference for waves travelling to one telescope compared to those travelling to the other and this introduces a phase difference in the two signals. When the signals are superposed this phase difference determines how they will combine. If they are in phase they add to create a larger signal if they are exactly out of phase they will cancel. As the source tracks across the sky the phase difference changes so the combined signal goes through a series of maxima and minima. Sources with an angular size greater than about d/λ (where d is the length of the baseline) will not be resolved because waves from different parts of the source will be received at the same time and no constant phase relation will be maintained. This allows large detail to be rejected and fine detail to be measured from the way the fringes vary with time. In effect two dishes separated by a distance d have a resolving power along that axis equal to about λ/d. Of course, they do not collect as much radiation as a dish of that diameter, but the long baseline interferometer does represent a cheap and simple solution to a serious problem. In a similar way a 2D array of radio telescopes can synthesise a 2D dish equal to the area of the array. A related but more sophisticated idea is to use the Earth's rotation to move a line of radio

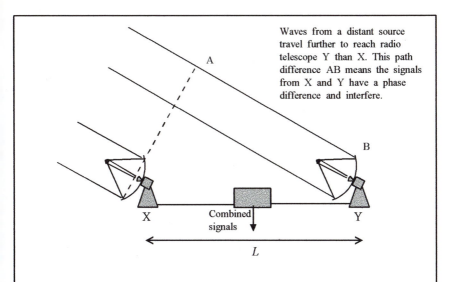

Waves from a distant source travel further to reach radio telescope Y than X. This path difference AB means the signals from X and Y have a phase difference and interfere.

Figure 13.6 A radio interferometer on a baseline of length L has a resolving power comparable to that of a telescope with a single dish of diameter L.

telescopes so that they sweep out a large aperture and then combine their signals as if they originated from separate sections of a single dish. This idea of *aperture synthesis* was pioneered by Martin Ryle at Cambridge in the 1960s.

The Merlin (multi-element radio-linked interferometer network) array is a group of radio telescopes (including Jodrell bank) straddling the Anglo-Welsh border that synthesises a single telescope several hundred kilometres across. All the telescopes are pointed at the same radio source and their signals are sent back to Jodrell Bank by microwave links where they are recombined. The Very Large Array in New Mexico has 27 telescopes linked in a Y-shaped array providing greater sensitivity to faint sources but less resolving power than Merlin. Very Long Baseline Interferometry (VLBI) combines the signals from radio telescopes separated by enormous distances, even on different continents to provide incredible resolving power. To do this they record their signals electronically and these are later combined (with appropriate timings) by computer. This is only possible by using accurately synchronised atomic clocks, but when it works properly it is capable of synthesising an aperture with a diameter equal to that of the Earth (although its collecting power and sensitivity to faint objects is far worse than for a single telescope of that size). This idea has been taken even further. In the mid-1990s Japan launched a radio telescope into space. Its signal, when combined with the signal from a terrestrial telescope, synthesises the aperture of a telescope several times the diameter of the Earth!

13.4 OTHER WAVELENGTHS

13.4.1 Ultra-violet Astronomy

The Hubble Space Telescope has ultra-violet (UV) cameras and spectrometers, but the groundbreaking UV space telescope was the International UV Explorer (IUE) launched in 1978. UV radiation is blocked by the atmosphere so ground-based telescopes are not practical. The IUE contained a 45 cm aluminium reflector and was sensitive to wavelengths from about 100 to 300 nm and sensitive to magnitude 16 (about 10 000 times fainter than optical sources visible to the naked eye). The satellite was placed in a geosynchronous orbit over the Atlantic, making it continually visible from Europe and America and giving astronomers direct control over its targets and exposure times. This was the first time this had been possible and it opened up space astronomy to a much larger population of researchers.

UV astronomy is important because it picks up radiation from very hot sources. In particular stars with surface temperatures above 40 000 K, and the IUE was important in identifying and studying these sources. Another advantage of the UV region is that it contains many of the resonance lines associated with common elements such as hydrogen, carbon, oxygen and nitrogen. These can be readily identified and used to measure elemental ratios in stars and galaxies. One of the early discoveries was that the ratio of elements in some of our local galaxies are not the same as in the Milky Way, an idea predicted by theory but not previously tested.

13.4.2 Infra-red

Infra-red (IR) radiation is partly absorbed by the atmosphere but can be detected at ground level and many of the new and planned optical telescopes are also designed to observe in the infrared. However, the most important observations so far have been made by space telescopes.

Thermal radiation is a problem for all infra-red telescopes and detectors. All objects radiate as a result of their temperature and this thermal radiation peaks in the IR part of the spectrum and increases as the fourth power of the temperature (Stefan's law). This means that an IR telescope at moderate temperatures (e.g. room temperature) will produce its own background 'glow' that will affect its detectors. For this reason IR detectors and their surrounding spacecraft must be refrigerated. The Infrared Astronomical Satellite (IRAS) was launched in 1983 and was kept at less than 16 K by evaporating on-board liquid helium into space. Its main aim was to survey the IR sky and produce an IR map of sources. It operated successfully for a year (until its supply of liquid helium ran out) and surveyed the entire sky three times in this time, and at four different wavelengths (12, 25, 60 and 100 μm). It also compiled IR spectra and provided astronomers with an enormous amount of information about cool stars. One of the advantages of making observations in the IR is that IR radiation can penetrate dust and gas clouds that are opaque at other wavelengths. Stars usually form from these clouds and IR astronomy has provided many images of protostars forming in vast stellar 'nurseries' where galaxies interact with one another.

The successor to IRAS, the Infrared Space Observatory (ISO), was launched in November 1995. It carried a reflecting telescope 0.6 m in diameter and could be pointed with an accuracy of 2.7 arcseconds. IR collected by the primary mirror could be reflected to four separate detectors recording images and spectra across a range of wavelengths from 3 μm to 200 μm. The spacecraft was placed in an elliptical transfer orbit with a perigee of 1000 km and apogee of 70 500 km so that the instruments (which were cooled to 3 K by an on-board liquid helium cryostat) were carried out of range of the Earth's belt of trapped protons and electrons for about 16 hours of each 24 hour orbit. The spacecraft was controlled from ESA's tracking station at Villafranca, near Madrid, from where it was 'visible' for about 14 hours of every day. The observatory continued to send back results for over two years during which time it recorded over 26 000 observations.

In November 1989 NASA launched its first satellite dedicated to cosmology. The Cosmic Background Explorer (COBE) was equipped with several instruments including FIRAS (Far Infra-red Spectrophotometer) designed to measure the cosmic background radiation over a range of wavelengths from 100 μm to 1 cm. This covers the microwave background discovered by Arno Penzias and Robert Wilson in 1964, thought to be an 'electromagnetic echo' of the Big Bang. This radiation began as high-energy gamma rays created when most of the matter and antimatter created in the Big Bang annihilated. Up to about 300 000 years after the Big Bang the radiation interacted strongly with matter, causing repeated ionisations and recombinations that brought the radiation to thermal equilibrium with matter. This model implied that the radiation would have a characteristic black-body spectrum. After 300 000 years the radiation decoupled from matter and retained its black-body spectrum. Since that time the universe has been expanding and the radiation has been red-shifted. In the early 1960s the Big Bang model predicted a cosmic background of thermal radiation corresponding to the thermal spectrum from a perfect radiator at a few degrees above absolute zero. Penzias and Wilson discovered that the radiation did have a black-body spectrum and corresponded to a temperature of about 3 K. This quickly became one of the most important pieces of evidence for the Big Bang theory and more precise measurements of the radiation spectrum and intensity can be used to test aspects of cosmological theory. FIRAS made a very accurate measurement of the spectrum, confirming its black-body characteristic and showing that it corresponded to a temperature of 2.730 ± 0.060 K and is more intense in the direction of the Virgo cluster than in the opposite direction. This dipolar distribution is because of our galaxies local motion toward Virgo (and perhaps toward a 'Great Attractor' beyond). A second instrument, the Differential Microwave Radiometer (DMR), measured the intensity of the background radiation very accurately at four frequencies: 9.5 mm, 5.7 mm and 3.3 mm. In 1992 George Smoot announced the results of the first COBE measurements. They showed that the background radiation is not perfectly smooth. It has fluctuations corresponding to temperature differences of a few tens of millionths of a degree over large regions of the sky. These fluctuations were expected, since there must have been some inhomogenities in the early universe from which the galaxies would eventually form. Smoot compared the event to 'seeing the face of God' which seems a little over the top (especially in the light of later criticisms levelled at the accuracy of the results) but it was very important for cosmology. Also the large distances over which these fluctuations spread suggested that quite distant parts of the early universe had actually been in some kind of contact at an even earlier stage. This suggested that these parts had moved rapidly apart in a period of exponential expansion (usually called inflation) that immediately followed the Big Bang and preceded the later expansion of the conventional theory. Later ground based observations confirmed the fluctuations, but the inflationary theory is by no means accepted by all cosmologists.

13.4.3 X-ray Astronomy

The Röntgen Satellite (ROSAT) has been the most successful space telescope observing in the X-ray and extreme UV regions. Previous X-ray space telescopes (such as the High Energy Astronomical Observatory 2, or Einstein Observatory) suffered from an inability to focus the radiation. This is because materials capable of refracting or reflecting X-rays are also strong X-ray absorbers. This rules out refracting or reflecting telescopes with similar designs to optical telescopes. However, it is possible to reflect X-rays if they strike the reflector at a glancing angle. In ROSAT the main instrument is the German X-ray telescope (XRT), a set of four nested slightly tapering metal cylinders. X-rays entering the wider aperture at glancing incidence reflect from the polished inner surfaces of each cylinder and come to a common focus some way beyond the end of the 'tube'. Proportional detectors in the focal plane respond like Geiger counters to the focused X-rays and are tuned to particular wavelength ranges. The telescope can respond in the range 0.6 –10 nm. Outside the spacecraft is a second British telescope designed to focus hard UV (5-25 nm). This has a similar structure, consisting of three nested tapering cylinders, but it is shorter than the X-ray telescope because the longer wavelength waves can be reflected at larger incident angles. Extreme UV (EUV) is absorbed strongly by gas in the interstellar medium, so EUV measurements are a good way to probe the density of the local interstellar medium.

X-rays carry information about the most violent objects in space, including X-ray galaxies, black holes, white dwarfs, binary systems and quasars. One of the main aims of ROSAT was to carry out a full sky X-ray survey, but it has also made detailed observatons of many individual X-ray sources. It also identified the source of the X-ray background, something about which astronomers had speculated since its discovery in 1962 during a NASA mission to look for X-rays from the Moon. Over 50% of the background is coming from individual quasars and further observations of faint sources seems likely to account for the rest in a similar way. X-rays are also expected to be emitted from matter falling into black holes, so X-ray images have been important in linking large black holes to the centre of active galaxies.

Figure 13.7 Observational astronomy has also revolutionised our view of our own planet. This spaceborne radar image shows Southern California's San Fernando Valley with its dense network of streets and freeways in the center. The Santa Monica Mountain range is shown along the bottom of the image and it separates the valley from the city of Los Angeles. Urban planners can use images like this one to study land-use patterns in cities and surrounding areas. The image was acquired by the Spaceborne Imaging Radar-C/X-band Synthetic Aperture Radar (SIR-C/X-SAR) onboard the space shuttle Endeavour on October 3, 1994. SIR-C/X-SAR, a joint mission of the German, Italian and the United States space agencies.   Photo credit: NASA

Figure 13.8 We have even begun to explore beyond our home planet. Astronaut Jim Irwin setting up the Lunar roving Vehicle during the Apollo 15 mission.   Photo credit: NASA

14
STARS AND DISTANCES

14.1 THE SCALE OF THE UNIVERSE

One of the most difficult problems in astronomy is working out the distance to astronomical objects. As our understanding of stars and galaxies has grown through the twentieth century a number of techniques have been developed and compared so that an approximate scale can be used to position everything from the objects within our own solar system to the galaxies at the far limit of the Hubble deep field image. At the end of the century our best guess of whether the universe will expand forever or recollapse depends on our ability to measure distances and hence determine the Hubble constant. Unfortunately this is still a very difficult and controversial area and even in the 1990s newspaper and magazine headlines could ask questions such as, 'Are stars older than the universe?', exploiting the uncertainties in measurements to draw a rather paradoxical conclusion! In this chapter we shall look at some of the techniques used to fix these astronomical distance scales.

The metre is a convenient unit of length for objects of approximately human size. In this unit the distance to the Moon is about 4×10^8 m and the distance to the Sun is 1.5×10^{11} m. The distance to Proxima Centauri, the closest star to the Sun, is 4×10^{16} m and the distance across our galaxy is about 10^{21} m. The distances to the most distant galaxies in the Hubble deep field survey is more like 10^{26} m. These large numbers make it convenient for us to invent new units for astronomical distances.

The distance from the Earth to the Sun varies slightly as the Earth moves in a near circular ellipse around it. The mean distance from the Earth to the Sun is used as a unit of distance for local measurements. It is called the Astronomical Unit (AU) and calculated using Newton's law of gravitation and the measured distance to some of the inner planets.

$$1 \text{ AU} = 1.459 \times 10^{11} \text{ m}$$

Between 1609 and 1620 Johannes Kepler (1571-1630) published three laws for planetary motion derived from careful observations made mainly by the great

Danish astronomer Tycho Brahe and partly by himself. They were:

- The orbits of planets are ellipses with the Sun at one focus.
- A line from the Sun to a planet sweeps out equal areas in equal times.
- The squares of planetary periods in years are equal to the cubes of their mean distances from the sun in AU (or T^2/r^3 = constant in any units).

In the 17th century Newton showed that Kepler's laws could be derived from the inverse square law of gravitation which led to the following equation:

$$T^2 = \frac{4\pi^2 a^3}{G(M_S + M_P)}$$

T = the orbital period;
a = the semi-major axis of the elliptical orbit;
M_S = the mass of the Sun;
M_P = the mass of the planet.

Kepler's third law drops out of this expression when it is realised that the mass of any of the planets is much less than the mass of the Sun. Furthermore, if the Earth is taken as the planet and T and a are measured in years and AU respectively, then the remaining constant is unity so that $T^2 = a^3$ and the distance (in AU) to any planet is simply equal to $T^{2/3}$.

In practice it is difficult to make a direct measurement of the AU by measuring the distance to the Sun. However, the equation above can be used to calculate the AU if the orbital period and orbit for any planet is known. The first accurate calculations were made by Giovanni Cassini using the terrestrial parallax of Mars – its change in apparent position when viewed from opposite sides of the Earth. More recently the value has been refined using radar reflection measurements from the surface of Venus.

Once the AU is known the distances to nearby stars can be calculated using parallax as the Earth moves around the Sun. Friedrich Wilhelm Bessel made the first successful measurement of stellar distance using parallax in 1838. He chose 61 Cygni because it had the greatest known proper motion and so is likely to be nearby (proper motion is a systematic change in position with time against the background of 'fixed stars' and is only visible for stars that are relatively close to us – the effect was first observed by Edmund Halley in 1718). 61 Cygni has a proper motion that amounts to 5 seconds of arc per year. When this had been subtracted there was an annual to and fro motion of about 0.58 arcseconds that must be due to the change in parallax brought about by the Earth's own orbital motion. The astronomical parallax is defined as one half of the angular shift (i.e. equal to the angle subtended by the radius of the Earth's orbit (1 AU) as seen from the star). For 61 Cygni the astronomical parallax p is therefore about 0.29 arcseconds. Using simple trigonometry (and a small angle approximation) the distance to 61 Cygni is 1/tan p astronomical units. This is about 7×10^5 AU. The nearest star to the Earth is Proxima Centauri, which has an astronomical parallax of 0.76 arcseconds and a distance of 2.7×10^5 AU.

Even for measuring distances to local stars the AU is beginning to seem too small, so another astronomical unit is introduced, the parsec. This is the reciprocal of the astronomical parallax p (p in arcseconds). A star at 1 parsec from Earth would have an astronomical parallax of 1 arcsecond and a distance of 1/tan p = 206 265 AU. For even greater distances the appropriate unit is the megaparsec. Our galaxy has a diameter of about 10^5 light years which converts to about 30 000 parsec or 0.03 megaparsec.

Table: Comparison of various astronomical units					
Units	Metre	Au	Parsec	Light year	Megaparsec
Metre	1	6.9×10^{-12}	3.3×10^{-17}	1.1×10^{-16}	3.3×10^{-23}
AU	1.5×10^{11}	1	4.9×10^{-6}	1.5×10^{-6}	4.9×10^{-12}
Parsec	3.0×10^{16}	2.1×10^5	1	3.3	10^{-6}
Light year	9.5×10^{15}	6.4×10^4	0.31	1	3.1×10^{-7}
Megaparsec	3.0×10^{22}	2.1×10^{11}	10^6	3.3×10^6	1

Table: Some astronomical distances					
	m	AU	Light years	pc	Mpc
Atoms	10^{-15}				
Humans	2				
Diameter of Earth	1.3×10^7				
Diameter of Sun	1.4×10^9				
Mean radius Earth's orbit	1.5×10^{11}	1	1.6×10^{-5}	4.8×10^{-6}	
To nearest star Proxima Centauri	4.1×10^{16}	2.7×10^5	4.3	1.3	
To Sirius	8.2×10^{16}	5.4×10^5	8.6	2.6	
To Polaris	3.4×10^{18}		360	110	
To Rigel	8.3×10^{18}		880	570	
Milky Way 'diameter'	9.5×10^{20}		10^5	3×10^4	0.03
To Andromeda galaxy	2.4×10^{21}		2.5×10^5	7.5×10^4	0.075
To Virgo galaxy	5.0×10^{23}		5.3×10^7	1.6×10^7	16
To Ursa Major	6×10^{24}		7×10^8	2×10^8	200
To Hydra galaxy	3×10^{25}		3×10^9	8×10^8	800
Furthest HDF image	10^{26}		10^{10}	3×10^9	3000

14.2 STARS

14.2.1 Stellar Magnitudes

When you look up at the night sky you see thousands of stars, but they do not all look the same. Some are brighter than others and, if you look carefully on a clear night, they have a range of different colours. Astronomers cannot set up controlled experiments, vary the conditions and see what happens. The only evidence they can use to test their theories about stars and galaxies is contained in the radiation that reaches us through vast distances of almost empty space. Every clue has to be weighed and considered and brightness and colour are two of the most important clues we have. Brightness depends mainly on two factors – distance and luminosity. Colour depends on surface temperature. It is amazing how much we can work out from such limited information.

The first person to classify stars by their apparent magnitude (brightness) was Hipparchus in about 130 BC. He divided the stars into classes based on how bright they appeared in the night sky. The brightest stars were classified as magnitude 1, those that were just visible to the naked eye as magnitude 6. In practice the intensity of a magnitude 1 star is 100 times that of a magnitude 6 star, so the 5 magnitude steps correspond to a multiple of 100. For a geometric series of magnitudes each magnitude must be a times the intensity of the previous one with $a^5 = 100$. This means that going up one magnitude increases the intensity by a factor of $a = 2.51$. So magnitude 3 is 2.51 times as intense as magnitude 4 and so on.

In Hipparchus's scheme all the stars were thought to reside on the same celestial sphere, so a comparison of apparent magnitudes would be a true comparison of stellar luminosities. We now know that the stars are at an enormous range of different distances, so apparent magnitude alone is not very useful. To allow for this a simple convention has been adopted. The apparent magnitude of stars is adjusted to the value it would have if the star were 1 pc from Earth. This adjusted value is called the absolute magnitude (M). A comparison of absolute magnitudes is as good as a comparison of luminosities.

Luminosity is the gross power output of a star. This is not linked directly to the apparent magnitude but is related to the absolute magnitude. The smaller the numerical value of the absolute magnitude, the *greater* the star's luminosity. The link is:

$$\frac{L_1}{L_2} = a^{M_2 - M_1}$$

Stellar luminosities vary over 30 magnitudes giving a range of a^{30} in luminosity. The Sun, a star with absolute magnitude +4.8 has a luminosity of 4×10^{26} W and an apparent magnitude of −27. Sirius, at a distance of 2.65 pc has an apparent magnitude −1.46, and absolute magnitude +1.42 and a luminosity 9×10^{27} W.

Maths Box: Absolute and Apparent Magnitudes

Let the intensity from a star at distance d be I_d (all distances measured in pc).
This will obey an inverse square law with increasing distance.
The absolute magnitude is M, and apparent magnitude m, so:

$$\frac{I_d}{I_{10}} = \frac{10^2}{d^2} = \frac{100}{d^2}$$

This ratio of intensities can also be expressed in terms of m and M.

$$\frac{I_d}{I_{10}} = a^{M-m} \qquad \text{where } a = 100^{1/5}$$

(If the star was at 10 pc then $M - m = 0$ and the ratio is 1)

$$a^{M-m} = \frac{d^2}{100}$$

Taking logarithms (to base 10) of both sides:

$$(M - m)\log a = 2\log\frac{d}{10}$$

But,
$$\log a = \log 100^{1/5} = \frac{1}{5}\log 100 = \frac{2}{5}$$

$$M - m = 5\log\frac{d}{10}$$

$$m = M - 5\log\frac{d}{10}$$

These relationships between absolute and apparent magnitudes, distance and luminosity can be used to measure distances. This is possible when a star can be identified as belonging to a particular type. Then it is possible to estimate its luminosity, calculate its absolute magnitude and use the apparent magnitude to measure its distance. The difficulty is in knowing what type of star it is. One way to do this is to use the spectrum of light emitted by the star, and this will be discussed in a later section (the method is called spectroscopic parallax). Another way is to pick unusual types of star whose behaviour is well understood, or stars at particular points in their life cycle when their behaviour is to some extent predictable. Two examples are:

- **Variable stars** (e.g. Cepheid variables) whose luminosity varies periodically. The period of variation is linked to luminosity, so it can be used to work out the absolute magnitude. These allow distances out to about 4 Mpc to be measured using terrestrial observations. The HST extends this to about 40 Mpc, but beyond that it is too difficult to resolve them.
- **Supernovae**. Massive stars explode at the end of their lives and emit an enormous amount of energy in a very short time. These stellar flashlights have particular characteristics that can again be used to estimate their luminosity and hence, from their apparent magnitude, their distance. This method has been very important for measurements to very distant galaxies (since the great luminosity of supernovae makes them visible even at very great distance). This method has been used to measure distances over hundreds of Mpc.

To understand how stars can be used to estimate distances we need to understand what makes them shine. This is discussed below, and the application of physical theories to this area of astronomy is one of the great success stories of twentieth century physics.

14.2.2 Stellar Fusion

The problem of how stars shine was not solved until the twentieth century and it became intimately linked to arguments concerning the age of the Earth. Two hundred years ago the prevailing view was that the age of the Earth was likely to be measured in thousands of years rather than millions (or billions). This was partly based on theological arguments, the most famous of which was put forward by Bishop Ussher in 1654. He added the ages of the biblical patriarchs to the length of recorded history and concluded that the Earth came into being in 4004 BC. During the nineteenth century this idea was challenged by two scientific theories:

- James Hutton, an eighteenth century geologist, suggested that hills and mountains have been shaped by the eroding action of wind and water over an unimaginably long period of time.
- In 1859 Charles Darwin published the *Origin of Species* in which he proposed evolution by natural selection over a very long time – his own estimate of the age of the earth was 300 million years.

It was not only the theologians who were worried by these ideas. A commonly held theory about the Sun was that it continued to shine because of chemical combustion, a process that would consume the entire mass of the Sun in no more than 10 000 years. In the 1850s William Thomson, later Lord Kelvin, analysed a different theory, first put forward by James Waterson. He calculated how long the Sun could shine if its energy derived from the gravitational potential energy released by infalling matter (he imagined a whirling vortex of meteors tumbling into the Sun and heating it). The rate of infall was fixed by assuming it must also account for the Sun's rotation. This led to an age of about 30 000 years. A few years later he replaced the idea of meteors falling onto a pre-existing Sun and assumed that the Sun itself is formed by gravitational collapse and gradually converts gravitational potential energy to heat (which is radiated away) as it continues to collapse. This dramatically extended the theoretical age of the Sun to about 20 million years. This seemed to solve the problems posed by geology and evolution neither of which could take place on Earth without some external energy source, in this case solar radiation. The flux of radiation from the Sun would require a rate of collapse of just 20 m per year, a figure that seemed reasonable at the time and implied that the Sun must be very old (although not, as it turned out, old enough). Lord Kelvin also proposed another way to calculate the age of the Earth. He assumed it formed with the rest of the solar system in a process of gravitational collapse and was once a ball of molten rock. This seemed consistent with the increase in temperature noticed in coal mines, an observation that suggested that the inside of the Earth must be warmer than the outside and so the planet was still cooling. By estimating the rate of heat loss during the age of the Earth Kelvin calculated that it had been cooling for about 25 million years, in good agreement with his estimate of the age of the Sun.

Kelvin's ideas were challenged by another geologist, John Joly, who worked out how long it would take for the oceans to become salty. His results suggested that around 100 million years would be required – much more than Kelvin's estimate. Other geologists used the rate of mountain formation and valley erosion and came up with even longer timescales. As the century turned Kelvin and Joly argued their cases at the British Association and the Royal Society. But by then radioactivity had been discovered and this led ultimately to a much longer physical estimate of the age of the Earth and indirectly to a mechanism capable of keeping the Sun shining for 10 billion years.

In 1904 Rutherford spoke at the Royal Institution of Great Britain. He had been investigating the newly discovered radioactivity and realised that it represented a major source of energy (although Rutherford did not think it was one that could be usefully harnessed). Kelvin had assumed that there had been no additional sources of heat since the Earth formed. Rutherford adjusted Kelvin's calculation to allow for the presence of radioactive isotopes in the Earth which would provide an extra heat source as they decayed over geological time. He showed that this might extend the age of the Earth to billions of years rather than tens of millions. Kelvin never liked the radioactivity, but it has provided a powerful way to date rocks on Earth and rocky materials recovered from the Moon or meteorites. One method relies on potassium-40 which decays in two ways:

$$^{40}_{19}K \rightarrow \, ^{40}_{20}Ca + \, ^{0}_{-1}\beta + \, ^{0}_{0}\bar{\nu} \quad \text{in 89\% of decays}$$

$$^{40}_{19}K \rightarrow \, ^{40}_{18}Ar + \, ^{0}_{1}\beta^{+} + \, ^{0}_{0}\nu \quad \text{in 11\% of decays}$$

The second decay is interesting because argon, an inert gas, is not likely to have been present in solidifying rocks. This means any argon present now must have been formed from the decay of potassium. As time goes on the ratio of potassium-40 to argon-40 falls (as potassium decays and argon forms) so the age of a rock can be calculated from this ratio. There are several other isotopic methods to determine the age of rocks. The oldest rocks on Earth are about 4.6 billion years old. This agrees pretty well with the age of Moon rock and the ages of rocks that fall to Earth as meteorites suggesting the solar system itself formed a little under 5 billion years ago. Of course, this implies that the Sun was also formed at that time, and has been shining ever since. Its power output is about 4×10^{26} W. None of the mechanisms discussed above could account for such an enormous power output over such an incredibly long period of time. George Darwin (second son of the famous naturalist, and a prominent physicist) suggested it might be powered by decaying radioactive isotopes, but calculations showed that a Sun made entirely of uranium could not account for the observed radiation flux!

The problem of the Sun's energy source was solved in the 1930s. Arthur Eddington was one of the first physicists to apply Einstein's theory of relativity to real physical problems. He realised that the mass of a helium-4 nucleus is slightly less than the mass of four hydrogen nuclei and wondered whether the intense temperatures and pressures in the core of a star would be sufficient to fuse hydrogen to helium. If so the reaction should release an enormous amount of energy. Using Einstein's mass-energy equation we can see that the Sun's total power output of about 4×10^{26} W can be accounted for by a relatively tiny rate of transfer of mass to energy.

$$E = mc^2$$

$$P = c^2 \frac{\mathrm{d}m}{\mathrm{d}t} = 4 \times 10^{26}$$

$$\frac{\mathrm{d}m}{\mathrm{d}t} = 4 \times 10^9 \, \mathrm{kgs}^{-1}$$

This represents only 2 parts in 10^{21} per second and if all the Sun could be consumed in this way (which it cannot) the Sun would last for about 10^{13} years, more than enough to account for all the geological changes on Earth and for the radioactive dating of the rocks. However, there was one big problem. For two hydrogen nuclei to fuse they must get close enough to one another for the strong nuclear force to overcome their electrostatic repulsion. This needs a closest approach of about 10^{-15} m. Eddington used Kelvin's gravitational collapse model to estimate the core temperature and found that, high though it was, it was insufficient to initiate fusion reactions.

At this stage another of the new theories came to the rescue, quantum theory. In classical physics a particle has a definite position and momentum at each point on its path. In quantum theory the uncertainty principle implies that these quantities are not fixed, but lie with in a range of values linked to Planck's constant. In effect two protons separated by, say 10^{-14} m, have a chance of being closer or farther away than this. This means that in some collisions that should not (classically) bring them close enough to fuse they do indeed fuse. The process is called 'tunnelling' since they effectively tunnel through the Coulomb potential barrier. The net effect is that fusion reactions which combine protons can and do take place in the core of the Sun. George Gamow, Fritz Houtermans and Robert Atkinson worked this out in 1928. However, fusing two protons does not explain how hydrogen turns to helium and powers the Sun, but it is a step along the way.

The mechanism that dominates helium and energy production in the Sun is called the proton cycle and was discovered by Hans Bethe in 1936. It proceeds via three distinct fusion steps, the first of which (because the tunnel effect has a low probability for each individual proton-proton collision) is terribly slow, with each proton having a 'lifetime' of about 10^{10} years before it combines with another one to form a deuteron!

$$\text{Step 1} \quad {}^1_1\text{H} + {}^1_1\text{H} \rightarrow {}^2_1\text{H} + {}^0_1\beta + {}^0_0\nu$$

$$\text{Step 2} \quad {}^1_1\text{H} + {}^2_1\text{H} \rightarrow {}^3_2\text{He} + {}^0_0\gamma$$

$$\text{Step 3} \quad {}^3_2\text{He} + {}^3_2\text{He} \rightarrow {}^4_2\text{He} + 2{}^1_1\text{H}$$

The overall effect is to convert 4 protons to a helium nucleus and release approximately 28 MeV. The low probability of step 1 is the reason the Sun will continue to shine for billions of years. If that step was much quicker then the rate of reaction would be greater and the Sun would use up its nuclear fuel much more rapidly – so the very impediment to fusion is also the reason the Sun has been around long enough for us to evolve. It generates about 80% of its energy from the proton-proton cycle and its lifetime is calculated to be about 10 billion years. Assuming the age of the Sun is similar to that of the planets, it has been around for about 5 billion years and so is almost exactly half way through its life.

Bethe worked out an alternative mechanism by which stars can convert their hydrogen to helium, the carbon cycle. In a star like the Sun less than 0.1% of atoms are carbon and the probability that protons can break through the potential barrier surrounding the carbon nucleus rises rapidly with temperature, so this will occur frequently in hot large mass stars. Helium can then be created using carbon nuclei as a kind of catalyst:

$$ {}^1_1\text{H} + {}^{12}_6\text{C} \rightarrow {}^{13}_7\text{N}$$

$$ {}^{13}_7\text{N} \rightarrow {}^{13}_6\text{C} + {}^0_1\beta + {}^0_0\nu$$

$$ {}^{13}_6\text{C} + {}^1_1\text{H} \rightarrow {}^{14}_7\text{N}$$

$$ {}^{14}_7\text{N} + {}^1_1\text{H} \rightarrow {}^{15}_8\text{O}$$

$$ {}^{15}_8\text{O} \rightarrow {}^{15}_7\text{N} + {}^0_1\beta + {}^0_0\nu$$

$$ {}^{15}_7\text{N} + {}^1_1\text{H} \rightarrow {}^{12}_6\text{C} + {}^4_2\text{He}$$

There are also gamma rays emitted, some of which are created when the two positrons annihilate with electrons. The net effect has been to convert four protons to a helium nucleus and emit neutrinos and energy.

The rate of energy generation by both proton-proton fusion and the carbon cycle depends on the temperature, but much more sharply for the latter process (proportional to T^{17} compared to T^4). The Sun has a core temperature of about 15 million kelvin and generates about 80% of its energy from proton-proton fusion and 10% from the carbon cycle. For stars over about two solar masses the carbon cycle is the dominant process. Hans Bethe was awarded the Nobel Prize for Physics in 1967 *'for his contributions to the theory of nuclear reactions, especially his discoveries concerning the energy production in stars'*.

For a star to be stable it must balance gravitational forces tending to make it collapse against the internal radiation pressure generated by fusion reactions. This radiation pressure will increase with the mass of the star. Below about 0.08 solar masses the core temperature is too low to support fusion so the star never ignites (Jupiter is not a star for this reason; it is about 50 times too light). Above about 120 solar masses the radiation pressure is so great that it blows excess mass away, so stars lie within a range of masses from about 0.08 to 120 times the mass of our Sun.

14.2.3 Nucleosynthesis

One of the most exciting aspects of physics is its ability to give a new slant on old philosophical questions. The gradual realisation that the stars shine because they fuse hydrogen to helium and the success of atomic theory in explaining the Periodic Table of the elements led to the startling idea that all the elements we find in our bodies (apart perhaps from some of the hydrogen) were probably formed by fusion reactions in stars. We are, quite literally, stardust.

The early universe was dominated by hydrogen, the simplest element and the one that would have formed most easily as the hot dense bubble of matter and radiation formed in the Big Bang expanded and cooled. In the extreme conditions of the first few hundredths of a second some of this hydrogen would fuse to form deuterium, helium-3 and helium-4. After a few hundred seconds the temperature had fallen so far that fusion reactions could no longer occur. By this time about 25% of matter (4 out of every 16 nucleons) had been converted to helium-4. Stars and galaxies began to form by gravitational collapse in regions of above average density about 100 million years later, and the stars began to manufacture more helium from hydrogen and small quantities of heavier elements such as carbon, oxygen and nitrogen. However, when we look at the spectra of stars we see spectral lines characteristic of many elements, where do these come from? The key to this problem was provided in 1957 in a joint paper published by Margaret and Geoffrey Burbage, William Fowler and Fred Hoyle. In this paper the essential steps of nucleosynthesis are worked out in detail.

The idea that the elements were synthesised from hydrogen is quite straightforward, The detailed mechanism is not. There is a problem because no stable isotopes exist with mass numbers 5 or 8, so it is not possible to build up the heavy nuclei simply by adding protons one at a time. The solution was proposed by E. Öpik and E. Salpeter in 1951. They showed that a triple-alpha reaction can occur when the core temperature exceeds about 4×10^8 K:

$$_2^4\text{He} + {}_2^4\text{He} + {}_2^4\text{He} \rightarrow {}_6^{12}\text{C}$$

But even this did not solve the problem because the probability for the process (its cross-section) was still far too small to produce enough carbon. In 1953 Fred Hoyle suggested that the probability might be greatly enhanced (by a factor of about 10^7) by a nuclear resonance at about 7.7 MeV. This reduced the temperature required for helium burning to about 10^8 K, in agreement with estimates of the core temperatures at which the process must be important if it is to account for the luminosity and abundance of carbon in massive stars. Once carbon has formed heavier nuclei up to iron-56 can be formed by fusion with helium nuclei or absorption of neutrons. This was supported by a survey of the cosmic abundance of the elements published by Hans Suess and Harold Urey in 1956 which emphasised three things:

• Abundances drop rapidly with increasing mass number.
• Nuclei which are simple multiples of the helium nucleus are particularly abundant.
• There are peaks of stability corresponding to the 'magic numbers'.

Above iron-56 the creation of nuclei becomes endothermic so they will not be produced in any significant quantity in the cores of stars during their normal lifetimes. In 1957 A. Cameron pointed out that heavy nuclei could be synthesised explosively in a supernova. Later analysis confirmed that this model could fit the measured abundances of heavy nuclei. The presence of significant quantities of heavy elements in our solar system is evidence that it is not a first generation system but formed from the debris of ancient supernovae.

This gives us a powerful picture. Stars fuse hydrogen to helium by the proton or carbon cycles depending on their mass and temperature. When the hydrogen in the core runs out gravitational collapse increases the core temperature and may ignite helium-burning. This will depend on the mass of the star. In fact heavier stars will exhaust the helium in their cores, collapse further and fuse carbon-12 to magnesium-24, or oxygen-16 to silicon-32 and so on up to iron-56. For a massive star the final reactions that create nickel and iron release very little energy and only postpone the inevitable gravitational collapse. At this stage the flux of neutrinos from nuclear reactions in the core is far greater than the flux of photons from the surface. In fact one of the most important recent supernovae (supernova 1987A) was heralded by an increase in neutrinos in terrestrial detectors many hours before the increase in its luminosity was seen. As the energy generated by fusion reactions falls off the star begins a rapid collapse. Gravitational energy pays for the formation of elements heavier than iron at the same time as it breaks up core elements formed during millions of years of fusion. An enormous increase in nuclear reactions in the core creates an explosive blast of neutrinos that hurls the outer layers of the star including significant quantities of the heavy nuclei into space.

14.3 STELLAR SPECTRA

14.3.1 Analysing Starlight

The high-energy gamma rays created in the core of a star are scattered, absorbed and re-emitted very many times before they escape from the photosphere and radiate into space. The opaque hot gas that absorbs and re-emits the light brings it into thermal equilibrium with the gas on the outer surface of the star. This means that the radiation spectrum from a star resembles that of an ideal black body (although this is far from exact). Black body radiation obeys Wien's Law and Stefan's Law:

- **Wien's Law**:

$$\lambda_p T = \text{constant} = 2.898 \times 10^{-3}\,\text{mK}$$

λ_p is the peak wavelength in the spectrum

T is the temperature of the black body

Wien's law tells us that the colour of the star is related to its surface temperature. Red stars are relatively cool at about 3000 K, yellow ones (such as the Sun) are closer to 6000 K and white ones around 10 000 K. Some blue stars have surface temperatures in excess of 20 000 K.

- **Stefan's Law**:

$$I = \sigma T^4$$

I is the energy radiated per unit area per second

σ is Stefan's constant $= 5.7 \times 10^{-8}\,\text{Wm}^{-2}\text{K}^{-4}$

T is the temperature of the source

Stefan's Law gives the star's luminosity, the total power radiated: $L = \sigma A T^4$. The intensity of the star's radiation at a distance r from the star can then be related to its luminosity by:

$$I = \frac{L}{4\pi r^2}$$

If the type of star is identifiable so that its approximate size is known, then L can be estimated using Stefan's law and its apparent intensity can then be used to calculate its distance.

In 1815 Joseph von Fraunhofer noticed that the continuous spectrum of light from the Sun is crossed by thousands of fine dark lines. These are absorption lines formed when radiation passes through cooler low-density gases in the Sun's atmosphere. Some photons excite electrons inside atoms to make quantum jumps to higher levels. These excited atoms then re-radiate photons at the same frequency but in random directions. The net effect is that the intensity at these absorption frequencies is significantly reduced and a dark line (an absorption line) is seen in the solar spectrum.

Absorption lines occur at wavelengths and frequencies that correspond to allowed energy jumps in the atoms that happen to be present in the stellar atmosphere, so the pattern of dark absorption lines tells us about the composition of the star. Hotter stars will be able to excite higher energy transitions and induce ionisation, so the spectrum of a star gives detail about both composition and temperature. From the middle of the nineteenth century to the 1920s astronomers made many attempts to classify stars by their spectra. However, the crucial work was carried out by a group at Harvard, and Annie Jump Cannon in particular. She adapted existing schemes to form a basic sequence of seven spectral classes labelled O, B, A, F, G, K and M (see table) representing stars of decreasing temperature.

Table : Some important spectral lines	
Wavelength /nm	Absorber (symbol)
656.3	Hydrogen (H)
589.3	Neutral sodium (Na I)
587.6	Neutral helium (He I)
527.0	Neutral iron (Fe I)
516.7, 517.3, 518.4	Neutral magnesium (Mg I)
495.5	Titanium oxide (TiO)
486.1	Hydrogen (H)
468.6	Ionised helium (He II)
438.4	Neutral iron Fe I
430.0	CH molecule
434.0	Hydrogen (H)
422.7	Neutral calcium (Ca I)
410.1	Hydrogen (H)
396.8	Ionised calcium (Ca II)
393.4	Ionised calcium (Ca II)

Table: The seven main spectral classes			
Class	Effective temperature /K	Colour	Characteristic absorption lines
O	28 000- 50 000	Blue	He II, He I
B	9900 - 28 000	Blue-white	He I, H
A	7400 -9900	White	H
F	6000 - 7400	Yellow-white	Metals, H
G	4900 -6000	Yellow	Ca II, metals
K	3500 - 4900	Orange	Ca II, Ca I, molecules
M	2000 - 3500	Orange - red	TiO, molecules, Ca I

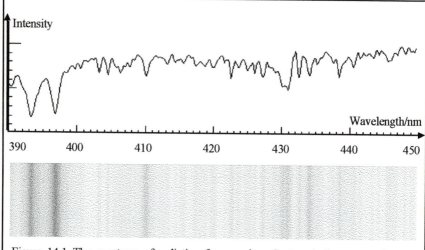

Figure 14.1 The spectrum of radiation from a class G star similar to our Sun. The two prominent dark bands at the short wavelength end of the spectrum are the K and H calcium lines. These are also prominent in galactic spectra and can be used to calculate galactic red-shifts.

14.3.2 The Hertzsprung-Russell Diagram

In 1914 the Danish astronomer Ejnar Hertzsprung and the American astronomer Henry Russell Norris plotted a graph of absolute magnitude against spectral class. For the large majority of stars here is a very clear correlation – a large absolute magnitude corresponds to a high temperature and a low absolute magnitude corresponds to a low temperature. This graph of luminosity versus spectral class is called the Hertzsprung-Russell diagram and this relationship defines the 'main sequence' of stars. Theory suggested that more massive stars would be hotter and more luminous, so the main sequence was also thought to represent a mass sequence (higher masses corresponding to stars at the top left of the diagram).

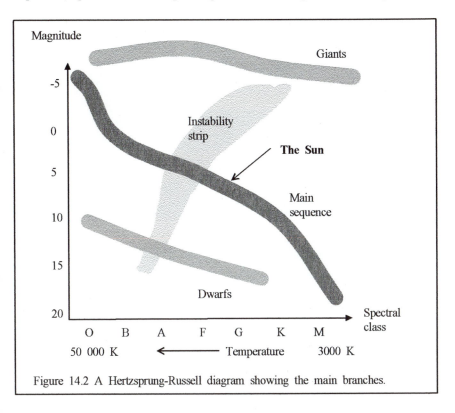

Figure 14.2 A Hertzsprung-Russell diagram showing the main branches.

A significant minority of stars do not lie in the main sequence. The first of these to be investigated were the red giant stars, such as Betelgeuse, which are relatively cool (hence red) and yet have a high luminosity. Eddington was convinced they must be much larger than red stars on the main sequence (hence 'giants') so he persuaded Albert Michelson to attempt to measure the diameter of Betelgeuse using an optical interferometer. Michelson set up a 6-metre interferometer across the disc of the 100-inch telescope at Mount Wilson and was able to confirm that the diameter of Betelgeuse was enormous, about equal to the diameter of the orbit of Mars around the Sun. At first the idea of a large cool star seemed to contradict the theory that mass and luminosity are linked. However, whilst this is true for main sequence stars burning hydrogen to helium in their cores, it is not the case for older stars that have exhausted their supplies of hydrogen and turned their cores into helium. In 1942 Mario Schönberg and Subrahmanyan Chandrasekhar showed that stars cannot remain stable burning hydrogen once their helium core exceeds about 10% of the star's mass. The core itself then begins to collapse and

Figure 14.3 Hubble space telescope images of stars. The top photograph is the first direct image of the disc of a star other than the Sun. It is of Alpha Orionis or Betelgeuse, a red supergiant. This photograph was taken in ultra-violet light using the Faint Object Camera. The lower image is the first direct look, in visible light, at alone neutron star. It was identified by its high temperature, small size and low luminosity. It is about 28 km in diameter.

Photo credit: NASA

hydrogen burning continues in a vastly expanded shell around the core. It is this expansion that blows the star up into a red giant. It also tells us that red giants are stars at the end of their lives. Their high luminosity implies that they spend far less time in this phase than on the main sequence.

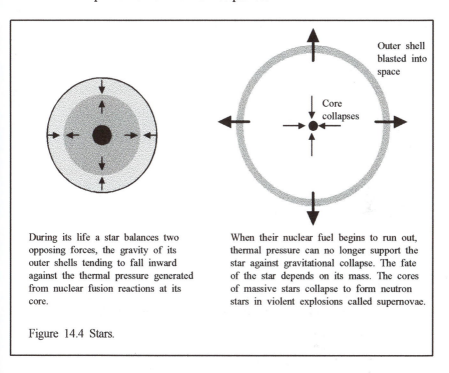

During its life a star balances two opposing forces, the gravity of its outer shells tending to fall inward against the thermal pressure generated from nuclear fusion reactions at its core.

When their nuclear fuel begins to run out, thermal pressure can no longer support the star against gravitational collapse. The fate of the star depends on its mass. The cores of massive stars collapse to form neutron stars in violent explosions called supernovae.

Figure 14.4 Stars.

There were also other anomalies – faint (low luminosity) stars with spectra characteristic of very hot stars. These 'white dwarfs' turned out to be the collapsed cores of old stars supported against further gravitational collapse by the electron degeneracy pressure (a consequence of Pauli's Exclusion Principle, which prevents two electrons from falling into the same quantum state in the same system). The theory of white dwarf stars was also an early example of the application of quantum theory to astrophysics. Eddington argued that such a massive dense star should cause a significant red shift to the radiation leaving its surface. This was confirmed in a series of careful measurements made by Adams in 1925, effectively testing general relativity and stellar theory in the same experiment. Chandrasekhar showed that cores with a mass greater than about 1.46 times the mass of the Sun (this is known as the Chandrasekhar mass) cannot even be supported by electron degeneracy pressure, so heavier cores must continue to collapse.

Walter Baade and Fritz Zwicky suggested that the core of a massive supernova might continue to collapse until all its electrons and positrons combine to form a smaller denser object called a neutron star supported by neutron degeneracy pressure, with a density of about 10^{18} kgm^{-3}. Such compact objects, a few kilometres across would not re-radiate significantly and were not directly observed until the 1960s when pulsars (rapidly rotating neutron stars that radiate energy in a 'beacon' that flashes across our field of view creating regular pulses) were discovered by Jocelyn Bell-Burnell and Martin Hewish. But even neutron degeneracy pressure has a limit and for cores above 2.5 solar masses there are no known physical forces that can prevent ultimate gravitational collapse to a singularity or black hole.

A star like our Sun will spend most of its life fusing hydrogen to helium on the main sequence. Toward the end of its life its helium core will collapse and the outer layers will expand beyond the orbit of the Earth as it becomes, briefly, a red giant star. Gradually the outer layers will drift away and the core will remain as a white dwarf, growing cooler and dimmer as time passes.

In 1983 the Nobel Prize for Physics was awarded jointly to Subramanyan Chanrasekhar *"for his theoretical studies of the physical processes of importance to the structure and evolution of the stars"* and to William Fowler *"for his theoretical and experimental studies of the nuclear reactions of importance in the formation of the chemical elements in the universe"*.

14.3.3 Supernovae

Novae are quite common. They announce their appearance as a sudden brightening of a star (hence 'novae' or 'new') which disappears again in a few weeks. They are caused by an explosive thermo-nuclear interaction between tightly bound members of a binary star system (e.g. as a white dwarf steals hydrogen from a larger companion sparking off nuclear explosions on the surface of the white dwarf).

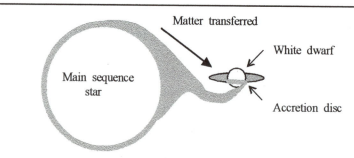

Figure 14.5 A nova (new star) flares up when matter transferred from an evolving main sequence star falls onto a white dwarf and initiates fusion in the surface layers. The light curve peaks in a few weeks and the whole process repeats with a period of the order of a hundred thousand years.

Supernovae are far more violent events involving an incredible increase in luminosity and only about half a dozen have been observed within our galaxy in the last thousand years. Three were identified by Japanese and Chinese astronomers in 1006, 1054 and 1181, another was noted by Tycho Brahe ('Tycho's star') in 1574 and another by Kepler in 1604. But it was not until the 1920s, when astronomers realised that the Milky Way is just one of an enormous number of galaxies, that the incredible power of a supernova explosion was fully appreciated. Their unprecedented brightness and (as we shall see) predictable spectra and power makes them ideal 'standard candles' with which to measure the size and rate of expansion of the universe. Unfortunately they are also quite rare and, since they are only bright for a short time, they are difficult to 'catch' and study. To see how this is possible, we must first consider how a massive star dies.

In 1926 Cambridge University Press published a book by Sir Arthur Eddington entitled *On the Internal Constitution of Stars*. In it he described a new type of star that has an average density way beyond that of ordinary matter. These stars are the white dwarfs. An example is Sirius B (companion to Sirius). Its mass is 1.05 times the mass of the Sun and yet its radius is only 5000 km (less than 1% of the radius of the Sun). This makes its density more than a million times greater than that of the Sun:

- average density of the Sun: 1.4×10^3 kgm^{-3}
- average density of the Earth: 5.5×10^3 kgm^{-3}
- average density of Sirius B: 4×10^9 kgm^{-3}.

Eddington was worried by these stars. If they are supported by thermal pressure then what will happen to them when they radiate energy away and begin to cool? Once the thermal pressure has gone the star must have some other means of support against gravitational collapse, but what could it be? Eddington did not accept the idea that stars may collapse indefinitely and could only suppose that a cold dead star must support itself against collapse in the same way that ordinary matter does, by the resistance of the atoms themselves to compression. But this implied that the star must expand to normal densities as it cools, and there was no way then to account for the energy source that would be needed to 're-inflate' it against gravity. This problem was called 'Eddington's Paradox' and seemed to contradict the measured density of stars like Sirius B. Another consequence of a star being small and dense is that light leaving its surface has a relatively large gravitational red-shift and this might be able to be measured in the spectra of light reaching the Earth. These measurements were carried out in 1925 by W.S. Adams at the Mount Wilson Observatory. He confirmed the existence of the red-shift, a result that prompted Eddington to remark that:

"Professor Adams has killed two birds with one stone, he has carried out a new test of Einstein's General Theory of Relativity and he has confirmed our suspicion that matter 2000 times denser than platinum is not only possible, but is actually present in the Universe."

(Quoted in *Black Holes and Time Warps*, Kip Thorne, Picador, 1994)

So what does prevent collapse? The solution was developed by R.H. Fowler and S. Chandrasekhar between 1926 and 1930. Fowler had published a paper entitled *On Dense Matter* in which he applied the laws of quantum theory (the Copenhagen version hot off the press!) to the behaviour of matter under extreme pressure. To understand the new idea we must take into account the wave-nature of the electrons. If atoms are compressed into a smaller and smaller volume it follows that the electrons within them are also constrained to fit that volume. This puts an upper limit on their allowed de Broglie wavelength. Now, the de Broglie relation (see below) shows that electron wavelength is inversely proportional to momentum, so electrons in a small box will have much greater momentum than electrons in a large box. They will also have greater kinetic energy. This can be interpreted in one of two equivalent ways.

- Compressing the atom must involve a pressure large enough to oppose the pressure (acting outwards) of the electrons bashing on the walls of their enclosures – like the pressure of a classical gas).

- Compressing the atom must pay for the extra kinetic energy of the confined electrons, so that an increasing force is needed to produce further compression. Once again there is an outward pressure from the confined electrons.

This pressure is called electron degeneracy pressure and is negligible at the density of ordinary matter but dominant at the pressures inside a white dwarf star. In fact it is this pressure that holds the star up against gravitational collapse.

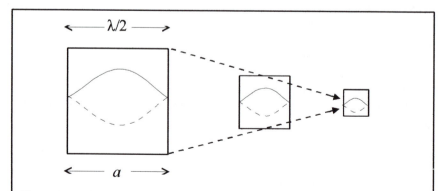

Figure 14.6 Electron degeneracy pressure. As electrons are confined in a smaller and smaller volume their de Broglie wavelength is constrained to fit the box. This results in greater momentum and a high pressure.

Maths Box: De Broglie and Degeneracy

Imagine an electron trapped in a box of side a.
If the electron waves must fall to zero at each boundary then the maximum wavelength is $2a$.

$$mv = \frac{h}{\lambda} = \frac{h}{2a}$$

$$KE = \frac{h^2}{4a^2 m}$$

So the electron momentum is inversely proportional to the box side and its kinetic energy is inversely proportional to the box side squared. As far as the degeneracy pressure itself is concerned, reducing the size of the box means the electrons have less distance to travel before hitting it and are also moving faster. In addition to this, the area they can strike is reduced. All of these effects act in the same direction – to increase pressure. It is quite clear that the degeneracy pressure will become increasingly important as the atom is squeezed into a smaller and smaller volume.

In 1930, Chandrasekhar took this analysis one step further. He realised that the compression would make the electrons go faster and faster and at some point relativistic corrections must be made. The electrons cannot continue to increase their velocity indefinitely. There is a finite limit, the speed of light. And there is a second effect, the mass of the moving electrons should increase. When both effects were incorporated Chandrasekhar had shown that relativity reduces the effectiveness of the degeneracy pressure compared to the classical model. Relativistic stars are easier to compress than classical stars. This has startling consequences: it means that a star of mass greater then 1.4 times the mass of the Sun will *not* be stopped from collapsing by electron degeneracy pressure. The value of 1.4 times the mass of the Sun is the largest mass possible for a white dwarf and is called the Chandrasekhar limit. Subsequent observations confirmed that none of the then known white dwarf had a mass above this limit.

Eddington was not impressed by this analysis, and he had a personal reason to oppose it. He realised that the Chandrasekhar limit implies that stars with mass above the Chandrasekhar limit cannot end their lives as white dwarfs. They must either throw off enough matter during their lifetime so that they drop below the limit or, and this was much more of a problem, continue to collapse beyond the white dwarf stage to even more extreme densities.

It is at this point that the story of dense matter links to supernovae. In the 1930s Fritz Zwicky began to classify supernovae. He divided them into two main classes:

- TYPE I: These are found mainly in the galactic halos and are associated with the explosion of low mass old stars. Their defining spectral characteristic is an absence of hydrogen lines. They increase in brightness by up to 20 magnitudes and explode outwards at high velocity, perhaps 3% of the speed of light. Later Type I was subdivided to Type Ia and Type Ib.

- TYPE II: These are less luminous than Type I supernovae, and increase in brightness by 16-17 magnitudes. Their spectra do contain hydrogen lines and their light curve has a plateau region. They expand outwards at about half the speed of a Type I supernova and they are confined mainly to the galactic disc and spiral arms. All of these observations suggest they are associated with the explosion of higher mass younger stars with an intact hydrogen envelope.

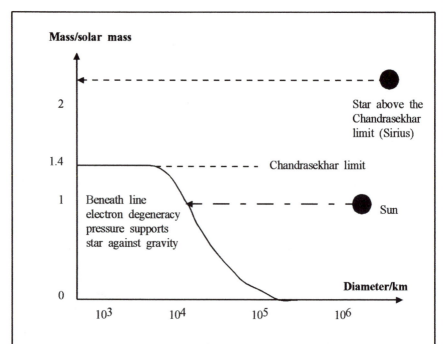

Figure 14.7 During most of a star's life it is supported against gravitational collapse by thermal pressure. Once its nuclear fuel begins to run out it shrinks. If its mass is less than the Chandrasekhar limit it will eventually support itself by electron degeneracy pressure. More massive stars cannot be supported in this way.

Zwicky knew that an enormous energy source must be found to explain the amazing power of supernovae. He proposed that, under extreme high pressure conditions, the core of a star might be compressed so hard that the electrons are literally forced into protons to create a gas of neutrons. This would mean that the core could suddenly collapse to a density comparable to that of nuclear matter and release an enormous amount of energy in the process. Very few astronomers took him seriously, and the details were not worked out until the 1960s, but Zwicky's hunch proved correct and the first neutron star was observed in 1968. This gives the following picture of the life and death of a heavy star (about 20 times the mass of the Sun):

Approx. 10^7 years – *nuclear burning of hydrogen to helium;*
Approx. 10^6 years – *nuclear burning of helium to carbon and oxygen;*
Approx. 10^5 years – *nuclear burning of carbon to Neon and magnesium;*
Approx. 20 years – *nuclear burning of oxygen to silicon and sulphur;*
Approx. 1 week – *nuclear burning of silicon and sulphur to iron.*

This leaves a supergiant layered star with an iron core of about 1.4 solar masses. The core is supported for awhile by electron degeneracy pressure as the core loses energy electrons attack the iron nuclei reversing the fusion process to reform lighter and lighter nuclei as the core begins a rapid collapse. Then,

Approx. 0.1 seconds – *collapse of core to form a neutron star – about 10^{46} J are radiated (more than 99% going to neutrinos). The origin of this energy (which is about 100 times as much as the Sun will radiate in its entire 10 billion-year lifetime!) is the gravitational potential energy of the collapsing matter. The inward collapse involves speeds of about 0.25 c. The core stops collapsing when it is supported by neutron degeneracy pressure at a diameter of 10-20 km.*

Shockwaves from the collapsed core bounce outwards blasting the outer layers of the star with neutrons and neutrinos, creating heavy nuclei and sending the debris out into pace to form the material out of which new stars (such as our Sun) may form.

This description of a Type II supernova was confirmed in many respects by the discovery of supernova 1987a in the Large Magellanic Cloud, a small companion galaxy to the Milky Way. One of the most dramatic feature of this discovery was the detection of neutrinos some 20 hours ahead of visual confirmation.

Type I supernovae are even more violent and their spectra, being devoid of hydrogen, point to the collapse of a white dwarf star. What could provoke such an event? Ordinary novae are thought to be caused by thermonuclear fusion reactions set off in hydrogen captured by a white dwarf from its companion in a binary system. What if the white dwarf were already poised on the Chandrasekhar limit so that extra matter could tip it over? This might cause a sudden collapse that ignites fusion reactions like a gigantic thermo-nuclear bomb. This is the characteristic Type Ia explosion. Since the mass of the object that collapses is fixed, so is its energy output. This makes it ideal to act as a standard candle.

Type Ib supernovae are probably formed in a similar way to Type IIs, but in stars that have somehow lost their hydrogen during their lifetime, perhaps because it has been stripped away by a companion star.

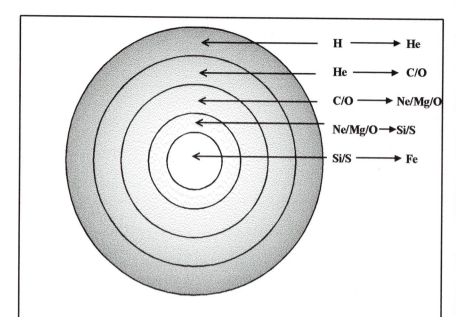

Figure 14.8 The final stage before a massive star explodes as a supernova. The star has developed a layered structure (not drawn to scale) and fusion reactions take place towards the inner edge of each layer. Beyond iron, fusion reactions are endothermic, so once an iron core forms no more energy is released from fusion reactions and it undergoes a catastrophic collapse sending shock waves out through the star blasting the outer layers off into space.

14.3.4 Using Supernovae

The energy output, light curve and spectrum of a Type Ia supernova is well known, so once they are observed they can be used to measure distances in the usual way (by comparing apparent magnitude against absolute magnitude). Their brightness increases to about magnitude −19 at its peak, making them almost as bright as an entire galaxy and thus visible at extreme distances. The difficulty is in seeing them in the first place, since their light curve peaks in a matter of just a few weeks and there are only 2 or 3 per millennium per galaxy. During the 1990s two groups managed to get around this problem. The Supernova Cosmology Project led by Saul Perlmutter at Berkeley was first to announce its results, but the High-z Supernova Search led by Alex Filippenko and Brian Schmidt of Australia's Mount Stromlo and Siding Springs Observatories soon confirmed their results.

To catch a supernova the sky is surveyed in about 50-100 separate regions, each containing about 1000 visible galaxies. This survey is carried out when the seeing is best, in the dark skies just after the new Moon. Three weeks later the same

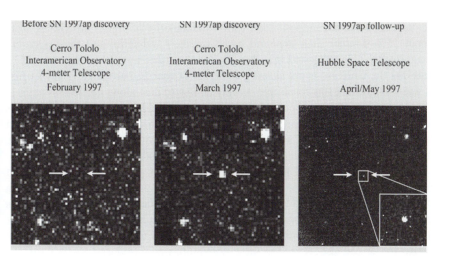

Before SN 1997ap discovery	SN 1997ap discovery	SN 1997ap follow-up
Cerro Tololo Interamerican Observatory 4-meter Telescope February 1997	Cerro Tololo Interamerican Observatory 4-meter Telescope March 1997	Hubble Space Telescope April/May 1997

Figure 14.9 Three images from observations made by the Supernova Cosmology Project of one of the most distant supernovae ever recorded. From the left: the first two images (from the Cerro Tololo Interamer Ican Observatory 4-metre telescope) show a region of sky just before and just after the appearance of a Type Ia supernova. The third image shows the same supernova as observed by the Hubble Space Telescope. This much sharper image allows a much better measurement of the apparent brightness and hence distance of the supernova. The brightness of these supernovae can be predicted from theory, so accurate measurements can be used to measure the deceleration or acceleration of the universe and so help predict its eventual fate.

Photo credits: S.Perlmutter et al, The Supernova Cosmology Project and NASA

patches of sky are re-surveyed and the two sets of images are carefully scanned to detect any significant differences. Any sudden increases in brightness may be supernovae. This information is then used to direct the giant Keck telescope or the HST onto the bright regions just before the next new Moon (again taking advantage of the dark skies). Keck or HST is used to measure the spectrum of the bright object to see if it is a Type Ia supernova. If so its red-shift is measured and its distance calculated. In January 1998 Perlmutter's group announced the discovery of a supernova with red-shift z = 0.83 at a distance of about 7 billion light years (light left this supernova when the universe was half its present age), then the most distant individual star to be measured. Each survey produces about a dozen supernovae and the number of high-z (large red-shift) supernovae measured so far (mid-1999) is approaching 100.

When the surveys started the idea was to use distant supernovae to calculate the recession velocities of their associated galaxies to work out the rate at which gravitational attraction is slowing down the expansion of the universe. This would help us to work out what the ultimate fate of the universe is likely to be. If the deceleration parameter is large then the universe will eventually stop expanding and recollapse, if it is small enough the universe will expand forever. However, the actual results did not support either idea – the majority of high-z supernovae were about 15% dimmer than expected based on their red-shifts and their apparent magnitudes. This suggested that they are much further away than they should be and that the Universe is not slowing down as it expands, but speeding up. At first it was thought that this result may simply be due to dust along our line of sight to the supernovae making them appear dimmer and redder (since the dust scatters short wavelengths more than long wavelengths). To check this, the results thought to be most affected by dust were discarded and the remainder analysed in isolation. They told the same story, and this was confirmed by both groups.

If this result is correct it has major implications for physics and cosmology. It means the universe must be open and will not recollapse. But it also resurrects a debate started back in 1915 by Einstein himself. When Einstein discovered his equations for the gravitational field (general relativity) he realised that they do not allow for a static stable universe, which he assumed to be the case, but lead to expansion or collapse. Rejecting these possibilities he introduced a new term, the cosmological constant, which acted like a long range gravitational repulsion and which might balance the tendency for gravitational collapse in the universe at large. After Hubble's discoveries Einstein described the cosmological constant as his biggest blunder, but he may have been a little premature in this. If the expansion of the universe is really accelerating then we will need to add something similar to the cosmological constant back into the equations. Interpreting this extra term will be very interesting. It is already being suggested that the energy source for the accelerating expansion may come from the vacuum itself and might involve new quantum effects.

Previously it was pointed out that the Universe (without a cosmological constant) will just stop expanding if it has a certain critical density giving it a density parameter $\Omega = 1$ and making the overall geometry of the Universe flat. So far observations have failed to find enough matter to reach this figure. However, if a cosmological constant Λ is introduced it may be possible for this to combine with the density parameter to make the universe expand forever and yet have overall flat space-time geometry. For this to be the case we need:

$$\Omega + \Lambda = 1$$

So far the evidence available puts the following limits on these parameters:

$$\Lambda > 0.3$$
$$\Omega < 0.7$$

These are consistent with the possibility that the Universe is actually flat.

Early in the 21st century two missions will make detailed measurements of the background radiation and attempt to pin down these parameters more precisely. The European Space Agency will launch the Planck Mission and NASA will launch the Microwave Anisotropy Probe. These, together with further measurements of distant, high-z supernovae, should reveal the ultimate fate of the universe.

14.3.5 Black Holes

If the core of a supernova is above about 2.5 solar masses neutron degeneracy pressure is insufficient to prevent gravitational collapse. The intense gravitational field close to the collapsing core eventually raises the escape velocity to the speed of light. Once this limit is reached nothing can escape. Although a thorough treatment of black holes really needs the full general theory of relativity, many important results come out of Newton's theory too. For example, it is simple to derive an expression for the radius at which the escape velocity equals the speed of light:

$$v_{esc} = \sqrt{\frac{2GM}{R}} \geq c \quad \text{from Newtonian gravitation theory}$$

$$R \leq \frac{2GM}{c^2}$$

The critical radius at which the escape velocity equals the speed of light is called the Schwarzschild radius. This depends on mass, so a larger mass black hole has a larger Schwarzschild radius, R_S. However, density is mass per unit volume, so a larger black hole can form at lower density and if the total mass is large enough a black hole could be formed from material at ordinary densities.

If you were to watch something (from a safe distance) collapse to form a black hole you would notice time running slower and slower as the objects you observe get closer and closer to the Schwarzschild radius. This is because of gravitational time dilation and would make time stop when the objects reach R_S. In one sense this implies that the inside of the black hole is beyond the end of time in the external universe!

14.3.6 Using Stars to Measure Distances

Once the mechanism by which stars shine was reasonably well understood it became possible to infer a star's luminosity and magnitude from its spectrum. Its estimated magnitude and apparent magnitude can then be used to work out its distance using:

$$m - M = 5\log\frac{d}{10}$$

$$d = 10^{\frac{m-M+5}{5}} \quad \text{(in parsecs)}$$

This method is called spectroscopic parallax and such methods are useful out to about 50 kpc. A modified version of this method can be applied to star clusters.

The stars within a particular cluster are all about the same age having formed together from the same gas cloud. If the apparent brightness of stars in the cluster is plotted against their surface temperatures (or spectral classes) on an H-R diagram they form a line parallel to the main sequence but some way below it (because their apparent magnitudes are less than their absolute magnitudes, assuming they are more than 10 pc away). The average vertical distance between the two lines of the HR diagram is the distance modulus $m - M$ for the cluster and can be used in the equation above to work out their distance. This method is called main sequence fitting and can be calibrated by comparing the results for distant clusters with those of nearby clusters (such as the Hyades) for which geometric methods can be used.

In 1912 Henrietta Leavitt at Harvard published a paper giving details of variable stars, called Cepheid variables, in the Magellanic Clouds. Although their absolute distance was not then known they were all at about the same distance from the Earth so their apparent magnitudes could be used to compare their absolute luminosities. She discovered that their periods of variation were related to their luminosity by a fixed rule. This provided a powerful new method for estimating distances. Once the period luminosity rule was calibrated using nearby variable stars whose distance can be measured by other means it could be used to predict their absolute magnitude and then to work out distance using the familiar equation. Since variable stars are usually very luminous they are also visible at great distances so distance measurements based on variable stars (particularly Cepheid variables) have become extremely important. Variable stars observed from the HST allow distance measurements out to about 40 Mpc.

Beyond this distance even a variable star cannot be resolved so a brighter 'standard candle' is required. One possibility is to use supernovae (see section 14.3.4), exploding stars that, for a short time, become more luminous than an entire galaxy (increasing by about 20 magnitudes). If it is assumed that supernovae of a similar type have a similar maximum luminosity then their apparent magnitude can be used in the same way as for known types of star to estimate their distance. This extends the distance scale to hundreds of megaparsec. Beyond that we use the red shifts of distant galaxies or quasars, an approach explained in more detail in the next chapter.

15

COSMOLOGY

15.1 GALAXIES IN SPACE AND TIME

"Cosmology used to be considered a pseudoscience and the preserve of physicists who might have done useful work in their earlier years, but who had gone mystic in their dotage. There were two reasons for this. The first was that there was an almost total absence of reliable observations. Indeed, until the 1920s about the only important cosmological observation was that the sky at night is dark. But people didn't appreciate the significance of this ... There is ... a second and more serious objection. Cosmology cannot predict anything about the universe unless it makes some assumption about the initial conditions. Without such an assumption, all one can say is that things are as they are now because they were as they were at an earlier stage."

(Stephen Hawking, *The Nature of Space and Time*, Princeton, 1995)

15.1.1 Beyond the Milky Way

When you look up at the night sky on a clear night you should be able to make out a band of stars crossing your field of view. This is the Milky Way, a view through the disc of our own elliptical galaxy from a point in one of the spiral arms. There are about 10^{11} stars in our galaxy and our Sun is just one of them. We now know that there is a similar number of galaxies spread through the universe, but this is a twentieth century discovery based on the work of one of the most important modern astronomers, Edwin Hubble.

In the 1920s Hubble used the new 100-inch reflecting telescope at Mount Wilson to study the structure of some nearby nebulae (star clusters). The resolving power of this telescope was good enough to pick out individual stars in the Andromeda spiral nebula. Among these were some Cepheids, so Hubble used Shapley's period-luminosity law to estimate their luminosities and hence their magnitudes. From this he could calculate distances and discovered that they were much farther away than the majority of stars in the Milky Way. He concluded that Andromeda

Figure 15.1 Galaxy M51. Seen from a distance our own galaxy might look something like this.
                                                                                    Photo credit: NASA

and other similar objects are probably galaxies in their own right rather than nebulae within the Milky Way.

Hubble identified a large number of galaxies and suggested a system of classification that is usually referred to as the 'tuning fork diagram'. In Hubble's day this was regarded as a picture of the way galaxies evolve. Now we know this is not the case, but it is still a convenient classification scheme. The main classes are spiral, elliptical and irregular, but many other sub-divisions are possible and active galaxies and quasars must be added to the list.

While Hubble refined his measurements of galactic distances, one of his colleagues, Vesto Slipher was busy measuring the red-shifts of galactic spectra. When the source of radiation moves relative to an observer the waves received by

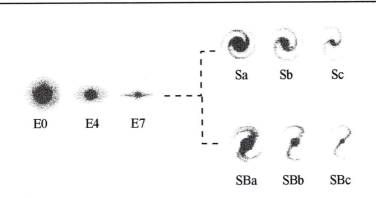

Figure 15.2 Hubble's 'tuning fork diagram'. Hubble divided the galaxies into three main classes, elliptical (E), spiral (S) and spiral barred (SB). He thought the diagram above showed the sequence of galactic evolution. This idea has now been abandoned, but the classification scheme is still in use.

the observer are shifted to longer or shorter wavelengths. If the source approaches, the wavelength is shortened as the waves are 'squashed' between the source and receiver, this is called a 'blue-shift'. If the source is moving away, the wavelength is 'stretched' or 'red-shifted'. This is similar to the way that the sound of a moving vehicle seems to rise and then fall as it approaches, passes and then moves away from you and was analysed mathematically by Christian Doppler in the first half of the nineteenth century.

Maths Box: Red-shift

Shift in wavelength = (change in wavelength) / (unshifted wavelength)

or
$$z = \frac{\lambda' - \lambda}{\lambda} = \frac{\Delta\lambda}{\lambda}$$

A positive shift means the wavelength has increased – this is called a red-shift (since red light is at the long wavelength of the visible spectrum).

For velocities small compared to c this gives a simple link between red-shifts and recession velocities:

red-shift $\propto v$
$$z = \frac{\Delta\lambda}{\lambda} = \frac{v}{c}$$

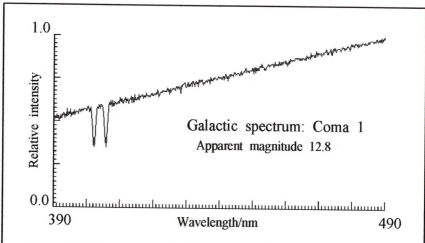

Figure 15.3 The spectrum of radiation from Coma 1. Notice the two prominent absorption bands. These are the H and K lines from ionized calcium. For a source at rest they have values 396.8 nm and 393.4 nm. If the galaxy is moving they will be shifted to longer or shorter wavelengths (red or blue shifts). The red shift of Coma 1 can be measured from the graph.

Slipher expected galaxies to have random motions like the proper motions of stars in our own galaxies, so he expected some of the galactic spectra to be red-shifted and some blue-shifted. He was surprised to discover that this is not in fact the case. By 1923 36 of the 41 galaxies he had surveyed had red-shifts. The only ones with blue-shifts were all nearby. This implied that the majority of galaxies are moving away from us. Howard Robertson combined Hubble's distances with Slipher's red-shifts and realised that the speed of recession increases with distance. Hubble refined the values for red-shift and distance in the 1930s and proposed the 'Hubble Law' which has become a cornerstone of cosmology:

Maths Box: Hubble's Law

Galactic red-shift is directly proportional to galactic distance
or
recession velocity is directly proportional to distance:

$$z = v/c \qquad (v \ll c)$$
$$z \propto d$$

So: $v \propto d$

giving: $v = H_0 d$ where H_0 is the Hubble constant.

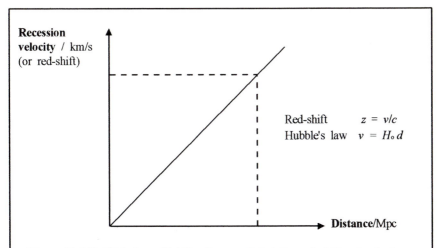

Figure 15.4 Hubble's Law. Hubble discovered that the red shifts of galaxies are directly proportional to their distances from Earth. This led to Hubble's Law and the discovery that the universe is expanding.

15.1.2 The Expanding Universe

"The discovery that the universe is expanding was one of the great intellectual revolutions of the twentieth century. With hindsight, it is easy to wonder why no one had thought of it before. Newton, and others, should have realised that a static universe would soon start to collapse under the influence of gravity. But suppose instead that the universe is expanding. If it was expanding fairly slowly, the force of gravity would cause it eventually to stop expanding and then start contracting. However, if it was expanding at more than a certain critical rate, gravity would never be strong enough to stop it, and the universe would continue to expand forever."

(Stephen Hawking, *A Brief History of Time*, Bantam, 1988)

When Einstein formulated general relativity he realised that it suffered from a similar problem to Newton's theory of gravitation. If the universe contains a finite amount of matter and is static it must collapse. This seemed unrealistic to Einstein (as it had to Newton) so in 1917 he added an extra term to his equations which he called the cosmological constant. This acts like a long distance repulsion and can be adjusted to ensure a static universe. Ironically, and also in 1917, Willem de Sitter demonstrated that a universe (albeit one containing no matter!) in which space-time continually expands is a possible solution to Einstein's equations. If particles of matter are sprinkled into the de Sitter universe they act as if they are moving away from one another and for a while Slipher's results were called the 'de Sitter effect'. In the 1920s the Russian astrophysicist and mathematician Alexander Friedman and the Belgian cosmologist, Georges Lemaître derived a variety of different possible universes from Einstein's equations. Some expanded forever, some expanded and recollapsed, some stopped expanding after an infinite time, none needed the cosmological constant.

Einstein came to regard the cosmological constant as 'the greatest mistake' he ever made. If he had followed his theory he could well have predicted the expanding universe before Hubble's discovery. As it was it took a long time before the idea that the universe was in a state of expansion really caught on. The overall expansion follows from Hubble's Law and the assumption that the universe is more or less homogeneous. This leads to an interesting interpretation. Whereas the standard Doppler effect arises from the relative motions of source and observers within a fixed reference frame of absolute space, this reference frame does not exist in general relativity. The relativistic interpretation of Hubble's law is that space itself is expanding so that the distance between galaxies increases with time because the space between them 'stretches'. This leads to a number of visual analogies, the most common of which are related to expanding rubber sheets or balloons. Imagine a very large rubber sheet whose dimensions are expanding in all directions at a uniform rate. Any two marked points on the surface will move away from one another as the sheet expands. Furthermore, the further apart the two points happen to be, the more rapidly they recede from one another. This simple model reproduces Hubble's Law, but it can be misleading. Imagine the galaxies are represented by small circles drawn onto the rubber sheet. These too would expand with the sheet, so would any observers and their measuring instruments within these galaxies. This would make the expansion unobservable! A more realistic model would have galaxies represented by coins riding on the surface of the sheet. Their separation would still increase, but the strong internal forces in the coin material will prevent them from expanding. Our model of the expanding universe is more like this, galaxies, stars, planets and people are in bound states held together by strong internal forces, which do not yield to cosmic expansion.

This model of the expanding universe assumes that the universe looks the same from any point, so there is no centre. The expansion itself is not an expansion into pre-existing empty space, it is the expansion of space itself. In this respect there is no 'outside' and it is possible that the universe is finite and unbounded. The overall geometry of the universe will then depend on the mass density within it. If the density is low the universe will expand forever, if it is high the universe will stop expanding and collapse. The cosmological density parameter is given the symbol Ω and is equal to the ratio of the actual density to the critical density. Its value determines the nature of the universe:

- $\Omega > 1$ – the universe will stop expanding and collapse, its geometry is spherical and it is a closed universe;
- $\Omega = 1$ – the universe has critical density and would stop expanding after an infinite time, its geometry is flat;

- $\Omega < 1$ – the universe continues to expand forever, its geometry is hyperbolic and it is an open universe.

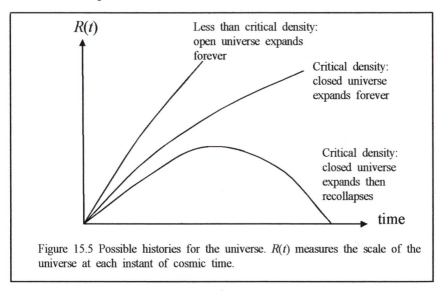

Figure 15.5 Possible histories for the universe. $R(t)$ measures the scale of the universe at each instant of cosmic time.

All surveys of visible matter lead to values for Ω significantly less than 1. However, there are good reasons to believe that the universe contains a considerable amount of dark matter, although the nature of this dark matter is not yet known. We shall return to this idea of 'missing mass' later in the chapter.

Maths Box: The Critical Density

The critical density ρ_c can be derived from Newton's equations (which lead, surprisingly, to many of the same results as in general relativity). Imagine the universe is filled with a uniform dust of matter at density ρ and is in a state of expansion. The critical density will be that for which the expansion eventually stops. Imagine a spherical shell of matter of radius x as it expands outwards. The uniformly distributed matter at radii $r > x$ will have no effect on it, but it will be pulled back by all matter within a sphere of radius x. it will escape (i.e. expand forever) if its total energy is greater than 0.

Mass of shell of thickness δx: $\quad \delta m = 4\pi x^2 \rho \delta x$

Mass of inner sphere: $\quad M = \dfrac{4}{3}\pi x^3 \rho$

Shell velocity relative to centre: $\quad v = H_0 x$

Kinetic energy of shell: $\quad KE = 2\pi x^4 \rho {H_0}^2 \delta x$

GPE of shell:
$$GPE = -\frac{GM\delta m}{x} = -\frac{4G\pi x^3 \rho \delta m}{3x}$$

Total energy:
$$TE = GPE + KE \geq 0$$

Therefore to escape:
$$2\pi x^4 \rho H_0^2 \delta x \geq \frac{16 G\pi^2 x^4 \rho^2 \delta x}{3}$$

giving
$$\rho \leq \frac{3H_0^2}{8\pi}$$

The critical density is therefore:
$$\rho_c = \frac{3H_0^2}{8\pi G}$$

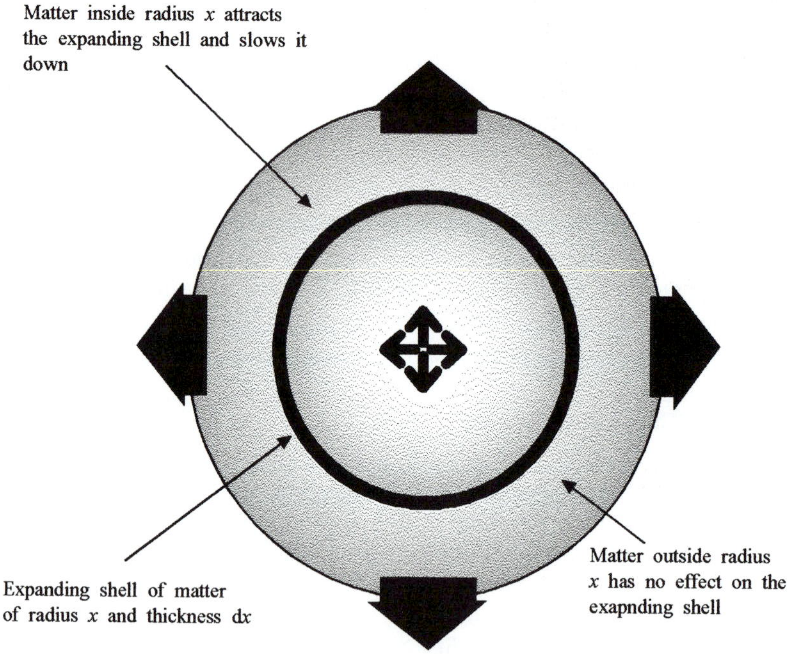

Matter inside radius x attracts the expanding shell and slows it down

Expanding shell of matter of radius x and thickness dx

Matter outside radius x has no effect on the exapnding shell

Figure 15.6 The expansion of a dust-filled universe. If the density of matter in this universe is less than a critical value, the universe will expand forever.

Another way to deal with the evolution and history of the universe is in terms of a deceleration parameter q. The net effect of mass in the universe should be to slow the rate of expansion in all cases, so the value of q will be related to the mass density. The greater the mass density the larger the deceleration parameter. The deceleration parameter is defined in terms of the global geometry of the universe by defining a scale factor R which changes with time. If the expansion rate can be measured at different times then an estimate of q can be made and related back to the critical density. It should be possible to do this because observations of very distant galaxies see them as they were when the radiation left them and this might be billions of years in the past. In 1998 a series of observations using supernovae in distant galaxies to gauge their distance seemed to show that, rather than decelerating the expansion of the universe is continuing to accelerate (see Chapter 14)! This raises the possibility of repulsive gravitational forces once again, but distance measurements for the farthest galaxies are notoriously unreliable so this question is unlikely to be settled until the twenty-first century.

We have already used the analogy of an expanding rubber sheet to describe the expanding universe. This can be taken further. Imagine the universe is filled with a grid from which we can work out the co-ordinates of any object. We can also use these co-ordinates to calculate the co-ordinate separation l between any pair of objects, perhaps between the Earth and a distant galaxy. Expansion will not affect the co-ordinates or co-ordinate separations of objects in the universe. However, their actual separation d, as measured by us, will increase with time. The link between co-ordinate distance and measured distance is determined by a scaling factor $R(t)$. This scale factor changes with time. We seem to live in a universe in which the scale factor is increasing uniformly and homogeneously. This description emphasises the idea that it is space itself that is stretching rather than galaxies moving through some fixed absolute space. It also gives a 'feel' for the central idea of general relativity, that the presence of matter in the universe distorts space-time geometry. The maths box below shows how the idea of a scale factor can be related to the Hubble Law.

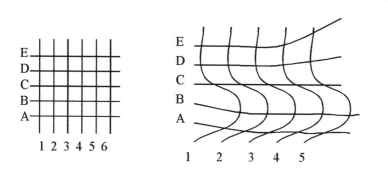

Figure 15.7 The expansion and curvature of space-time can be compared to the expansion and distortion of a large (4D) sheet of graph paper. If the co-ordinates of points are embedded in the sheet then the co-ordinate separation of two points never changes. However, the actual distance measured through the universe does increase. Uniform expansion can be described by a scale factor $R(t)$ that was zero at the Big Bang and has been increasing ever since.

Maths Box: The Scale of the Universe

If the scaling factor is $R(t)$ and the co-ordinate distance between two points is l then the measured distance between them is given by:

$$d = R(t)l$$

The recession velocity is just the rate of change of d:

$$v = \frac{d}{dt}(d) = l\frac{dR(t)}{dt}$$

If l is now replaced using the first equation we get:

$$v = \left[\frac{dR(t)/dt}{R(t)}\right]d$$

The bracketed expression is the Hubble constant, which will in general be time dependent:

$$H_0 = \left[\frac{dR(t)/dt}{R(t)}\right]$$

Maths Box: The Radiation and Matter Eras

As the scale of the universe $R(t)$ increased, the total volume increased too. This reduced the density of matter and energy in the universe. However, these two things do not change in the same way.

Volume available $\propto R(t)^3$ mass density $\propto 1/R(t)^3$

Radiation however gets red-shifted and since photon energy is given by:

$$E = hf = \frac{hc}{\lambda} \quad \text{with } \lambda \propto R(t)$$

This makes the energy density of radiation more sensitive to the scale factor:

Radiation energy density $\propto 1/R(t)^4$ (fourth power from Stefan's law)

In the early universe radiation dominated – this is referred to as the radiation era. After about 100 000 years matter dominated and the energy density of the universe today is totally dominated by matter.

15.2 THE SPACE-TIME OF THE UNIVERSE

15.2.1 The Geometry of the Universe

We have already remarked that Einstein reinterpreted gravity as a distortion of space-time geometry. Particles then follow geodesics through this curved space-time. The equations of general relativity relate space-time curvature to the distribution of mass energy and momentum (which all distort geometry). To solve the equations you have to put in the initial mass/energy/momentum distribution and then solve them to find the resulting space-time geometry. If the density of mass/energy/momentum is low then the equations reduce to those of Newtonian gravitation. The first exact solution was by Karl Schwarzschild who solved the equations for a simple uniform spherical mass. The Schwarzschild solution led to the idea of the Schwarzschild radius inside which a collapsing star will form a black hole. A rough rule of thumb to decide whether general relativistic corrections to Newton's equations must be taken into account is given by the following relation:

$$\text{Significance of general relativistic corrections } \Gamma \sim \left(\frac{v_{esc}}{c}\right)^2$$

For a spherical mass like the Sun or Earth:

$$v_{esc} = \sqrt{\frac{2GM}{r}}$$

$$\Gamma = \frac{2GM}{rc^2}$$

At the surface of the Earth $\Gamma \sim 10^{-9}$. At the surface of the Sun $\Gamma \sim 10^{-6}$.

In both cases Newtonian theory works brilliantly (the solar deflection of starlight predicted by Einstein and detected by Eddington in 1919 was only 1.76 arcseconds). However, the situation is different near the surface of a collapsed neutron star where typically $\Gamma \sim 0.5$, showing that general relativistic effects will be very important. Even white dwarfs (as we have seen) produce measurable gravitational red-shifts. Their values of Γ are more like 10^{-3}.

So far we have discussed the effect of individual masses on their local space-time geometry, but we can also discuss the overall geometry of the universe. The simplest way to approach this big question is to assume that matter is uniformly distributed as a fine dust throughout the entire universe and then work out the consequences. However, before we do this we should step back for a moment and review some ideas put forward by mathematicians long before Einstein created general relativity.

Our conventional ideas about geometry derive from Euclid in about 300 BC. He wrote the 'Elements', a treatise on geometry (it also covers some other aspects of mathematics, particularly number theory) in 13 books. It was partly his own discoveries and partly a summary of ancient mathematical ideas brought together in a systematic way that had a major influence on the development of Western culture (well beyond mathematics). Euclid's geometry was set out in an axiomatic structure so that all the theorems derive from a set of axioms which are taken to be self-evident. For nearly two millennia Euclidean geometry went unchallenged, but in the nineteenth century a number of mathematicians began to question its foundations. In an axiomatic system theorems derive their truth from the truth of the axioms on which they are based and one axiom in particular was being questioned. It is called the axiom of the parallels and states:

- *There is only one straight line that passes through a point and is parallel to another straight line.*

In the middle of the nineteenth century a Russian mathematician, Nicolai Lobachevski and a Hungarian mathematician, János Bolyai proved that non-Euclidean geometries are possible and are perfectly self-consistent. They replaced the axiom about the parallels with one that allows an infinite number of lines to pass through the point and all to be parallel to the original (in the sense that, if they are extended indefinitely, they never cross it). Later a German mathematician, Bernhard Riemann derived another form for non-Euclidean geometry in which there are *no* parallel lines (i.e. all extended lines intersect). The properties of Euclidean and non-Euclidean geometries can be illustrated by two-dimensional surfaces.

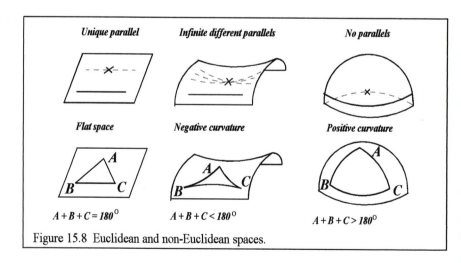

Figure 15.8 Euclidean and non-Euclidean spaces.

- Euclidean: a flat sheet of paper.
- Bolyai and Lobachevski: the surface of a saddle – 'negative curvature'.
- Riemann: the surface of a sphere – 'positive curvature'.

Now it might be objected that the last two surfaces are actually curved in a third spatial dimension and so are not really examples of two-dimensional geometry. However, every point on each surface can be uniquely defined by just two co-ordinates (such as latitude and longitude on Earth or polar co-ordinates) so the relation between lines and points within the surface is entirely contained in a two dimensional co-ordinate system. The third dimension is there (in our examples) but is not essential to the argument (or to the geometry). Non-Euclidean 2D surfaces could exist self-consistently even if no higher dimensions exist. This idea of *intrinsic* geometry or curvature is important in general relativity. The curvature of four-dimensional space-time does not necessarily imply the existence of a fifth dimension into which it curves.

This raises the important question of what kind of geometry best describes the universe. Einstein's attempt to create a static universe by introducing the cosmological constant forced the universe to have positive curvature and to be closed, so that space is finite and unbounded like the two-dimensional surface of a sphere. The time axis, however, escaped this curvature and ran on eternally, so that there would be no cosmic recurrence. The closed spherical space curves around the linear time axis to form a universe whose best 2D space-time representation would be a cylinder. Willem de Sitter, a Dutch mathematician produced another static model based on the cosmological constant. In de Sitter's universe both time and space are curved to form a four-dimensional spherical universe in which recurrence is inevitable.

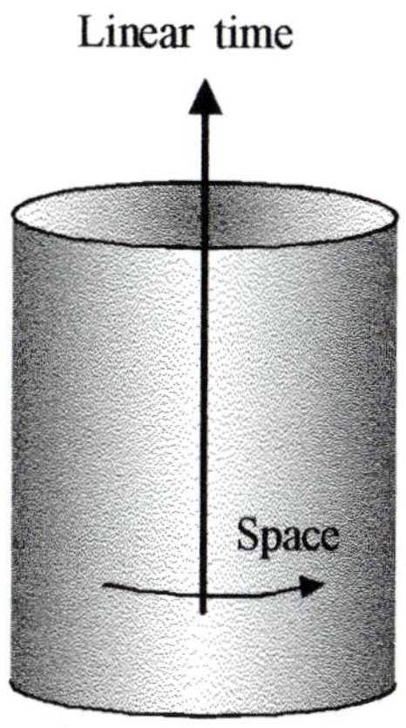

The discovery of Hubble's Law showed that these static models do not represent the universe in which we live. Before this the brilliant Russian mathematician Alexander Friedman had discovered an error in Einstein's calculations and derived two dynamic models for the universe without introducing a cosmological constant, one that expands forever and one that

Figure 15.9 A two-dimensional picture of Einstein's static eternal spherical universe. Space is closed but the time axis is infinitely long.

expands and then contracts, possibly repeating this cycle over and over again. Georges Lemaître also tackled Einstein's equations and came to the conclusion that the universe began in a highly compressed state and expanded outwards increasing its scale as time passed. He called the original state the 'primeval atom'. We now refer to this as the 'Big Bang'.

15.2.2 Cosmic Background Radiation

In 1948 George Gamow analysed the state of the universe immediately after the Big Bang and came to the conclusion that, for the first few hundred thousand years, the universe was opaque. Most of the matter and antimatter created in the Big Bang would have been annihilated, creating intense gamma radiation that would continually ionise any hydrogen atoms formed when protons captured electrons. This continual absorption and re-emission brought the radiation into thermal equilibrium with matter as the universe continued to expand and cool. Eventually, perhaps one million years after the Big Bang (only 0.01% of the estimated present age of the universe), the universe had cooled sufficiently that the photons no longer had enough energy to ionise atoms, so matter and radiation decoupled. Through the billions of years that followed the universe continued to expand and this thermal radiation got 'stretched' or 'red-shifted'. Gamow calculated that it should now have a spectrum characteristic of the present temperature of the universe, about five degrees above absolute zero. The black-body spectrum for a 5K source has a peak in the microwave region of the electromagnetic spectrum, so Gamow's prediction was that there should be a ubiquitous cosmic microwave background radiation.

In 1964 Arno Penzias and Robert Wilson were working at the Bell Laboratories in Holmdel New Jersey using a 6 m directional antenna designed for satellite communications. They were disturbed to find a consistent background noise that seemed to come from all directions. Further analysis showed that it had a black-body spectrum characteristic of a source temperature of about 3.5 K. Robert Dicke at Princeton realised the importance of their discovery and refined Gamow's calculations to show that the background radiation from the Big Bang should indeed have a temperature of around 3 K. In 1989 the Cosmic background Explorer satellite made very accurate measurements of this spectrum using the Far Infra-Red Spectrophotometer and determined that the temperature of the cosmic background radiation is 2.725 ± 0.01 K and fits a black-body almost perfectly. It also showed that (once Doppler effects due to our local motion are corrected for) the radiation is almost perfectly uniform. The 'almost' is important here. Detailed analysis of the COBE data suggests there are slight inhomogeneities in the background radiation corresponding to temperature variations of a few tens of micro-kelvin across substantial areas of the sky. This is a very important result because the background radiation gives us a kind of image of the universe as it was about a million years after the Big Bang. If it was perfectly uniform then it is hard to account for the non-uniformities in the distributions of galaxies, clusters of galaxies and super-clusters that we see now. On the large scale the universe seems to have strings of matter connected by long filamentary structures and separated by enormous voids. The 'bumps' in the uniformity of the background radiation are the origin of these structures. Arno Penzias and Robert Wilson shared one half of the 1978 Nobel Prize for Physics *"for their discovery of the cosmic microwave background radiation"*.

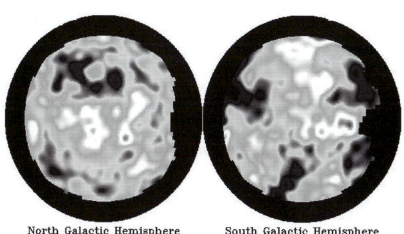

North Galactic Hemisphere   South Galactic Hemisphere

Figure 15.10 These are the oldest large-scale structures ever observed. They show the intensity distribution of the cosmic background radiation. Dark patches are above average, light patches below average. These 'wrinkles' are thought to correspond to the density variations in the early universe that eventually led to galaxy formation. The data was gathered from a 4 year microwave sky survey carried out using the COBE satellite.

Image credit: NASA

The cosmic background radiation and galactic red-shifts are both consistent with the Big Bang model, but they are not the only evidence. One of the other pillars on which Big Bang theory rests concerns the abundance of light nuclei in the universe.

Helium is not easy to observe because it has high excitation energy and so is only obvious in the spectra of hot stars. However, by the 1960s it was quite well established that helium-4 accounted for about 25% of all baryonic matter (i.e. matter made out of protons and neutrons). This was in agreement with calculations based on nucleosynthesis following the Big Bang which also predicted 25% helium-4. The calculations also showed that this ratio was pretty well independent of the present density of baryonic matter, which was a pity because it prevented an estimate of the present matter density which could then help answer questions about the overall geometry of the universe. Fortunately this is not the case for some other light nuclei created in primordial nucleosynthesis. Deuterium-2, helium-3 and lithium-7 abundances are all sensitive to the density of baryonic matter. In the 1970s measurements of deuterium abundances were used to put an upper limit on the baryonic density of the universe of about 1.5×10^{-28} kgm^{-3}. This is way below the critical density (by a factor of over 100). If baryonic matter dominates the universe then we must live in an open universe with negative curvature that will continue to expand forever. But there are good reasons for suspecting that a lot of mass may not be baryonic and may not be 'visible' to our telescopes because it does not radiate. This missing mass or dark matter has been a subject of great discussion in the late twentieth century, with all kinds of exotic and not so exotic candidates being proposed.

Nucleosynthesis in stars, operating for about 10 billion years, has converted about 2% of their hydrogen to helium – a small fraction of the 25% helium abundance in the universe at large. Most of the helium was created in the *first few minutes* after the Big Bang when the entire universe was similar to one great hydrogen bomb. The raw materials were protons and neutrons in a ratio of about 10 to 2 (more protons because neutrons, being slightly more massive, are unstable) and these combine readily to form deuterons. But deuterons are also easily broken apart by the intense radiation filling the universe. About 100 seconds after the Big Bang the temperature is low enough for deuterons to survive and the ratio of protons to neutrons has risen to about 14:2. For every 12 nucleons there are 2 deuterons and 12 protons. However, the temperature is still high enough to fuse deuterons to form helium nuclei and by about 200 seconds the vast majority of deuterons have disappeared, forming the 25% helium we see now. Nonetheless a small fraction of deuterons survive and this fraction is very sensitive to the density of matter in the universe when it is hot enough for fusion of deuterons to take place. If the density of matter is high the probability of deuteron collisions is also high and so more is converted to helium. If it is low then the probability is smaller and more survives. A similar argument holds for the abundance of helium-3 and lithium-7 which are also created in small amounts at this time. The observed abundances of these nuclides are in good agreement with standard Big Bang models.

15.2.3 The Hubble Constant and the Age of the Universe

It is important to realise that the Hubble 'constant' is not constant with time. It has already been mentioned that (notwithstanding very recent observations) the gravitational effect of matter in the universe should cause the expansion to slow down so that H_0 falls with time. However, a simple estimate of the age of the universe can be made by assuming that the present expansion rate is fixed and using the inverse of the Hubble constant as a time scale.

$$v = H_0 d$$

If these galaxies had been separating at the same speed since the Big Bang then:

$$T_H = \frac{d}{v} = \frac{1}{H_0} \quad \text{(beware units!)}$$

However, this still requires us to measure H_0, and our ability to do so depends critically on the accuracy of the cosmic distance scale. Hubble's original distance measurements led to a Hubble period of about 1.8 billion years, less than half the age of the Earth's crust (calculated from radioactive dating of rocks)! In the 1950s Walter Baade corrected the distance scale and since that time estimates have usually fallen in the range 80 to 40 $kms^{-1} Mpc^{-1}$ leading to Hubble periods in the range 10 to 20 billion years. The lower values (favoured by recent measurements) cause some problems because they are shorter than the calculated ages of the oldest stars, but few cosmologists think this will lead to the downfall of the Big Bang theory. Most assume that a combination of distance corrections and refined calculations of how stars work will reconcile the values. The Hubble period gives an idea of the age of the universe, but the actual age depends on the deceleration parameter, and that varies from model to model. At the present time there is no clear value for this parameter, so most discussions use the Hubble period.

15.3 INFLATION

15.3.1 The Very Early Universe

There are two major problems with the simple Big Bang/expanding universe model. One concerns the overall geometry of the universe and the other concerns its homogeneity.

Present estimates of the density parameter Ω place it below 1 but within striking distance of that value. This is quite remarkable, because all models in which the overall geometry is not flat predict that Ω will change with time. If the universe has positive curvature (a closed universe) then Ω starts above 1 and increases rapidly as the universe expands. If the universe has negative curvature (open universe) then Ω starts below 1 and falls rapidly toward zero as the universe expands. Observational astronomy places the value of Ω around 0.25, which is quite close to 1. If the universe is run backwards from now toward the Big Bang this would imply that Ω was *extremely* close to 1 in the very early universe (closer than 1 part in 10^{50}), something that theorists find hard to accept as just a coincidence. Many would say that Ω *is* 1 and the shortfall in observed mass is simply because the 'missing mass' is cold dark and/or exotic. This would imply that the universe as a whole is flat. Why?

The other problem concerns the microwave background radiation. COBE showed that this is remarkably uniform from all directions but does show tiny fluctuations in intensity and temperature that are just big enough to account for galaxy formation. This is fine, but some of these variations have an angular width of several degrees, showing that the corresponding parts of the early universe must have been in some kind of thermal contact soon after the Big Bang (in order to reach thermal equilibrium). However, for this to be possible they must have been close enough together for light emitted from one part to have reached the other before they expanded 'out of reach'. This problem would still remain even if the background radiation were perfectly uniform – that would imply that the whole universe was in thermal equilibrium when the radiation decoupled from matter, but the original expanding universe scenario does not allow this. Any particular point in the universe would only be causally connected to a part of the universe, not all of it, so it is hard to see how a universal equilibrium temperature could come about. At any moment after the Big Bang there is a definite 'horizon distance' equal to the maximum distance light could have travelled since the Big bang. Any regions separated by a distance greater than this cannot be in causal contact. The problem with the conventional expansion theory is that the horizon distance has been considerably less than the size of the universe for most of its history. This makes it very hard to explain the apparent large-scale homogeneity of the universe.

In the early 1980s Alan Guth proposed a new theory about how the very early universe evolved. This idea was developed by Andrei Linde, Andreas Albrecht and Paul Steinhardt. The theory proposes that there was an early explosive inflation prior to conventional expansion, and if true it would mean that the very early universe was much smaller than previously thought, in fact much smaller than its own horizon distance. This would allow all parts of the early universe to be in causal contact and explains how the observable universe can be so homogeneous now – it is just a small part embedded in a much larger region that inflated incredibly rapidly following the initial Big Bang. This solves the problem of large-scale uniformity and also solves the flatness problem. In inflation theory small regions increase in size by a very large factor, perhaps 10^{50} times in a time of order 10^{-32} s. Imagine small section on the surface of a balloon as it inflates – the curvature of the surface diminishes as the balloon grows in size. If it were expanded 10^{50} times the curvature would be indistinguishable from a flat surface. The observable universe represents a tiny section of a much larger universe that inflated in an analogous way, so the value of Ω was necessarily driven toward unity.

Inflation offers a way to resolve the problems of homogeneity and flatness, but what drives inflation? The mechanism is linked to the grand unified theories that link all the fundamental forces together at high energies, and leads to the intriguing possibility that the entire universe came into being from absolutely nothing. As Guth described it, the universe itself may be the 'ultimate free lunch'.

15.3.2 Broken Symmetry and the Higgs Mechanism

Think of a liquid on the point of freezing to form a crystalline solid as it cools. Before it freezes the liquid possesses rotational symmetry – it looks the same in all directions. Once it has frozen particular regions form crystalline structures with properties that depend on direction – there are preferred axes and the rotational symmetry has been broken. Different regions within the liquid may crystallise in different orientations and discontinuities will form along the boundaries of these microcrystals. Furthermore, the phase change releases energy (the latent heat of crystallisation). This is an example of a process in which symmetry is spontaneously broken as a system cools. The model of inflation describes the universe as a whole in an analogous way.

The very early universe, immediately after the Big Bang, was symmetric and homogenoues. In the extreme conditions of high temperature the forces of nature were not distinguished from one another but appeared as aspects of the same superforce, as predicted by some present day grand unified theories. However, as the universe expanded it cooled and the forces separated out to form electromagnetism, weak and strong nuclear forces and gravitation as we see them today. This transition was like a phase change of the entire universe, a spontaneous symmetry breaking which released an enormous amount of energy. It is this energy that condensed to form the particles of matter and antimatter in the early universe. The rapid inflation required to solve the homogeneity and flatness problems came about because the universe expanded beyond its critical temperature for the phase change and became effectively super-cooled. In these conditions small regions of symmetry would suddenly change state and grow at a phenomenal rate, increasing in size by a factor of perhaps 10^{50} before settling into the more leisurely expansion of the conventional model. After about 10^{-30} s, the rapid inflation slowed and produced exactly the same conditions as in the standard model of the expanding universe. This means inflation theory is in complete agreement with all the successful predictions about background radiation, abundance of light elements, etc. that follow from the traditional model of the expanding universe.

In the 1960s Peter Higgs at Edinburgh was trying to understand how the weak and electromagnetic fields can be symmetric at high temperatures (electroweak unification) and yet break this symmetry and appear as distinct forces at low temperatures. Somehow the massless photon and the very massive W-plus, W-minus and Z-nought particles (called vector bosons) are indistinguishable at high temperatures. At lower temperatures the symmetry breaking process must endow the vector bosons with mass. Higgs suggested that this might come about if all of the particles interact with a set of new fields which are now called Higgs fields. If this idea is correct there would have to be new quanta associated with the Higgs fields that would appear as new massive particles – the Higgs bosons. So far these have not been detected, but the Higgs bosons associated with electroweak symmetry breaking are predicted to have a mass within reach of the LHC which should come on line at CERN early in the 21st century.

But what has this got to do with inflation? The answer lies in the nature of the Higgs fields and was first discussed by Alan Guth in the early 1980s. In the very early energy-dense universe all the forces of nature would be unified and the Higgs fields would be in a state of maximum symmetry. However, this symmetric state does not correspond to the minimum energy density in the Higgs field and is a kind of 'false vacuum'. As the universe expanded and cooled it made a transition to a lower energy state of the Higgs field, the 'true vacuum', in which the symmetry between the fundamental forces is broken. In fact, if the transition was slow compared to the rate at which the universe cooled it might go into a 'supercooled state' before 'changing state'. When a particular region of the universe then makes its transition to this lower energy state in the Higgs field it does so very rapidly, growing like a crystal at a phenomenal rate. Furthermore, the energy lost from the Higgs field as it 'jumps' to a lower energy state is fed back into the universe to create a vast amount of matter and antimatter – the precursors

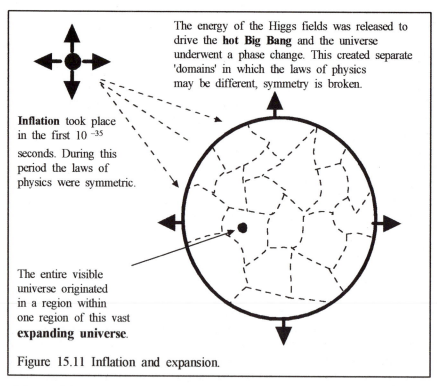

The energy of the Higgs fields was released to drive the **hot Big Bang** and the universe underwent a phase change. This created separate 'domains' in which the laws of physics may be different, symmetry is broken.

Inflation took place in the first 10^{-35} seconds. During this period the laws of physics were symmetric.

The entire visible universe originated in a region within one region of this vast **expanding universe**.

Figure 15.11 Inflation and expansion.

of the traditional expanding universe scenario. Of course, different regions of the very early universe might 'crystallise' differently, resulting in discontinuities that should appear as very massive objects in the universe. However, the incredible rate of inflation means that the entire visible universe is much much smaller than the typical size of one 'crystal'. This explains why we have not observed these massive structures. In the traditional model the horizon distance is much greater than the typical size of a 'crystal' and so we should see many discontinuities (in fact their enormous mass should have a profound effect on the way our part of the universe has evolved).

   One of the interesting consequences of the inflationary model is that the final state of the expanding universe is pretty well independent of the details of its initial state (information about this is lost in the rapid inflationary stage). This is not the case in the traditional model. It has led some theorists to speculate that the initial conditions are completely irrelevant and the creation and evolution of the universe as a whole proceeds as a direct consequence of the laws of physics and not from some material 'seed' in a particular state). The idea that the entire universe leapt into being as a result of a quantum fluctuation is interesting but had always been dismissed because it seemed to violate fundamental conservation laws. One of these is the conservation of energy and another is the conservation of baryon number. Other conservation laws such as the conservation of charge are not a problem because they are consistent with the idea that the total charge in the universe could be zero (for example, whenever a positively charged particle is created in a particle collision it must be accompanied by a negatively charged particle as well). On the other hand the universe appears to contain a large amount of energy and a large number of baryons and all experiments seem to conserve these quantities. Inflation offers a way round this. If all the forces are unified at high energies then there should be a symmetry between quarks and leptons which would allow the lightest baryons (the protons) to decay. This solves the problem of baryon number. The problem of energy may also disappear (literally). It is true that the matter present in the universe represents a large amount of positive energy, but gravitational potential energy is negative and the sum of positive and negative energies may actually cancel. If it does then the universe as a whole may have zero energy, charge, mass, etc. and so no conservation laws would have been violated if it came into being spontaneously – out of absolutely nothing!

15.4 PUZZLES

15.4.1 The Missing Mass

If the inflationary model is correct it implies that the universe must have a critical density (Ω). We have already pointed out that the density inferred from observation is considerably less than this (although of the same order of magnitude). This suggests that estimates based on observable matter are underestimates – there must a lot of 'dark matter' (matter that we do not detect from Earth because it does not radiate). Even without the inflationary theory there are good reasons to believe that a lot of the mass in the universe is invisible. In particular, the calculated masses of galaxies and clusters of galaxies seem to be much greater than the masses of the stars they contain.

The first indication that this might be the case was discovered by Fritz Zwicky in the 1930s. He surveyed the Virgo and Coma Berenices galaxy clusters and used the virial theorem to estimate their total masses. This theorem states that, for a gravitationally bound system, the mean kinetic energy is half the magnitude of the gravitational potential energy (as it is for a planet in a circular orbit around the Sun). This leads to the equation:

$$\overline{v^2} = \frac{GM}{R}$$

where v is the velocity of a galaxy within the cluster, M is the mass of the cluster and R is its radius. Values of v and R can be obtained from observation and then used to calculate M. The mass of the cluster can also be estimated from its luminosity (e.g. by comparing its total luminosity with that of the Sun and assuming its mass is in the same proportion). When this was done the calculated mass turned out to be about 20 times greater than could be accounted for by its visible luminosity. More recently some of this missing mass has been accounted for in ionised gas that reveals itself by X-ray emissions, but about 80% is still 'missing'.

On a smaller scale it is possible to estimate the mass of a galaxy from the motions of stars within it. This has been carried out for our galaxy and others and always results in a calculated mass significantly greater than the mass we would expect from estimates based on luminosity. The technique is really quite simple. Stars orbiting at radius r from the centre of a galaxy must be held in orbit by a centripetal force provided by gravitational attraction to the centre of the galaxy. This attraction is to the mass (M_r) contained inside radius r:

$$\frac{GM_r}{r^2} = \frac{v^2}{r}$$

leading to an equation for the orbital speed v:

$$v = \sqrt{\frac{GM_r}{r}}$$

Observations suggest that the orbital speed (measured using the Doppler shifts from stars at different radii from the centre of a galaxy) is roughly constant and independent of r. This is only possible if M_r is directly proportional to r. On the other hand, estimates of how mass varies with radius based on luminosity suggest that it falls exponentially away from the centre of the galaxy. This implies that the ratio of dark matter to visible matter increases rapidly toward the outer edges of a galaxy and leads to the idea of a galactic halo dominated by dark matter. This idea is consistent with the hypothesis that the missing mass is mainly some form of weakly interacting massive particles. If galaxies formed from the gravitational collapse of large clouds of baryonic and dark non-baryonic matter then both types would have gained a large amount of kinetic energy in the collapse. The baryonic matter would have been able to radiate a lot of this away and so sink toward the centre of the galaxy. The dark matter does not radiate, so its mean kinetic energy remains higher and it stays concentrated in the outer parts of the galaxies.

The nature of dark matter has been a subject for intense speculation. Some physicists think it must be in the form of exotic and as yet undiscovered particles. Others have suggested that a great deal of it may simply be in the form of cool low mass stars and planetoids that are too dim to be seen. Recent observations of gravitational microlenses in our own galaxy suggests that some of the dark matter is almost certainly in the form of cool baryonic matter, but if too much of it turned out to be in this form there would be serious problems for inflation theory which, when combined with conventional expansion, limits the amount of baryonic matter in the universe. The processes of nuclear fusion which took place in the young universe and which takes place in stars are pretty well understood and allow physicists to work back from the present abundances of isotopes such as deuterium to the density of baryonic matter in the early universe. These calculations set an upper limit on the density of baryonic matter of about 10% of the critical density.

If this is coupled with the inflationary prediction that $\Omega = 1$ then at least 90% of the universe must be contained in non-baryonic matter. If it is confirmed that neutrinos have mass they will reduce (but not solve) the dark matter problem. To account for all the dark matter they would need a mass of about 10 eV/c^2. The results from super Kamiokande suggest a value closer to 0.1 eV/c^2. Of course, it is possible that there are some other masssive neutrino-like particles that we have not yet discovered. In the end the two general possibilities are some form of cold dark matter (baryonic or non-baryonic) or hot dark matter (particles moving close to the speed of light that would have evaded the gravitational collapse that formed the galaxies and so are more likely to be found between the galaxies than within them). The answer to the problem may well be a combination of both possibilities. Having said that, it still seems possible that some modification in the inflationary scenario, or even a new model of the very early universe, may remove the need for some or all of the dark matter, or at least change the apparent constraints on the theory.

15.4.2 The Mysterious Vacuum

Aristotle claimed that nature 'abhors a vacuum'. By 'vacuum' he meant a completely empty space. In some senses modern views of the vacuum tend to agree – there is no such thing as absolutely empty space and the vacuum itself is the seat of some of the most violent and exotic physical events. We have seen that the collapse of the 'false vacuum' released the enormous amount of energy that fuelled the hot Big Bang and ultimately 'paid' for the creation of the stars and galaxies. On the smallest scales the vacuum is teeming with creation and annihilation as pairs of virtual particles burst into existence as a consequence of Heisenberg's Uncertaity Principle. This links energy and time so that the energy in a small region of space has an uncertainty ΔE over short time periods Δt obeying the relation:

$$\Delta E \Delta t \leq h$$

In practice this means that the energy of the vacuum continually fluctuates and these quantum fluctuations manifest themselves as virtual particles. A pair of virtual particles of rest mass m could come into being as long as they annihilate within a short time Δt for which:

$$2mc^2 \Delta t \leq h$$

In other words the pair of virtual particles can exist for a time of order:

$$\Delta t \leq \frac{h}{2mc^2}$$

For an electron positron pair this is about 4×10^{-21} s.

The presence of quantum fluctuations means that the vacuum itself now has a small mass and energy density. In the very early universe the vacuum energy was enormous and provided a kind of repulsive gravitational force (a negative pressure) that caused the inflation of the universe. This early vacuum state was actually a 'false vacuum' in the sense that there were lower energy states into which it could decay (the 'true vacuum' of space as we see it today). When it did so the incredible release of energy was the Bang in the Big Bang.

The ideas that the vacuum is not empty, that matter can come into being in a spontaneous quantum fluctuation, and a tiny seed can inflate by a factor of 10^{50} in a time of 10^{-32} s open the way to the suggestion that the entire universe may have leapt into being as a quantum fluctuation out of absolutely nothing. This tiny seed would then inflate, ignite the hot Big Bang and continue to expand for fifteen billion years up to the present day. This theory of quantum genesis is taken seriously by a number of cosmologists, but for it to be tenable the universe as a whole must contain exactly nothing! That is, all conserved quantities must have a total value of zero. This seems reasonable for electric charge and linear and angular momentum but there are problems, as we have already pointed out. Energy and baryon number are both conserved. It may be possible that the mass-energy of the universe is exactly balanced by its negative gravitational potential energy and this certainly seems reasonable on the basis of a closed universe with the critical density. If baryon number is always conserved then the only way it could be exactly zero is if there is an equal amount of matter and antimatter in the universe. It is still possible (but unlikely) that this will turn out to be the case. However, the observational evidence gives strong support to the idea that the universe is now dominated by matter, and we have no evidence for antimatter galaxies. Most models of the evolution of the universe assume that a small asymmetry in the production of matter and antimatter in the very early universe led to a slight excess of matter which remained to form the visible universe after matter and antimatter annihilated to create what is now the microwave background radiation. The only possibility that can save the idea of quantum genesis is that baryon number is not strictly conserved. Fortunately (for this theory) that is exactly what a number of GUTs predict, although there is, as yet, no experimental confirmation.

The vacuum becomes even more peculiar when we consider the effect of the Higgs fields. Most fields have zero energy density when they have zero magnitude. This is not the case for Higgs fields. In the early universe the Higgs fields were all zero but their energy density was positive. In fact, when the universe was incredibly tiny the energy density of the Higgs fields dominated. This non-zero energy density means that work must be done to increase the size of the region in which the Higgs field acts. This gives space filled with zero point Higgs fields an innate tendency to collapse. It is rather like a large block of rubber that has been stretched outwards in all directions and is attempting to spring back to its original shape and size. Another way of describing this inherent tendency to collapse is as a negative pressure. Now this negative pressure has another counter intuitive consequence. In general relativity the strength of gravity in a medium depends on two things – its mass-energy density (ρ) and its pressure (P) according to the equation:

$$\text{Strength of gravitational attraction} \propto \rho + \frac{3P}{c^2}$$

If the pressure is negative there is a possibility that this whole expression could become negative and gravity would then exert a repulsive force on masses! The energy density ρ_H in the 'false vacuum' (where the value of the Higgs fields is zero) is related to the pressure by:

$$P_H = -\rho_H c^2$$

So when these terms dominate the energy density and pressure of the universe we have:

$$\rho + \frac{3P}{c^2} = -2\rho_H$$

The value of ρ_H is around 10^{92} Jm^{-3} so the universe at this time (before about 10^{-35} s) is subjected to an incredible gravitational repulsion that drives the exponential inflation prior to the hot Big Bang.

Maths Box: Negative pressure

Imagine a spherical region of volume V in which the Higgs fields are zero and the energy density is ρ_H. Let the sphere expand by an amount ΔV then the energy increases by an amount:

$$\Delta E = \rho_H \Delta V$$

The force exerted on the surface is given by:

$$F = -\frac{dE}{dr} = -4\pi r^2 \frac{dE}{dV}$$

$$P_H = \frac{F}{A} = \frac{F}{4\pi r^2} = -\frac{dE}{dV} = -\rho_H$$

The negative pressure is equal to the energy density of the false vacuum.

Maths Box: Is the Total Energy of the Universe Zero?

A rough argument can be given to show that the negative gravitational potential energy of matter in the universe may well balance its positive rest energy.

$$\text{Hubble time} \quad T_H = \frac{1}{H_0}$$

$$\text{Hubble radius} \quad R_H = cT_H = \frac{c}{H_0}$$

$$\text{Critical density} \quad \rho_c = \frac{3H_0{}^2}{8\pi G}$$

At any particular time a mass m can only be affected by gravity from other masses within about one Hubble radius of it. If this distance is taken as the diameter of a sphere and mass m is assumed to be on the surface of this mass sphere then its gravitational potential energy is:

$$GPE = -\frac{GMm}{R_H}$$

where M is the mass within the hypothetical sphere.
Assume in addition that the universe is at the critical density. Then:

$$GPE = -\frac{GMm}{R_H} = -\frac{4G\pi r^2 \rho_c m}{3R_H} = -\frac{4G\pi R_H{}^2 3H_0{}^2 m}{3R_H \pi G} = -\frac{1}{2}mc^2$$

This is a very rough argument but it does show that the gravitational potential energy is of the same order of magnitude as the rest energy of masses in the universe (since the same argument would apply for any mass m). This makes it tempting to assume that the total energy of the universe is zero and the density of matter in the universe is the critical density. It is also consistent with speculations that the universe may have come into being from a quantum fluctuation.

Part 5
THERMODYNAMICS AND THE ARROW OF TIME

16 Time, Temperature and Chance
17 Toward Absolute Zero
18 CPT

16

The Second Law of Thermodynamics identifies an innate irreversibility in physical processes.
Entropy never decreases.
Boltzmann provided a controversial but powerful microscopic interpretation of entropy and irreversibility.
Thought experiments such as Maxwell's demon show that entropy and information are linked.
Hawking explores the thermodynamics of black holes.

17

Despite its apparent familiarity, the concept of temperature is a subtle one linked to entropy and energy.
Absolute zero is unattainable.
Very low temperatures can be reached by exploiting the laws of thermodynamics using transitions from order to disorder.
Kammerlingh Onnes's liquefaction of helium gives low temperature physics a kick-start and reveals some peculiar new properties – superconductivity and superfluidity.
This strange behaviour near absolute zero is linked to quantum statistics and the distinction between fermions and bosons.
Recent work has created a new state of macroscopic matter – the Bose-Einstein condensate.

18

Conservation laws are linked to symmetry principles.
Charge conjugation, parity and time reversal symmetries are all individually violated in some particle interactions but their combined effect (CPT) should be conserved.
The violation of time-reversal symmetry in some decays provides a fundamental microscopic arrow of time which is not obviously linked to the thermodynamic arrow.

16
TIME, TEMPERATURE AND CHANCE

16.1 THE SECOND LAW

"The law that entropy always increases – the second law of thermodynamics – holds, I think, the supreme position among the laws of Nature. If someone points out to you that your pet theory of the universe is in disagreement with Maxwell's equations – then so much the worse for Maxwell's equations. If it is found to be contradicted by observation – well, these experimentalists do bungle things sometimes. But if your theory is found to be against the second law of thermodynamics I can give you no hope; there is nothing for it but to collapse in deepest humiliation."

(A.S. Eddington, *The Nature of the Physical World*, New York, Macmillan, 1948)

"God has put a secret art into the form of Nature so as to enable it to fashion itself out of chaos into a perfect world system."

(Immanuel Kant, *Universal Natural History and Theory of the Heavens*)

The law of conservation of energy and the second law of thermodynamics are two of the most important discoveries in the history of physical science. Together they place constraints on what processes can occur and in what direction physical systems are allowed to evolve. A deep understanding of these ideas brings us face to face with one of the central problems of physics and philosophy, the nature of time. Whilst the laws themselves emerged from nineteenth century science, their interpretation and application in a wide range of theoretical and experimental processes has a major bearing on twentieth century physics, leading to a new understanding of irreversible processes and a link between entropy and information. In this section we shall briefly review the macroscopic observations that led to the second law and move quickly on to its microscopic interpretation in terms of the work of Boltzmann, Gibbs and Einstein.

16.1.1 Perpetual Motion Machines of the First Kind

"Indeed the phenomena of nature, whether mechanical, chemical, or vital, consist almost entirely in a continual conversion of attraction through space (he means 'potential energy'), *living force* (he means 'kinetic energy') *and heat into one another. Thus it is that order is maintained in the universe − nothing is deranged, nothing ever lost, but the entire machinery, complicated as it is, works smoothly and harmoniously."*

(James Joule, Matter, Living Force and Heat, *The Scientific Papers of James Prescott Joule*, Vol. 1, London, Taylor and Francis, 1884)

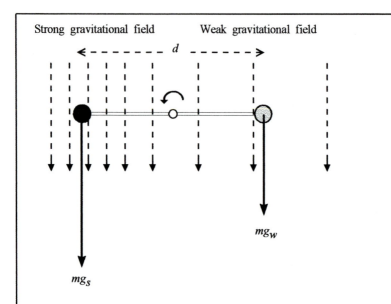

Figure 16.1 A perpetual motion machine of the first kind. This hypothetical machine uses the fact (true) that the strength of a gravitational field varies with position. While mass A is on the left of the pivot its weight is greater than that of mass B. It falls and drives the central axle which could be used to do external work. Once the rotor passes its vertical position B is heavier than A. This ensures that the resultant torque remains anti-clockwise. The motor continues to rotate and can in principle supply an unlimited amount of energy! The work available per rotation is $2md\,(g_s - g_w)$. Why is such a motor impossible?

For millennia scientists and inventors have tried to get around this principle and build a machine that can generate more energy than it consumes. Leonardo da

Vinci drew several versions, Robert Boyle suggested one and nowadays you can find a wide variety of designs published on the Internet. In the early days the machines were mechanical or gravitational, later they incorporated steam engines and electromagnetic devices, more modern versions rely on 'loopholes' in relativity or quantum theory or else tap into hidden energy sources such as the much-touted 'vacuum energy'.

  However, no known physical process violates the law of conservation of energy, so there is no way we can build a machine that operates according to the known laws of physics and generates energy from nothing. Any machine that is designed to do this is called a Perpetual Motion machine of the first kind. All such machines forbidden by the law of conservation of energy.

16.1.2 Heat Engines

The invention and development of the steam engine by Newcomen and Watt in the eighteenth century provided an incentive for theoretical thermodynamics but it was not until the nineteenth century that any real progress was made.

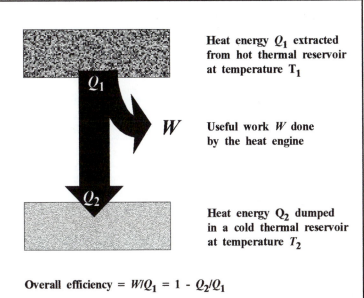

Heat energy Q_1 extracted from hot thermal reservoir at temperature T_1

Useful work W done by the heat engine

Heat energy Q_2 dumped in a cold thermal reservoir at temperature T_2

Overall efficiency = W/Q_1 = 1 - Q_2/Q_1

Figure 16.2 Energy flow in a heat engine. The net effect is to extract work W from the flow of heat from hot to cold reservoirs.

A steam engine (such as an internal combustion engine or a thermal power station) extracts work from the thermal energy released when its fuel is burnt. All heat engines operate between two temperatures, a hot source (provided by the burning fuel) and a cold sink (at a temperature at or above that of the surroundings). The overall effect is to divert some of the heat that flows from hot source to cold sink and make it do useful work. The big question concerned the efficiency of this process. How much of the energy extracted from the source can be used to do work? Is it possible to convert heat to work with 100% efficiency?

To illustrate the problem imagine braking hard and bringing a car to rest. In the stopping process the car does work against frictional forces in the braking system. This transfers kinetic energy to heat in the brake discs. Is it possible to build an engine that can transfer all of this thermal energy back to kinetic energy and return it to the car? In this form it is clearly a question about reversibility – is the transfer of work to heat reversible?

16.1.3 Perpetual Motion Machines of the Second Type

If it is possible to transfer heat to work with 100% efficiency then the atmosphere and oceans can be used as vast reservoirs of thermal energy that we can tap into to generate electricity. A machine that is designed to carry out the complete transfer of heat to work is called a Perpetual Motion Machine of the second kind. The diagram shows a plan for such a machine. It is important to realise that it does not violate the law of conservation of energy.

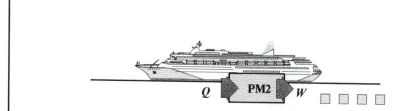

Figure 16.3 A perpetual motion machine of the second type would violate the second law of thermodynamics. For example it might extract heat from the ocean and use it to propel a boat. The PM2 above takes in water at about 280K and freezes it. The energy extracted is used to propel ice cubes backwards. The reaction propels the ship forwards.

16.1.4 Carnot, Kelvin and Clausius

"Notwithstanding the work on all kinds of steam engines, their theory is very little understood, and attempts to improve them are still directed almost by chance.

The question has often been raised whether the motive power of heat is unbounded, whether the possible improvements in steam engines have an assignable limit – a limit which the nature of things will not allow to be passed by any means whatever, or whether on the contrary, these improvements may be carried on indefinitely."

(Sadi Carnot, *Reflections on the Motive Power of Heat*, 1824)

Sadi Carnot (1796-1832) was a French engineer and physicist who was determined to put the theory of the steam engine onto a secure theoretical foundation. His major work, *Reflections on the Motive Power of Heat*, was published in 1824, eight years before his premature death from cholera and more than a decade before Joule's idea that heat is a form of energy replaced the incorrect caloric theory on which Carnot had based his work. Nonetheless, Carnot's analysis was so deep that his results turned out to be independent of the theory of heat and effectively established the second law of thermodynamics before the first law had been written down! Unfortunately he was little known outside Paris and his paper did not have much immediate impact. It was a quarter of a century later that Carnot's ideas were revived by Lord Kelvin in England and Rudolf Clausius in Germany.

Carnot was the first to realise that the working fluid in an engine (steam in a steam engine) is actually irrelevant to the underlying physics. This fluid is taken around a cycle of changes that return it to its initial state at the end of each cycle, so the only work that is extracted must be the difference between the heat that flowed into the fluid and the heat that flowed out around the cycle. This being the case he considered the most general ideal engine he could and worked out the maximum amount of work that could be extracted from it.

" ... *if the art of producing motive power from heat were to be elevated to the stature of a science, the whole phenomena must be studied from the most general point of view, without reference to any particular engine, machine, or operating fluid.*"

(Sadi Carnot, 1824, *Reflections on the Motive Power of Heat*)

The Carnot cycle proved that an ideal reversible heat engine has a maximum theoretical efficiency given by a simple equation:

$$\eta = 1 - \frac{T_2}{T_1}$$

where T_1 is the temperature of the hot source and T_2 the temperature of the cold sink.

His next step was to see what would happen if an engine could be built which beat the efficiency of the Carnot engine. He showed that this would make it possible to pump heat from a cold body to a hot body (something that never happens spontaneously in nature) and to combine the engines in such away that their net effect was to transfer heat entirely into work. In other words, if it is possible to build a heat engine that has a greater efficiency than an ideal reversible Carnot engine then it can be used to build a perpetual motion machine of the second kind. All experimental results and observational evidence suggested that this was not possible.

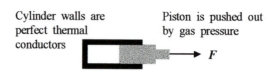

Cylinder walls are perfect thermal conductors

Piston is pushed out by gas pressure

An ideal reversible Carnot engine

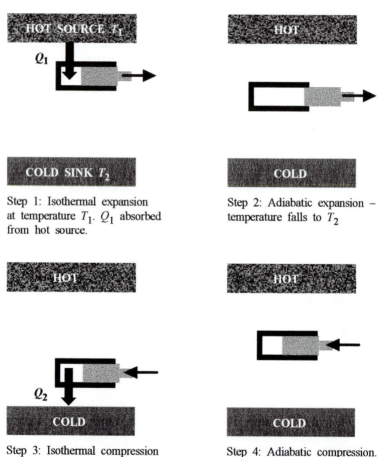

Step 1: Isothermal expansion at temperature T_1. Q_1 absorbed from hot source.

Step 2: Adiabatic expansion – temperature falls to T_2

Step 3: Isothermal compression at T_2. Q_2 dumped in cold sink.

Step 4: Adiabatic compression. Temperature rises to T_1 and system returns to original state.

Figure 16.4 The Carnot cycle. This extracts work reversibly from the heat flow between two thermal reservoirs at temperatures T_1 and T_2. Carnot showed that the maximum theoretical efficiency of such an engine is $1 - T_2/T_1$.

In 1849 William Thomson (later Lord Kelvin) rediscovered Carnot's work and in 1850 Clausius expressed the second law of thermodynamics in a very simple form.

The Second Law of Thermodynamics

- It is impossible for an engine, working in isolation, to transfer heat from a cooler to a hotter body.

In 1851 Thomson stated it in a slightly different way:

- It is impossible for an engine, working in isolation, to have the sole effect of transferring heat into work.

16.2 ENTROPY

16.2.1 Macroscopic Entropy

Clausius introduced a new quantity into physics – entropy. This can be understood quite simply in terms of Carnot's work. Imagine an ideal heat engine operating between temperatures T_1 and T_2 as before. The efficiency can be written down in terms of the energy transfers or the Carnot formula:

$$\eta = 1 - \frac{Q_2}{Q_1} = 1 - \frac{T_2}{T_1}$$

This leads directly to the relation:

$$\frac{Q_2}{T_2} = \frac{Q_1}{T_1}$$

Clausius treated the ratio of heat flow to temperature as a new quantity – the entropy change, ΔS for the process. For an ideal reversible heat engine the result above shows that the entropy lost by the hot reservoir is equal to the entropy gained by the cold reservoir. So ideal reversible heat engines 'conserve' entropy. What about real heat engines? These do not achieve the maximum theoretical efficiency so in general:

$$1 - \frac{T_2}{T_1} \geq 1 - \frac{Q_2}{Q_1}$$

leading to:
$$\frac{Q_2}{T_2} \geq \frac{Q_1}{T_1}$$
$$\Delta S_2 \geq \Delta S_1$$

This is an important result. It shows that the net effect of the operation of a real heat engine is to dump more entropy in the low temperature sink than it takes from the high temperature source. The overall effect on the universe at large is to increase entropy. Since processes that reduce entropy never occur this implies that the entropy of the universe as a whole tends toward a maximum value. It leads to a more general statement of the second law of thermodynamics:

• All natural processes increase the entropy of the universe.

Perpetual motion machines of the second kind would violate this law because they extract heat from a thermal reservoir (e.g. the ocean, thus reducing its entropy, but they do not dump any entropy in the environment, so their net effect would be to reduce the entropy of the universe – which the second law forbids.

HOT T_1 COLD T_2	**HOT T_1 COLD T_2**

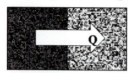 

POSSIBLE	**IMPOSSIBLE**
If heat Q is transferred from T_1 to a lower temperature T_2 then the entropy change is given by:	If heat Q is transferred from T_2 to a hotter temperature T_1 then the entropy change is given by
$\Delta S = -Q/T_1 + Q/T_2 > 0$	$\Delta S = -Q/T_2 + Q/T_1 < 0$
Entropy increases.	Entropy decreases.

Figure 16.5 Why heat flows from hot to cold.

The second law identifies an asymmetry in time. If all natural processes increase the entropy of the universe then the entropy must have been lower in the past than it will be in the future. This seems to associate an arrow of time with fundamental physical processes and this idea of a thermodynamic arrow of time soon became the focus of heated debates. The reason for this is easy to understand. All the other laws of physics – mechanics and electromagnetism – are time-reversible, i.e. the equations work in exactly the same way if the sign of time is inverted. This implies that fundamental physical processes do not distinguish between the past and the future, so where does the thermodynamic arrow of time come from?

16.2.2 Molecules in Motion

Kinetic theory developed as a microscopic model to explain the macroscopic behaviour of gases using Newton's laws of motion. The basic assumption was that a gas consists of an enormous number of tiny molecules in rapid random motion.

Each molecule is regarded as a spherical massive particle that only interacts with other particles and the walls of its container during collisions. Between collisions it travels in a straight line at constant velocity. This simple model proved to be incredibly powerful and soon provided an explanation of the gas laws and processes such as diffusion and chemical reaction rates. It also linked up nicely with the atomic hypothesis, supporting the ideas of those who claimed atoms were real rather than simply convenient mathematical constructs. More of this later.

Some ideas of kinetic theory were developed by Daniel Bernoulli in the eighteenth century, but the theory really took off in the nineteenth century under Clausius, Maxwell, Boltzmann and others. Clausius took the decisive step of identifying thermal energy as the total kinetic energy of random thermal motions and interpreted the first law of thermodynamics in microscopic terms.

In 1860 James Clerk Maxwell derived an expression for the distribution of molecular velocities in a gas at equilibrium. This was a major step forward but it did not describe the evolution of a system toward equilibrium, nor did it imply that a non-equilibrium gas would eventually reach this equilibrium state. That step was taken by the great Austrian physicist, Ludwig Boltzmann. Boltzmann represented the state of a gas containing N particles by a vector in $6N$-dimensional phase space and worked out an equation for the evolution of this vector from any starting 'position' toward equilibrium – which he identified with the distribution of velocities we now call the 'Maxwell-Boltzmann distribution'.

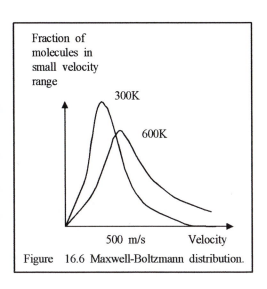

Figure   16.6 Maxwell-Boltzmann distribution.

This evolution toward equilibrium was contained in Boltzmann's 'H-theorem' where H is a function of the state of the gas and always decreases with time toward a minimum value which it achieves when the distribution of velocities within the gas satisfies the Maxwell-Boltzmann formula. The key to this time-evolution lay in probability. From the outside a gas state is determined by macroscopic parameters such as pressure, volume and temperature, whilst from the inside its state is determined by the microscopic distribution of velocities amongst particles. Boltzmann showed that the equilibrium state is the macro-state that can be realised in the maximum number of microscopic ways. Furthermore, when large numbers of particles are involved, the number of permutations representing the same macro-state increases so quickly toward the equilibrium state that the overwhelming probability is that non-equilibrium states will evolve spontaneously toward equilibrium states.

This idea can be illustrated very simply. Imagine having a container with two compartments. Place a large number (N) of identical objects in one of the compartments. Now select each object in turn and toss a coin. If the coin returns heads leave the object where it is. If it comes down tails move the object to the other compartment. What will the system look like if you return after along period of time? There are lots of ways in which the objects might be distributed so that there are n in one compartment and ($N - n$) in the other, but these micro-states are all indistinguishable from the macroscopic point of view (just as all the different molecular configurations that create uniform atmospheric pressure in a room are indistinguishable to a person in that room). In fact the number of ways in which an $n:(N - n)$ distribution might occur is given by:

$$W = \frac{N!}{(N-n)!}$$

This has a maximum value when $n = N/2$, when there are equal numbers of objects on either side of the barrier.

As N becomes large (remember there are about 10^{19} air molecules per cubic centimetre surrounding you) the maximum value of W becomes very large compared to its immediate surroundings. This means there are many more micro-states of the system close to this maximum value than far from it. The fundamental assumption of statistical mechanics is that all micro-states are equally accessible and so equally likely to occur. If there are Ω ways in which the system can be organised (Ω is the sum of all Ws) then the probability that the system is found (after a long period of time) in a particular macro-state X is given by:

$$p_X = \frac{W_X}{\Omega}$$

where W_X is the number of microscopic configurations which correspond to the macro-state X.

What does this mean? Simply that we are likely to find the system in a macro-state that can be realised in a large number of microscopic ways. This is no different to our expectation when we shuffle a pack of cards. We would be very surprised if the shuffled deck turned out to have all cards in suit and number order (a state achieved in only one of all possible card configurations). There are many more ways we would identify as 'shuffled' than as ordered even though every particular (microscopic) arrangement is equally likely. How does this relate to the kinetic theory of gases? The equilibrium macroscopic state is identified with the state that can be achieved in the maximum number of microscopic ways.

Imagine two objects at different temperatures placed in thermal contact. What will happen? Imagine transferring energy ΔQ by a flow of heat from one body to the other. The effect is to reduce the amount of energy in body A and increase it in body B. Each body has a finite number of particles which share the total thermal energy. The effect of transferring the energy from A to B is to reduce the number of possible micro-states in A and increase it in B. Which way will it flow? In the

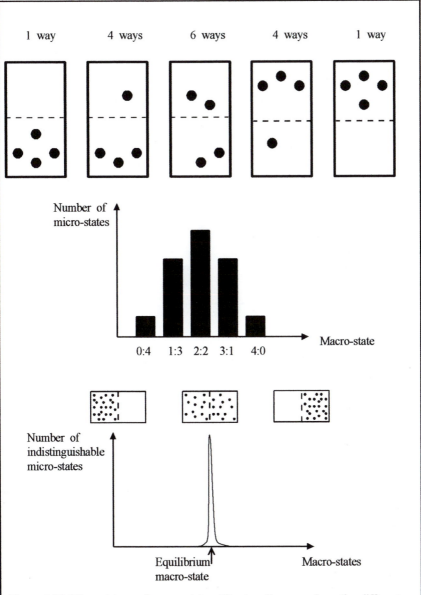

Figure 16.7 Micro-states and macro-states. The top diagrams show the different ways four molecules could be distributed in a partitioned container. From a macroscopic point of view the only observable is the ratio of particles in each half. An even split (2:2) is more likely than any other configuration, so this is the most likely macro-state. When the number of particles is large there are far more micro-states close to the equilibrium distribution (equal numbers on either side of the barrier) so the system will always evolve toward this macro-state, and stay there once it reaches it.

direction which increase the total number of micro-states available to the system as a whole. Why? Because this is more likely – if all micro-states are equally likely the original state will be adjacent to many other states, but far more of them will be closer to equilibrium than farther from it. With large numbers of molecules involved the evolution toward equilibrium is inevitable. But where is the equilibrium state? The two objects will eventually reach a combined state in which the transfer of heat from A to B reduces the number of micro-states in A by exactly the same amount that it increases the number of micro-states in B. This is thermal equilibrium and from the macroscopic viewpoint we would say that the two objects are now at the same temperature.

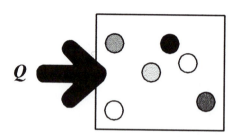

Figure 16.8 Temperature. If heat is supplied to something, its internal energy increases and there are more ways in which this energy can be distributed among available states. The number of micro-states increases. So does the entropy. The relative effect is greater when the object has little energy to start with. This links the entropy change to temperature:

Low temperature: adding Q increases entropy by a larger amount;
High temperature: adding Q increases entropy by a small amount.

Thermodynamic temperature is defined as the ratio of heat added to entropy increase. Thermal equilibrium is when the exchange of heat between two objects results in no overall change of entropy.

16.2.3 Microscopic Entropy

Boltzmann's work identifies an irreversibility in macroscopic physical processes, a tendency to evolve from non-equilibrium states toward equilibrium. Clausius's entropy principle must be linked to this. Boltzmann showed that entropy could be linked to the micro-states by the following equation:

$$S = k \ln W$$

where S is the entropy, k the 'Boltzmann constant' ($k = 1.38 \times 10^{-23}$ JK^{-1}) and W the number of microscopic configurations corresponding to the particular macroscopic physical state. Thermal equilibrium of two isolated bodies is achieved when they maximise their entropy – this corresponds to the macro-state realised in the maximum number of different micro-states.

16.3 IRREVERSIBILITY

16.3.1 A Thermodynamic Arrow of Time?

It is very tempting to identify Boltzmann's *H*-theorem and the time-oriented evolution of macroscopic systems toward more probable states with an arrow of time. But surely there has been some sleight of hand? Kinetic theory began with the assumption that molecules obey Newton's laws of motion and these laws are all time symmetric. Where did the asymmetry come from?

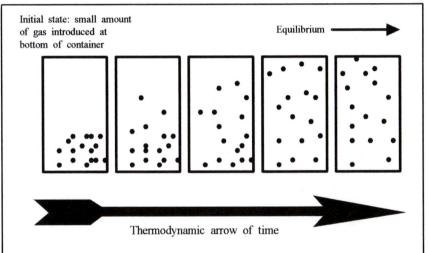

Figure 16.9 Shuffling micro-states results in macroscopic time evolution. Not that the states do not stop shuffling once equilibrium is reached, but they are unlikely to return to far-from equilibrium states because the vast majority of all micro-states correspond to the equilibrium macro-state.

This problem was highlighted by William Thomson and Josef Loschmidt in the 1870s in what has now become known as the Loschmidt paradox. Thomson asked what would happen if the instantaneous velocities of all molecules in a gas could be reversed at one instant. He pointed out that this should still represent a realisable physical state because the laws of motion which govern the behaviour of the individual particles are themselves time-reversible. He came to a remarkable conclusion:

"If then, the motion of every particle of matter in the universe were precisely

reversed at any instant, the course of nature would be simply reversed for ever after. The bursting bubble of foam at the foot of the waterfall would reunite and descend into the water; the thermal motions would reconcentrate their energy, and throw the mass up the fall in drops reforming into a close column of ascending water. Heat which had been generated by the friction of solids and dissipated ... would come again to the place of contact and throw the moving body back against the force against which it had previously yielded. Boulders would recover from the mud the materials required to build them into their previous jagged forms, and ... would become reunited to the mountain peak from which they had previously broken away. And if also the materialistic hypothesis of life were true, living creatures would grow backwards, with conscious knowledge of the future, but no memory of the past, and would become again unborn."

(W. Thomson 1874 *The Kinetic Theory of the Dissipation of Energy*, Proceedings of the Royal Society of Edinburgh, 8, 325-34)

The Loschmidt 'paradox' seemed to undermine the idea that the *H*-theorem gave a mechanical *derivation* of the irreversibility of the second law of thermodynamics. And there was a second attack, raised by the mathematician Zermelo in 1896 using an idea put forward by Henri Poncaré in 1893. Poincaré proved that any finite, bounded mechanical system (which might represent the entire universe) will evolve in phase space in such a way that states close to its initial state will recur periodically. This implies that evolution toward equilibrium will, sometime in the future result in evolution away from equilibrium. The system will pass through all of its available microstates repeatedly.

But Boltzmann's ideas were deeper than this. He was well aware that reversible laws by themselves cannot give rise to the irreversibility of thermodynamics. There is an extra factor to consider – the initial conditions. Given a far from equilibrium state, the probability that it evolves to a state closer to equilibrium is far greater than the possibility that it moves away from equilibrium. This does not mean that such a change is impossible, merely exceedingly unlikely. Time-reversed evolution is possible, so is Poincaré recurrence but given any particular universe in a non-equilibrium state (like ours) the probability that we observe even a tiny but macroscopic step against the thermodynamic arrow of time is almost zero. On the other hand, if we consider a microscopic system containing just a few particles the recurrence time may well be observable and the idea of an arrow of time defined within such a system becomes somewhat hazy. By the same token we should be surprised (and are!) to find any time-asymmetric behaviour in the interactions of sub-atomic particles.

16.3.2 The Heat Death of the Universe

"Thus general thermodynamics holds fast to the invariable irreversibility of all natural processes. It assumes a function (the entropy) whose value can only

change in one direction – through any occurrence in nature. Thus it distinguishes any later state of the world from any earlier state by its larger value of entropy. The difference of the entropy from its maximum value – which is the goal of all natural, processes – will always decrease. In spite of the invariance of the total energy, its transformability will therefore become even smaller, natural events will become ever more dull and uninteresting, and any return to a previous value of the entropy is excluded."

(Ludwig Boltzmann, 1898, *Gas Theory*, §89)

We appear to be living in a universe that is far from equilibrium. As time goes on all macroscopic processes increase the entropy of the universe, distributing energy among more and more states as dispersed low temperature heat. This led nineteenth century physicists to predict the forthcoming 'heat death' of the universe – a time in the far distant future when all useful sources of energy had been used and their energy transferred to large numbers of photons in low temperature thermal radiation. However, closer analysis and improved knowledge of the present state of the universe suggests that the heat death may not be quite so far off or quite so scary.

The universe came into being some fifteen billion years ago in a hot Big Bang and has been expanding and cooling ever since. The largest contribution to its entropy is from the radiation that fills the universe, most of which is tied up in the ubiquitous microwave cosmic background radiation. This is interesting, because the entropy of the background radiation is approximately constant. This follows because the universe cools as it expands. In cooling the number of photons per unit volume falls but the total volume increases and these two effects cancel each other out leaving a more or less constant number of photons. Of course stars and other astronomical objects radiate photons, but their effect compared to the background radiation is negligible. The entropy of baryonic matter is also negligible on the large scale because there are about 10^9 photons per nucleon (this ratio is called the specific entropy of the universe). This leads to the conclusion that the entropy of the entire universe is already close to its maximum (equilibrium) value and in some sense the heat death has already occurred. Here and there local hot spots (such as our Sun) still survive, burning nuclear fuel at extremely high temperature and radiating photons into space at a much lower temperature. These in turn are absorbed by nearby objects (such as the Earth) and the energy is eventually re-emitted at yet lower temperatures, ultimately adding to the background itself.

On the small scale life seems to buck the second law. How can complex 'ordered' systems such as human beings come into being as a result of a series of irreversible processes all of which must increase the entropy or disorder of the universe? The answer to this question is linked to the original idea of a heat engine. It is possible to extract some useful work from irreversible processes as long as the low temperature heat increases the entropy of the environment at the same time. In other words our 'order' is paid for by a much larger amount of disorder created in the universe at large. We live by consuming low entropy foods, extracting some useful work (to generate 'order') and dumping waste heat at low temperature back in the environment. This whole process is supported by the proximity of the Sun which acts as a high temperature thermal reservoir spewing out high frequency photons which pass through the terrestrial heat engine and are re-radiated into space at much lower temperatures.

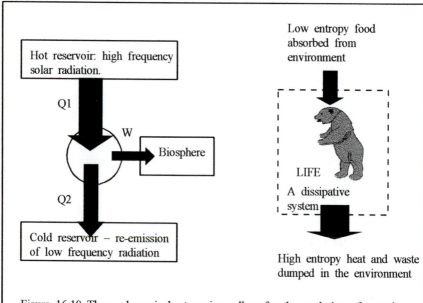

Figure 16.10 Thermodynamic heat engines allow for the evolution of complex systems without violating the second law of thermodynamics. Any temporary order trapped in a living thing is maintained by a through flow of energy and entropy that increases the total entropy of the universe.

16.4 BLACK HOLE THERMODYNAMICS

16.4.1 Black Holes as Heat Engines

One of the most surprising discoveries in twentieth century astrophysics was a fundamental link between the physics of black holes and thermodynamics. For this reason we will make a small digression at this point to review some of the ideas of black hole dynamics and relate them to the laws of thermodynamics and the history and fate of the universe.

Black holes are formed when the core of a supernova is more than about 2.5 times the mass of the Sun. This is the Landau-Oppenheimer-Volkov limit, above which neutron degeneracy pressure is insufficient to support the star against unlimited gravitational collapse. Such a core collapses to form a black hole. Schwarzschild was the first to solve the equations of general relativity in the vicinity of a static, spherically symmetric, uncharged, non-rotating black hole. His solution for the surrounding space-time is the Schwarzschild metric and it predicts the existence of a singularity at the centre of the black hole. However, most real black holes, if they formed at all, would be rotating, so many physicists were reluctant to assume that Schwarzschild's results implied that real black holes must have a space-time singularity at their centre.

In the mid-1960s R.P. Kerr and E.T. Newman solved Einstein's equations for rotating and charged black holes and in 1972 Stephen Hawking proved that whatever the detailed state of matter that collapses to form a black hole it is described by one of these Kerr-Newman solutions in the so-called 'Kerr-metric'. All that survives of the original object is its mass, charge and angular momentum – everything else is lost in the collapse. This consequence is often called the 'no-hair theorem'. It implies that when matter falls into a black hole information is lost from the universe. This is very suggestive, since information seems to be linked to the negative of entropy. If information is lost then entropy must increase, suggesting that there is an entropy associated with black holes and this increases when more matter falls in. The thermodynamic definition of temperature is linked to the rate at which the entropy of a body changes when heat is extracted or added reversibly. *If* black holes really do change their entropy when energy flows into them then they should (*if* they obey the laws of thermodynamics) change in temperature too. But *if* black holes have a particular temperature they should radiate like a black body, a conclusion that seemed to run counter to the very idea of a black hole. After all – not even light can escape from inside the black hole.

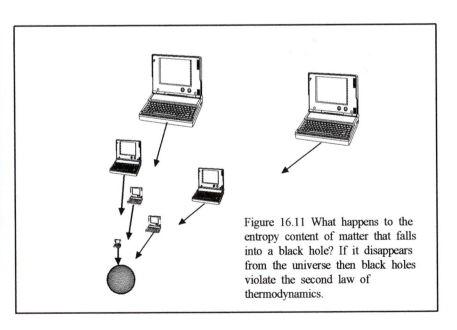

Figure 16.11 What happens to the entropy content of matter that falls into a black hole? If it disappears from the universe then black holes violate the second law of thermodynamics.

How could a black hole act like a black body? In the 1970s Bardeen, Brandon, Carter and Hawking had organised ideas about black holes into four 'laws':

- **Zeroth Law:** In equilibrium the surface gravity K of a black hole is constant over its entire surface.
- **First Law:** Black holes conserve energy – the energy change when a black hole changes state is given by:

$$\Delta E = \frac{c^2 K \Delta A}{8\pi G} + W$$

  where A is the surface area of the horizon and W is the work done on the black hole.
- **Second Law:** During any process the area of the horizon cannot decrease.
- **Third law:** It is impossible to reduce the surface gravity of a black hole to zero in any finite number of steps.

These laws bear an uncanny resemblance to the laws of thermodynamics. In particular, the second law implies that the surface area of the horizon behaves a bit like entropy and the first law links the surface gravity with temperature. The first person to connect these concepts in this way was Jacob Bekenstein in 1972 when he considered how much information is lost and entropy created when quanta are injected into a black hole. He and Hawking took this idea further producing the Bekenstein-Hawking formula for the entropy of a black hole in 1975:

$$S_{bh} = \text{constant} \times A$$

$$S_{bh} = \frac{A}{4}\left(\frac{kc^3}{G\hbar}\right)$$

where k is the Boltzmann constant.
This leads to a surface temperature given by:

$$T = \frac{\hbar k}{2\pi K c}$$

These results can be combined with Schwarzschild's solution to show that the entropy and surface area of a black hole are both proportional to the square of the black hole mass and the temperature is inversely proportional to mass:

$$S_{bh} \propto m^2$$

$$A \propto m^2$$

$$T \propto \frac{1}{m}$$

Putting numbers into the equation for temperature shows that massive black holes have very low temperature and so radiate their energy away at a low rate. For example, a black hole of about the same mass as the Sun would have a surface

Figure 16.12 **Stephen William Hawking** 1942-. Hawking's work in cosmology showed that the universe originated as a singularity. Combining general relativity, quantum theory and thermodynamics, he showed that black holes should be hot and radiate energy (Hawking radiation).

Artwork by Nick Adams

temperature of about 1 μK and a lifetime of about 10^{70} years. On the other hand, mini-black holes (which Hawking speculated may have been formed in large numbers in the Big bang) would be hotter, radiate more intensely and last less time. Furthermore, when a black hole reaches the final stages of 'evaporation' its mass gets very small, its temperature grows rapidly and it radiates its final energy in a burst of radiation that we should be able to detect (if such mini-black holes exist). So far we have not detected any of these exploding black holes.

This analogy is very interesting, but it still begs the question about radiation. If black holes are hot they must radiate. What is the mechanism? This was worked out by Stephen Hawking in 1974 in a brilliant synthesis of quantum theory, general relativity and thermodynamics. In quantum field theory the vacuum is not empty, it contains a seething mass of virtual particles which are continually created and annihilated. These vacuum fluctuations have a measurable effect on physical processes (e.g. the Lamb shift and the Casimir effect). Hawking considered what happens when virtual pairs are created close to the event horizon of a black hole. There are three possibilitities:

- The particles may be created and annihilated as they are in 'empty space'.
- The particles may both fall into the black hole.
- One of the pair may fall in and the other may escape.

The last of these is the process responsible for Hawking radiation. Hawking showed that the energy spectrum of particles leaving a black hole is that of an ideal black body radiator. This spectrum is determined by one thing – the temperature of the black body – given by the expression above. The laws of black hole physics are equivalent to the laws of thermodynamics, one of the most remarkable discoveries of theoretical physics.

Hawking went on to show that primordial black holes with masses around 10^{11} kg created in the Big Bang should have lifetimes comparable to the present age of the universe, so they should be exploding about now. So far they have not been discovered.

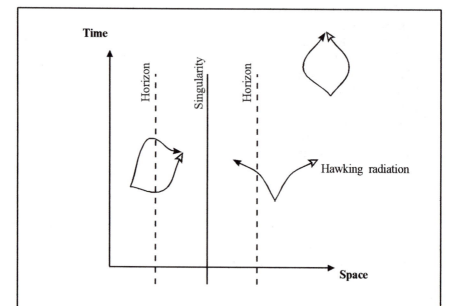

Figure 16.13 Hawking radiation. Virtual pairs of particles and anti-particles are created and annihilated all the time in otherwise empty space. If one member of the pair falls into a black hole, the other radiates away. The energy to create the pair came from the black hole, which decays. Hawking showed that the radiation has a black-body spectrum leading to the idea that black holes have a definable temperature.

16.4.2 Black Holes and the Universe

In the 1960s Hawking realised that the hot Big Bang theory behaves rather like a time-reversed gravitational collapse. In 1965 Roger Penrose showed that all black holes must have a singularity inside their event horizon, however the black hole formed. This raised a similar question about the origin of the expanding universe: does it follow that there must have been a singularity at the start of the beginning of time? In 1969 Hawking and Penrose proved that all hot Big Bang models must have a singularity at their origin.

This raises another interesting question. If the universe starts with a singularity and eventually collapses to a singularity does this mean that overall the history of the universe is symmetric in time – a Big Bang followed by expansion, collapse and then a Big Crunch? This would mean there is no real arrow of time associated with thermodynamics. Penrose argues against this. He claims that the singularity at the origin of the universe differs from the singularity at the end of the universe in the way in which space-time curvature forms the singularity. According to Penrose the original singularity, based on Friedmann's solutions to Einstein's equations, is a low entropy singularity. On the other hand the singularity that forms as matter clumps together through gravitational attraction forming black holes that eventually amalgamate to a Big Crunch is a high entropy singularity. Using the Bekenstein-Hawking formula the entropy of black holes soon dwarfs that of the microwave background radiation, providing about 10^{20} units of entropy per baryon in a solar mass black hole (as compared to 10^8 per baryon from the background radiation). By Penrose's estimation (in *The Emperor's New Mind*) the total entropy per baryon if all the matter in the universe forms one black hole is about 10^{43}. For 10^{80} baryons (an estimate based on the age, density and rate of expansion of the universe) this gives:

$$S_{universe} \approx 10^{123} \approx k \ln W$$

giving
$$W \approx 10^{10^{123}}$$

If this is correct it means there is an incredibly large number W alternative configurations in which matter and energy could have been arranged in the universe other than its present state. This seems to suggest that the universe originated in a very improbable state – *if* the low entropy Big Bang hypothesis is correct. Furthermore it links the existence of the second law of thermodynamics and the arrow of time with the constraints on space-time curvature in singularities at the origin and end of the universe. Of course, the universe may not end in a Big Crunch and recent evidence suggests that this is unlikely. What happens to entropy in an ever-expanding universe? It goes on increasing. The clumping together of mass due to gravity continues and this increases entropy, but not all the mass ends up bound into a single black hole. Eventually black holes are far from one another and the universe continues to become colder darker and more disordered (higher entropy) as predicted by the 'Heat Death' theories.

16.5 ENTROPY AND INFORMATION

16.5.1 Maxwell's Demon

The idea that information loss equates in some way to entropy increase did not originate in black hole thermodynamics. It came from statistical mechanics through a thought experiment put forward by Maxwell in his *Theory of Heat* (1872). Maxwell was responding to Thomson's idea that the mechanical universe would inevitably 'wind down' as energy is dissipated to heat and heat is radiated at ever-lower temperatures. Maxwell suggested that there is a way round the second law of thermodynamics if a microscopic intelligent being has sufficient information to follow the paths of individual molecules and manipulate them.

The second law forbids the isolated transfer of heat from cold to hot bodies and the complete conversion of heat into work in a heat engine. Maxwell imagined a system in which both of these processes might be achieved. Consider a container of gas in thermal equilibrium at uniform temperature and pressure divided in two by a partition containing a frictionless door. Close to this door is a tiny intelligent being ('Maxwell's demon') who can follow the motions of the molecules and open or close the door at high speed. Maxwell knew that the microscopic state of a gas in thermal equilibrium contains huge numbers of molecules moving with a broad distribution of velocities (the Maxwell-Boltzmann distribution) so some molecules will approach the partition at high speed whilst other approach relatively slowly. The demon's job is to open the partition when fast molecules approach from the left or slow ones from the right and keep it closed at all other times. In this way the average energy of molecules on the right increases whilst that on the left falls. Heat flows from cold to hot. This will convert a system in thermal equilibrium to a far-from equilibrium system in which one side is at high temperature and the other side at low temperature. The two sides of the system can now be used as hot and cold thermal reservoirs to drive a heat engine and extract work. Of course, the heat engine efficiency will be limited by the Carnot equation, but the waste heat is dumped into the cold reservoir and the demon can be brought into play to separate it out repeatedly. The overall efficiency of the system at converting heat into work can approach 100%. Maxwell's conclusion is worth repeating:

"The second law of thermodynamics has the same degree of truth as the statement that if you throw a tumblerful of water into the sea, you cannot get the same tumblerful of water out again."

(Letter from Maxwell to Strutt, 6 December 1870, quoted in R.J. Strutt, John William Strutt (London, 1924), 47.)

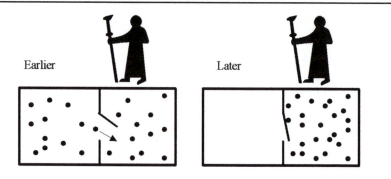

Figure 16.14 Maxwell's demon can sort molecules by opening and closing a delicate valve. In the example shown above, the demon has concentrated all the molecules on one side of the barrier, effectively reversing the arrow of time. He could also separate the gas into hot and cold parts so that these can be used to drive a heat engine and violate the second law.

What is the significance of Maxwell's demon? Is such a demonic device possible? As far as Maxwell was concerned, the second law operated in the macroscopic domain as a result of our lack of knowledge about microscopic configurations. Given complete knowledge we should be able to manipulate the microworld and make efficient use of thermal energy. I am not suggesting that Maxwell thought we could actually do this, but he did seem to think it was possible in principle. His main point was that the irreversibility cannot be derived from microscopic mechanics, it is a statistical result and not an intrinsic irreversibility. However, Maxwell's demon raised other issues too, and has been a source of debate over fundamental principles throughout the twentieth century.

16.5.2 The Szilard Engine

In 1929 Leo Szilard proposed a simplified version of the thought experiment involving a container with just one molecule. It worked like this.

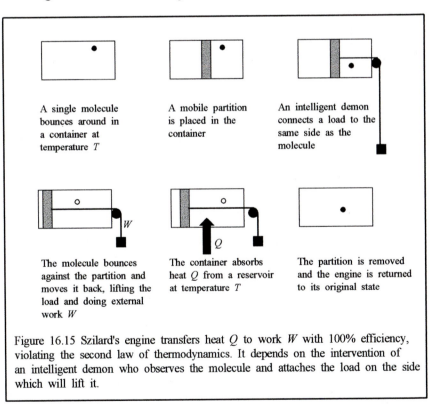

A single molecule bounces around in a container at temperature T

A mobile partition is placed in the container

An intelligent demon connects a load to the same side as the molecule

The molecule bounces against the partition and moves it back, lifting the load and doing external work W

The container absorbs heat Q from a reservoir at temperature T

The partition is removed and the engine is returned to its original state

Figure 16.15 Szilard's engine transfers heat Q to work W with 100% efficiency, violating the second law of thermodynamics. It depends on the intervention of an intelligent demon who observes the molecule and attaches the load on the side which will lift it.

1. Put a container (volume V) in contact with a thermal reservoir at temperature T.
2. Insert a light partition that divides the container in two equal halves.
3. Maxwell's demon observes the system and determines in which of the two halves the molecule is situated.

4. The demon attaches the partition via a pulley to a variable mass on the same side as the molecule (see diagram).
5. The molecule exerts a force on the partition and moves it back, lifting the mass and doing work W. At the same time heat $Q = W$ flows into the container to keep it in thermal equilibrium with the reservoir.
6. The mechanism is detached, returning the system to its original state.

The net effect of this cyclic process is to convert heat Q to work W and apparently reduce the entropy of the universe by an amount Q/T. This would violate the second law of thermodynamics. Szilard did not believe this conclusion. He felt that some change associated with the engine cycle must generate entropy, and focused his attention on the demon itself. The only additional process is the acquisition of information about the molecule by the demon and the storage of this information in the demon's memory. Szilard suggested that in gaining 1 bit of information (that the molecule is in the left or right half of the container) it must generate an amount of entropy equal to or greater than $k \ln 2$, the same as the decrease in entropy when the molecule is confined to half its original volume.

"One may reasonably assume that a measurement procedure is fundamentally associated with a certain definite average entropy production, and that this restores concordance with the second law. The amount of entropy generated by the measurement may, of course, always be greater than this fundamental amount, but not smaller."

(Leo Szilard, quoted in *Maxwell's Demon – Entropy, Information, Computing*, eds Harvey S. Leff and Andrew F. Rex, IOP/Adam Hilger, 1990)

Szilard's work drew attention to the links between entropy and information and the idea of binary bits long before the age of computing. But it was controversial too, the main problem being the nature of the connection between an objective physical property such as entropy and subjective knowledge about the state of a physical system. Szilard did not say exactly what it was about the measurement process that led to an increase in entropy – was it observation, storage in memory or erasure of memory? This provided fertile ground for further discussions.

Twenty years later the Szilard engine was given a fresh analysis by L. Brillouin and Dennis Gabor. They realised that the demon must interact with the gas molecule in order to determine its location. However, the system is in thermal equilibrium with a reservoir at temperature T, so the demon will be surrounded by thermal radiation with a black-body spectrum. In order to 'see' the molecule the demon will have to use a 'torch' that illuminates it against this background. They pointed out that this would require photons whose frequency is significantly above that of the peak in the black-body spectrum, otherwise they could not be distinguished from the background. This set a minimum energy transfer for the measurement process and led to an increase in entropy of the system that could balance the decrease when the engine transfers heat to work.

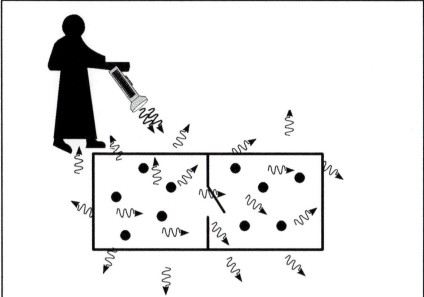

Figure 16.16 Maxwell's demon has to interact with molecules in order to see them. He could use photons. However, these must be distinguished from the black-body fog at temperature T. High-frequency photons have large momentum so they heat the gas and increase its entropy, balancing the reduction in entropy brought about by the demonic separation of molecules.

In 1949 Claude Shannon laid the foundations of information theory and used the ideas to give a fresh analysis of the Szilard engine. Imagine a physical system that could be in any one of N_0 different states that are all equally likely. Shannon argued that if we know nothing about the microscopic configuration then our information I about the system is zero. However, if we carry out measurements on the system this changes. We gain information and the system is then known to be in one of a smaller number $N_1 < N_0$ of states. The entropy of the system has reduced from:

$$S = k \ln N_0$$

to

$$S = k \ln N_1$$

so the entropy change of the system is given by:

$$\Delta S = k \ln \frac{N_1}{N_0}$$

However, our information about the system has increased. Shannon defined the change in information by analogy with the entropy change as:

$$\Delta I = -K \ln \frac{N_1}{N_0} = -\Delta S$$

If the constant K is set equal to k it is clear that the information gained is equal to the entropy lost and that information acts like a kind of negative entropy or 'negentropy'.

If we apply this to the Szilard engine it is clear that the net loss of entropy is equal to k ln2 (when volume available to the molecule is halved) and the information gained is also equal to k ln2, since it is binary information about which of the two sides of the box contains the molecule.

This seems pretty neat, but it still seems to relate subjective and objective aspects of reality. This problem was addressed by a number of people in the 1960s and 1970s including Rolf Landauer and Charles Bennet. They were particularly concerned about the way in which the demon stores and erases his information. Presumably the demon must have at least a binary physical system in which to store its bit of information. This implies the action of something akin to a Szilard engine inside the demon when he interacts with the apparatus. This system must be reset at the end of the engine cycle otherwise the whole system has not been returned to its initial state and the arguments levelled against the second law cannot hold. However, to erase memory the demon has to lose information. How does it do this? The details depend on the actual physical memory but the principle is rather like removing a partition from a Szilard engine (in fact a Szilard engine could act as a demon's memory!) And what is the effect of removing a partition – to increase the number of available memory states and increase the entropy of the system by at least k ln2. The problems raised by Maxwell's demon are removed by the need to erase information to return the system to its starting condition. Later this approach was given a full quantum treatment by Zurek and Lublin, who reached the same conclusion.

16.6 THE EXISTENCE OF ATOMS

"I, too, regard sense impressions as the source of all experience, but I do not believe that they should be consigned to oblivion again as soon as physical concepts have been formed; I ascribe a higher value to them, as a link between physics and the other sciences. I have attempted elsewhere to show how a unified physics can gradually be built up without artificial hypotheses ... "

(Ernst Mach, *The Guiding Principles of my Scientific Theory of Knowledge*, from *Physical Reality*, ed. Stephen Toulmin, Harper, 1970)

"I am always ready to be enlightened by facts. Mach doubts the reducibility of the second law to probability, he does not believe in the reality of atoms. Very well: perhaps he or one of his disciples will one day develop another theory more effective than the present one. We must wait and see."

(Max Planck, *On Mach's Theory of Physical Knowledge*, 1910, from *Physical Reality*, ed. Stephen Toulmin, Harper, 1970)

Statistical thermodynamics based on Boltzmann's kinetic theory convinced some physicists and chemists that atoms and molecules are real physical entities, not just convenient mathematical models useful for deriving macroscopic results. However, there was strong opposition to this view from scientists and philosophers who adopted Ernst Mach's theory of physical knowledge. Mach took sensory experience as the primary reality and refused to accept the existence of anything beyond the reach of human senses. This approach was taken up by the energeticists, particularly Ostwald, who tried to derive the main results of mechanics directly from the conservation of energy. There were several significant philosophical implications including:

- a refusal to accept that the irreversibility of the second law derives from probability and initial conditions;
- a refusal to accept the existence of atoms;
- a refusal to accept mechanical determinism as proposed by Laplace and others.

Boltzmann in particular was frustrated by the continual attacks of the energeticists and some have suggested that this contributed to his suicide at the age of 62 in 1906. If this is true it is a terribly irony – in 1905 Albert Einstein published a paper on Brownian motion which established a direct link between statistical mechanics and the experimental demonstration of the existence of atoms. Einstein's paper was published in the same volume of Annalen der Physik that included his famous papers on special relativity and light quanta (the photoelectric effect). Brownian motion is the random vibration of tiny visible particles as a result of collisions with invisible molecules. Einstein's main intention was to show where the atomic hypothesis and kinetic theory of matter would lead to observable and measurable consequences. Einstein's trick was to consider the average motion of the particle as a result of a large number of irregular impulses from the molecular motion. He showed that this gradual migration behaves like diffusion and can be modelled by a similar differential equation. The result was a link between the mean square displacement of the particle, its radius a, temperature T, viscosity η and the Boltzmann constant k.

$$\left\langle x^2 \right\rangle = \frac{kT}{6\pi\eta a}$$

"I think that these investigations of Einstein have done more than any other work to convince physicists of the reality of atoms and molecules, of the kinetic theory of heat, and of the fundamental part of probability in the natural laws."

(Max Born, Einstein's Statistical Theories, *Albert Einstein: Philosopher-Scientist, Library of Living Philosophers*, Vol. vii, 1949)

The influence of statistical thermodynamics was also apparent in Einstein's 1905 paper on light quanta and in Planck's earlier work on black-body radiation. The links between deterministic statistical mechanics and the emerging quantum theory are very strong indeed. Einstein himself continued to apply these ideas to the quantum theory of specific heats and (with Bose) to the statistics of integer spin particles.

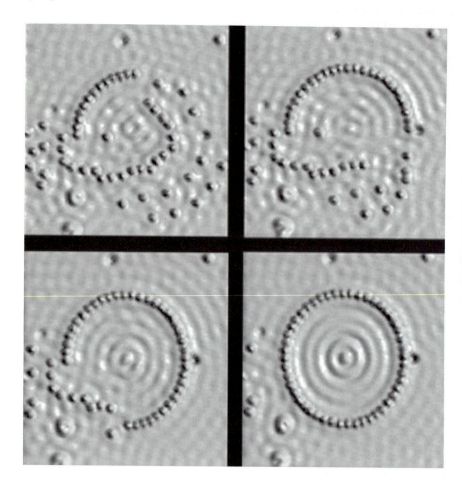

Figure 16.17 Scanning tunnelling electron microscope images of individual atoms. These remarkable images show how 48 iron atoms (the circular blobs) were arranged on the surface of copper to form a ring. Inside the ring the standing wave patterns show how the electrons are forced into quantum eigenstates. The image was formed by scanning the microscope tip over the metal surface at a height of just a few atomic diameters and measuring the tiny tunnel current between the surface and the tip. Variations in the current are used to map the topography of the surface. If you look carefully you can just see the hexagonal close-pack of copper atoms in the surface.

Image credit: IBM, Crommie, Lutz and Eigler

The idea that predictable physical behaviour might arise from a background of randomness in which details of individual processes are completely unknown was one of the most significant discoveries of the nineteenth century and has infiltrated virtually every branch of twentieth century physics. The discovery of quantum behaviour (by a direct application of statistical methods) changed the microscopic picture in a subtle way. In classical statistical mechanics the behaviour of individual particles was governed by the deterministic equations of Newtonian mechanics. In quantum theory the particles themselves do not possess well-defined properties of position and momentum, they are subject to the uncertainty principle. This means that identical initial states can evolve to different final states – the microscopic determinism disappears. However, this does not make any significant difference to the macroscopic consequences (at least not for systems with plenty of energy containing large numbers of particles).

16.7 LAPLACE'S DEMON AND DETERMINISTIC CHAOS

Statistical mechanics depends on two things, deterministic laws and initial conditions. This implies that if the present co-ordinates and velocities of all particles are known then all future and past co-ordinates and motions can in principle be calculated. This idea was first proposed by Roger Boscovich in 1763 but is better known from Laplace's writings.

"Thus we must consider the present state of the universe as the effect of its previous state and as the cause of its following one. An intelligence which could know, at a given instant, the forces by which nature is animated and the respective situation of the beings who compose it – and also was sufficiently vast to submit these data to analysis – would embrace in the same formula the movements of the greatest bodies of the universe and those of the lightest atom; for it, nothing would be uncertain and the future, like the past, would be present before its eyes."

(Pierre Laplace, *Essai philosophiae sur les probabilités*, Paris, 1814)

This 'intelligence' is possessed by a hypothetical creature known as Laplace's demon. If such a demon could exist and could possess this knowledge then everything is evolving in a fixed pattern and our impression of free-will is simply an illusion. We do what we must do because we too are part of this incredible but creatively barren Newtonian machine. Put in these terms it is hardly surprising that philosophers such as Mach rebelled against the consequences of atomism and statistical mechanics. In the twentieth century two discoveries have altered this perspective:

- Quantum theory describes physical processes in a dual way. An unobserved system evolves according to deterministic laws (e.g. the Schrödinger equation) whilst the effect of an interaction or observation is to cause the system to collapse at random into one or other of a number (possibly infinite) of alternative states. This destroys determinism. Quantum theory is discussed elsewhere in this book so we will not pursue it any further here.

- Chaos theory describes the behaviour of systems that are extremely sensitive to their initial conditions. A tiny change in initial conditions for a chaotic system results in future states that diverge exponentially. This puts severe restrictions on our ability to make long-term predictions for these systems.

The first person to study chaos was the French mathematician and physicist, Henri Poincaré, in the late nineteenth century. He noticed striking similarities between problems of prediction in celestial mechanics and problems in pure mathematics. However, the subject did not develop very far until the second half of the twentieth century. The reason for this was that the iterative methods required to attack the non-linear equations involved in chaotic dynamics need powerful computers and these were not available until then.

The difference between a non-chaotic system and a chaotic one can be illustrated quite simply. Imagine a mass suspended from a spring. If the mass is displaced and released its future positions can be predicted from deterministic equations. If there is a small error in the measurement of the initial displacement or release time (or of the mass suspended or the spring constant) then these errors go right through the calculation and result in a small error in our prediction of future positions or motions. Of course, the farther into the future we try to predict, the greater the discrepancy between our prediction and reality. But this error grows in a linear way with time. Now think of water flowing in a turbulent stream. Pick two adjacent particles in the flow and follow their motion (this could actually be carried out experimentally by dropping small pieces of paper onto the surface of a fast-flowing stream). They move rapidly apart even though the motion of the fluid is governed by deterministic laws derived from Newtonian mechanics. Similar initial conditions result in divergent outcomes. This implies that a small error in our knowledge of the intial co-ordinates and velocities of particles in turbulent flow will lead to rapidly (exponentially) growing errors in our predictions of their future motion. This is characteristic of chaotic systems – errors grow exponentially with time.

What is remarkable is that very simple dynamical laws can result in chaotic behaviour. For example, a simple pendulum free to oscillate in two directions becomes chaotic when driven by a small periodic force at the pivot – but only for some ranges of forcing frequency. At other frequencies its behaviour is non-chaotic and predictable.

What is the significance of this for determinism and predictability? Since the universe contains some deterministic and some non-deterministic systems it means that parts of it are in principle unpredictable (even without quantum theory). However, it does not mean they are not governed by deterministic processes – in fact this characteristic is often referred to as deterministic chaos. The point is that all measurements have some error associated with them, and no conceivable computer will ever be powerful enough to give a long-term accurate prediction for any chaotic system. Laplace's demon is not possible, even in principle. Does this mean that free-will and creative evolution can be reinstated? Some have argued that this is the case – chaos theory certainly shows that simple systems can behave in very complex ways, but the underlying processes are still deterministic, so any relation to consciousness and free-will will depend on our definitions. The future may be open in the sense that it is impossible to predict even though the processes by which it will come about are all predictable!

Ilya Prigogine, who won the Nobel Prize for Chemistry in 1977, sees the beginnings of a new philosophy of science in these discoveries. Classical thermodynamics was concerned with macroscopic equilibrium states ('being' as he would describe it), but most interesting situations involve flux, transitions between states and evolution (or 'becoming'). These involve far-from-equilibrium systems (such as living creatures, or working brains) which exist because they dissipate energy to the environment. For Prigogine one of the most important discoveries of the twentieth century is the realisation that there are fundamental limitations to the measurement process – we have been forced to take the observer and his apparatus into account. The idea that science is evolving toward a Platonic view with an objective reality fixed for all time seems to have failed. Relativity, quantum theory and chaotic systems have undermined it.

"Have we lost the essential elements of classical science in this recent evolution? The increased limitation of deterministic laws means that we go from a universe that is closed, in which all is given, to a new one that is open to fluctuations, to innovations.

For most of the founders of classical science – even for Einstein – science was an attempt to go beyond the world of appearances, to reach a timeless world of supreme rationality – the world of Spinoza. But perhaps there is a more subtle form of reality that involves both laws and games, time and eternity. Our century is a century of explorations: new forms of art, of music, of literature, and new forms of science. Now, nearly at the end of this century, we still cannot predict where this new chapter of human history will lead, but what is certain at this point is that it has generated a new dialogue between nature and man."

(Ilya Prigogine, *From Being to Becoming: Time and Complexity in the Physical Sciences*, W.H. Freeman, 1980)

17
TOWARD ABSOLUTE ZERO

17.1 TEMPERATURE

"My aim today is to introduce a course of lectures dealing with Low temperature Physics, which is now one of the most flourishing parts of modern science ... To get the proper perspective, let us first consider the position at the end of the last century ... Broadly speaking, a system is only influenced by a change in temperature if the energy differences associated with some possible change in the system are of the same order as its thermal energy; thus at low temperatures any temperature dependent phenomena are connected with small energy differences. Therefore it was thought that by decreasing the temperature still further one would be able to liquefy or solidify yet more gases with lower heats of evaporation or heats of melting, but that nothing unexpected would happen. Although this prospect was certainly important from the experimental point of view, it did not really produce much excitement among scientists. People busy with such activities were regarded in much the same way as someone who wanted to be first at the North Pole or run faster than anyone else. It is difficult now to realise that only a few years before the discovery of X-rays and radioactivity and the advent of quantum theory and relativity theory, it was still firmly believed that there was nothing essentially new to be found out. Even scientists such as Helmholtz took this view, although there was no dearth of signs that all was not well with classical theory.

In addition there is one other point which explained the relative indifference of scientists toward low temperature research. According to classical theory, absolute zero was not only unattainable, but was without much thermodynamic significance."

(F.E.Simon, Low Temperature Problems, A General Survey, in *Low Temperature Physics – Four Lectures*, Pergamon Press, London, 1952)

17.1.1 The Concept of Temperature

The macroscopic idea of temperature is linked to other macroscopic changes, such as the thermal expansion of mercury, the increase of pressure in a constant volume of gas or the increase in resistance of a metal wire. It is linked to microscopic behaviour in a loose way through the mean kinetic energy of molecules, but this is by no means an adequate definition of temperature. A proper understanding of temperature requires a definition of thermal equilibrium and an understanding of the link between entropy change and heat flow.

It is surprising to realise that the logic of temperature was only worked out in the twentieth century. It depends on the crucial idea of thermal equilibrium, but what does this mean? When are two objects in thermal equilibrium? In practice this is when they reach a state where there is no net heat flow between them. It is a steady state (if they are isolated from other influences) that we describe as having equal temperatures. This idea is so fundamental to thermodynamics that an additional law was tagged on to the other three in the 1920s to define it:

The Zeroth Law of Thermodynamics: *If A is in thermal equilibrium with B and C separately, then B and C are also in thermal equilibrium.*

This is linked to the second law of thermodynamics. Heat will flow spontaneously from one body to another if this flow causes an increase in entropy. Thermal equilibrium occurs when there is no net tendency for flow in either direction, absorbing heat from one body and dumping it in the other leaves the entropy of the universe unchanged.

Maths Box: Defining Temperature

For thermal equilibrium there is no change in total entropy when a small amount of heat, Q flows in either direction:

$$\delta S_{AB} \text{ (when heat } Q \text{ is transferred)} = \left(\frac{dS_A}{dE_B}\right)\delta Q - \left(\frac{dS_B}{dE_B}\right)\delta Q = 0$$

where the E's are the internal energies of A and B and the S's are their entropies. The equation states the condition for thermal equilibrium of A and B. It is quite clear that the condition reduces to:

$$\frac{dS_A}{dE_A} = \frac{dS_B}{dE_B}$$

This gives us a new thermodynamic quantity that is the same for A and B when they are in thermal equilibrium. The zeroth law extends this to all bodies so that a quantity called temperature can be defined.

In some ways it would be convenient if we defined temperature to be the rate of change of entropy with energy. In this scheme 'absolute zero' would be replaced by infinite temperature, and the difficulty of cooling toward this value would be reflected in the values achieved. But this would not correspond to existing ideas of temperature. Instead the thermodynamic temperature is defined by:

$$\frac{1}{T} = \frac{dS}{dE} \quad \text{or} \quad T = \frac{dE}{dS}$$

If the change in internal energy is brought about by a reversible heat flow in or out of the body then $\delta E = \delta Q$ so that:

$$T = \left| \frac{dS}{dQ} \right|_{rev}$$

This gives us a feel for the meaning of temperature. If a body is at a high temperature its entropy increases relatively slowly as it is heated. If it is at a low temperature its entropy increases rapidly. This also makes sense in terms of 'energy shuffling' – if there are already an enormous number of quanta in the system then the number of micro-states is huge and the addition of a few more quanta has a relatively small effect. On the other hand, a cold body containing few quanta has a much smaller number of possible micro-states. This number increases significantly with the addition of even a small amount of extra energy.

17.1.2 Absolute Zero

In classical physics absolute zero is the temperature at which molecular motion ceases so that each material forms a perfect crystal lattice. However, classical physics also predicts that the heat capacity of a solid remains constant to absolute zero. The reason for this is tied in with the idea of equipartition of energy. This is a powerful assumption that thermal energy will spread among all the available degrees of freedom (e.g. translation, rotation, vibration) of the particles so that each degree of freedom gets an average thermal energy of $\frac{1}{2}kT$. This leads to a rather awkward conclusion:

$$T = \frac{dE}{dS} = \frac{cdT}{dS} \quad \text{where } c \text{ is the heat capacity}$$

$$\delta S = \frac{c\delta T}{T}$$

$$\delta S \to \infty \quad \text{as } T \to 0$$

This means the entropy changes associated with small energy changes close to absolute zero blow up. If the entropy at any non-zero temperature is finite then the entropy at absolute zero must be minus infinity. One consequence of this would be

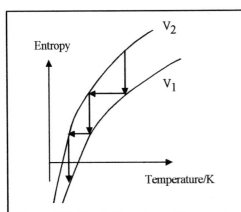

Figure 17.1 Classically entropy falls to minus infinity at absolute zero. The arrows show processes that manipulate the state of the system in order to lower its temperature. It is clear that absolute zero is unattainable.

that absolute zero is unattainable since we would have to dump an infinite amount of heat (and hence entropy) in the environment in order to pay for the infinite reduction in entropy of even the smallest sample as it was cooled to 0 K.

Quantum theory changes all this. In 1906 Einstein wrote a paper on the specific heats of solids in which he showed that specific heats should fall to zero as the temperature approaches absolute zero. This ground-breaking paper was the first ever to be written on the quantum theory of the solid state. Einstein considered a simple model solid in which the atoms oscillate independently about their equilibrium positions at lattice points with a frequency f, subject to similar quantum conditions to Planck's theory. The oscillators can have energies that are integer multiples of the quantum $E = hf$. He then assumed that the average energy per oscillator would fit the Planck law rather than getting kT as in classical equipartition of energy (this is kT rather than $\frac{1}{2}kT$ because each oscillator has both kinetic and potential energy). This had the effect of 'quenching' the oscillations at low temperatures (a result analogous to the solution to the ultra-violet catastrophe in black-body radiation) so that the specific heat of an Einstein solid would approach zero at absolute zero. This allows the entropy of the solid to approach a finite value at absolute zero.

While Einstein was working at the Patent Office in Bern and publishing papers on the specific heat of solids (along with relativity, quantum theory and statistical mechanics) in his spare time, Hermann Walther Nernst was also thinking about the behaviour of materials close to absolute zero. In 1905 he suggested that the specific heats of solids might approach a constant value at absolute zero and stressed how important it would be to gather more data about specific heats at low temperatures in order to test this idea. In fact he carried out much of the pioneering work himself and by 1910 he had results that supported the hypothesis. This work resulted in:

The Third Law of Thermodynamics: *as the temperature tends to zero the entropy of a substance also tends to zero.*

This is sometimes expressed in a rather different way, stating the impossibility of cooling something to absolute zero in a finite number of steps.

17.2 REACHING LOW TEMPERATURES

17.2.1 Catching Cold

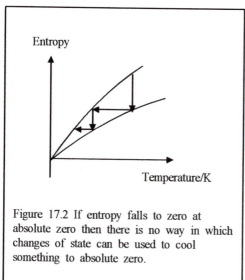

Entropy

Temperature/K

Figure 17.2 If entropy falls to zero at absolute zero then there is no way in which changes of state can be used to cool something to absolute zero.

Hot Big Bang cosmology assumes that the universe started off in a high temperature ordered state and has been cooling and increasing its entropy ever since. The expansion of the universe has reduced the temperature of the cosmic background radiation (formed when matter and anti-matter annihilated) to a uniform temperature of about 3 K above absolute zero. This is the temperature of 'empty space' so there are no natural thermal reservoirs at lower temperatures. This means that any attempt to cool things below 3 K will have to use apparatus that is entirely above that temperature and achieve the cooling effect by enabling some heat flow to the environment. It also means that we are creating environments that are colder than any that have ever existed in the history of the universe! How can this be done? In practice we manipulate the energy and entropy of a system in order to nudge it toward absolute zero. This can be made clearer if we consider the refrigerator – which is basically a heat engine run in reverse – as a heat pump.

The essential process in refrigeration is for the refrigerant to absorb heat from the inside of the refrigerator. A volatile refrigerant is pumped around a closed circuit passing in and out of the refrigerator. The cooling effect is obtained as the refrigerant vaporises and expands through a valve inside the refrigerator. This changes its state and increases the volume of gas, so its temperature falls and heat flows into it. However, to return the refrigerant to the refrigerator it must be recompressed and condensed. As this happens the energy it has absorbed will raise its temperature above that of its surroundings, so heat flows out into the environment. Since this energy is being dumped at a higher temperature than that at which it was absorbed it must be supplemented by extra energy (the electrical energy put in to run the refrigerator). In practice this supplement arrives by means of the compressor/pump that circulates the refrigerant. In this way heat flows from cold to hot without violating the second law of thermodynamics. Once the refrigerant has cooled to the temperature of the surroundings it is returned to the refrigerator and taken through the same cycle of changes repeatedly.

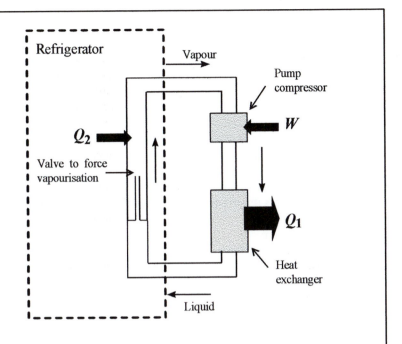

Refrigerator

Vapour

Pump
compressor

Q_2

W

Valve to force
vapourisation

Q_1

Heat
exchanger

Liquid

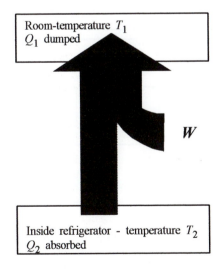

Room-temperature T_1
Q_1 dumped

W

Inside refrigerator - temperature T_2
Q_2 absorbed

Figure 17.3 A refrigerator is a heat pump. To satisfy the second law it must not reduce the entropy of the universe. This means that Q_1/T_1 must exceed Q_2/T_2. T_2 is less than T_1 so Q_1 is greater than Q_2. The extra heat comes from the work done in the compressor and pump.

The refrigeration effect is achieved when the refrigerant, in a liquid state, undergoes a dramatic, irreversible change of state to become a cold vapour. This transition forces the number of micro-states to increase so that the available energy spreads more thinly and the temperature drops. The heat absorbed from inside the refrigerator lowers the inside temperature. Refrigerants exploit a change of state to manipulate entropy and temperature. At very low temperatures more sensitive processes are needed. The essential characteristic is that the system can exist in two states with a significant difference in entropy. In addition there must be some external means of switching the system between these states. The refrigeration process then proceeds in four main steps:

1. Work W is done to put the substance that will absorb the heat into an ordered state. This increases its temperature (same energy, fewer micro-states).
2. Heat Q_2 flows from the refrigerant to the surroundings until it reaches the ambient temperature.
3. The refrigerant is brought into thermal contact with the object to be cooled and the 'force' which imposed the order upon it is switched off.
4. The refrigerant returns to its original disordered state absorbing heat Q_1 from the objects in contact with it (more micro-states, energy spreads more thinly). This lowers the temperature of the object.

Return to step 1.

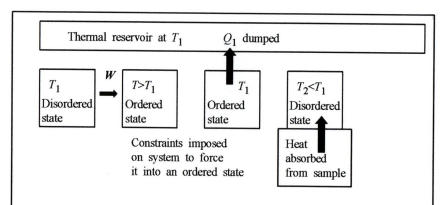

Figure 17.4 The essential principles of cooling. If the coolant is forced into an ordered state its thermal energy is shared by a reduced number of degrees of freedom and the temperature rises. Heat flows out to a thermal reservoir at T_1. The constraint is removed and the coolant temperature drops, allowing it to absorb heat from a sample in thermal contact with it.

Most cryostats make use of similar principles. The preliminary stages might involve liquid nitrogen, followed by liquid helium. The technique is simple (in principle): the liquid is allowed to evaporate at reduced pressure and the vapour is continually pumped away. Evaporation removes molecules with greater than average thermal energy so the average of the remaining molecules falls and so does the temperature. This method can get down to about 0.7 K with helium-4 and below 0.3 K if the less massive helium-3 isotope is used instead. An ingenious method to reach even lower temperatures was invented by Heinz London in 1951. He used a mixture of helium-4 and helium-3 which splits into two phases close to absolute zero. Both are mixtures but the concentration of helium-3 is low in one and high in the other. As far as the helium-3 is concerned the interface between these distinct phases acts like a liquid/gas boundary. The helium-3 'evaporates' from the 'liquid' (helium-3 rich) phase into the 'gas' (helium-3 depleted) phase, cooling the former. By pumping liquid helium away from the 'gas' phase the *dilution refrigerator* can achieve temperatures around 1 mK. Below this more subtle changes of state are used.

Two magnetic methods are very widely used. The first uses paramagnetic salts. These remain disordered down to very low temperatures. The technique of cooling by adiabatic demagnetisation uses the transition from this disorder to the order imposed by an external magnetic field:

1. Switch on the magnetic field – all the magnetic dipoles line up. The system becomes more ordered so the temperature rises. To prevent a reduction in entropy work must be done by the outside agent and this additional energy is also distributed as heat in the material.
2. Let the warmed material lose heat to its surroundings.
3. Bring it into contact with the object to be cooled and switch off the magnetic field.
4. The dipoles become disordered, the temperature drops and heat flows in from the object to be cooled.

If this method is used on samples that have already been cooled by a liquid helium cryostat it can cool them down below 0.1 K. The method eventually fails because it relies on thermal agitation to keep the dipoles disordered at the start of the process. At very low temperatures the thermal energy is too small, so even more sensitive methods are needed. One of these involves nuclear magnetic moments, which interact so weakly that they remain disordered down to extremely low temperatures. The nuclear demagnetisation technique has achieved temperatures below 10^{-5} K.

In the earl 1990s John Martinis invented a microelectronic refrigerator for use in electronic circuits. It is a superconducting quantum device that works by allowing only above average energy electrons to tunnel through a potential barrier. This has much the same effect as evaporation and the device can cool components to less than 1 K.

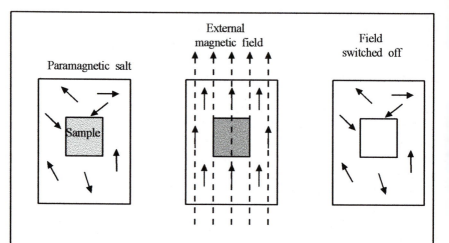

Figure 17.5 Paramagnetic cooling. A sample is placed in thermal contact with a paramagnetic salt. When an external magnetic field is applied the dipoles align and the temperature of the salt rises. Heat flows out of it until it has cooled back to its starting temeperature. Now the field is switched off and the temperature drops further.

17.2.2 Liquid Helium

Heike Kammerlingh Onnes, at Leiden University, led the team that liquefied helium-4 for the first time in July 1908. This was the first major result in the new field of low temperature physics, even so it was more than decade before other laboratories began to continue the work started by the Leiden group. Having liquefied helium (below 4.2 K) Onnes expected that they would be able to solidify it as well. However, repeated attempts at ever-lower temperatures were all unsuccessful – liquid helium would not solidify under its own vapour pressure (it was later successfully solidified at a pressure of about 30 atmospheres). This remarkable behaviour is a consequence of quantum theory. All atomic oscillators have a non-zero zero point energy and this is larger for light atoms. In helium the zero-point energy is large enough to prevent the atoms settling into fixed lattice positions in the solid state. Hence the failure to solidify.

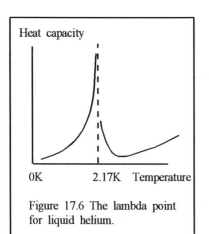

Figure 17.6 The lambda point for liquid helium.

Another strange property of liquid helium is its transition to a different distinct phase below 2.2 K (He-I above 2.2K, He-II below 2.2K). In the 1930s W. Keesom and A.P. Keesom investigated this phase change and discovered two odd things – there is no latent heat involved, but the specific heat is discontinuous. In fact a graph of specific heat versus temperature around 2.2 K has such an unusual shape that this temperature is called the lambda point. Zero latent heat means there is no entropy change at this phase change, a very surprising result. And there were other surprises in store. When rapidly boiling He-I is cooled through the λ-point it suddenly stops boiling – a fact linked to the sudden onset of extreme thermal conductivity. Yet another strange observation was the ability of He-II to leak from just about any container and to flow with zero viscosity through capillary tubes – a property described by Kapitsa as *superfluidity*.

These weird properties are related to macroscopic quantum states. In 1924 Einstein and Bose suggested that a collection of integer spin particles (bosons) should drop into a unique macroscopic quantum state as temperature approaches absolute zero. This behaviour is the exact opposite of that undertaken by half-integer spin fermions which obey the Pauli Exclusion Principle so that no two of them are allowed to have an identical set of quantum numbers.

17.3 QUANTUM STATISTICS

17.3.1 Fermions and Bosons

The explanation for the weird low temperature behaviour of liquid helium lies in the way the micro-states of a quantum system are counted. In classical physics identical particles are treated independently so that micro-states obtained by exchanging a pair of particles are counted as different micro-states – both must be counted. This amounts to the assumption that identical particles are nonetheless distinguishable. In quantum theory identical particles are *not* distinguishable, so a state in which two identical particles are exchanged is the same as the original state and must only be counted once when working out the entropy of the system. This indistinguishability means that the wave function representing a system of two particles must give the same observable consequences as one with those two particles exchanged:

$$|\Psi_{AB}|^2 = |\Psi_{BA}|^2$$

leading to two alternatives: $\Psi_{AB} = \pm\Psi_{BA}$

Both alternatives appear in nature:

- Bosons – integer-spin particles – have symmetric wavefunctions, there is no change when two particles are exchanged.
- Fermions – spin-half particles – have an asymmetric wavefunction, it changes sign when particles are exchanged.

A second classical assumption is that the presence of one particle in a particular energy state has no effect on the probability that another identical particle might also be in that energy state. The behaviour of quantum particles changes this. The chance of a boson occupying a particular energy state is enhanced if there are already other identical bosons in that state. The chance of a fermion entering a state that is already occupied is zero – this is the Pauli Exclusion Principle. This means that the distribution of particles among available energy states will be different for classical particles, bosons and fermions. (The reason we still need the classical distribution is because the situations we usually encounter in the macroscopic world are such that these quantum effects play an insignificant rôle – both the Bose-Einstein and Fermi-Dirac distributions are asymptotic to the Boltzmann distribution at high temperatures).

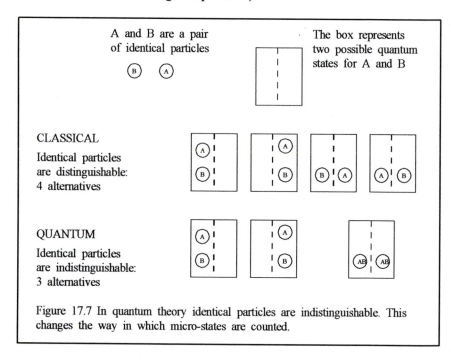

Figure 17.7 In quantum theory identical particles are indistinguishable. This changes the way in which micro-states are counted.

What does this imply? At low temperatures bosons will 'condense' into a single macroscopic quantum state in which the behaviour of all particles is correlated. This is called Bose-Einstein condensation. Above 2.2 K helium-4 behaves as a classical liquid (He-I). However, helium-4 nuclei are bosons and below 2.2 K (the lambda point) a rapidly increasing number of nuclei undergo a Bose condensation. The highly correlated behaviour in this state accounts for the increase (by more than 10^6) in thermal conductivity and the decrease (by a similar factor) in viscosity. At very low temperatures all the nuclei drop into the same low energy state. The tendency of bosons to cluster in the same energy state is also apparent in superconductivity and laser light.

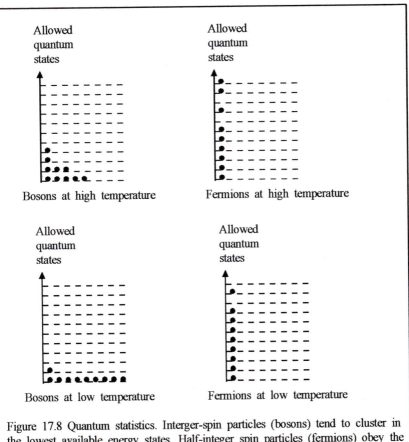

Figure 17.8 Quantum statistics. Interger-spin particles (bosons) tend to cluster in the lowest available energy states. Half-integer spin particles (fermions) obey the Pauli Exclusion Principle, so at low temperatures the lowest available states are all singly occupied.

For fermions the story is rather different. No two fermions are allowed to occupy the same quantum state so, at low temperatures all the fermions drop into the lowest available states filling each one singly. This behaviour is imporant for the explanation of conductivity in metals (electrons in metals behave like a fermi gas).

Box: Degeneracy Temperature

When is a fluid a quantum fluid? In other words, under what conditions do we have to take into account the wave nature of particles in a fluid? Presumably when the separation of the particles is comparable to or less than their de Broglie wavelength. This leads to a simple condition on temperature:

$$\lambda \geq a$$

$$\lambda = \frac{h}{p}$$

$$E = \frac{3}{2}kT = \frac{p^2}{2m} \quad \text{so} \quad p^2 = 3mkT$$

$$\lambda^2 = \frac{h^2}{3mkT} \geq a^2$$

So quantum effects will be important when

$$T \leq \frac{h^2}{3mka^2} = T_0$$

T_0 is called the degeneracy temperature and gives an indication of when we might effect the Bose condensation to occur. In practice this is just an indication. Other factors are also important but some general conclusions can be drawn:

- $T \propto \dfrac{1}{m}$ so lower mass particles will have higher degeneracy temperatures.

- $T \propto \dfrac{1}{a^2}$ so low density gases are not likely to exhibit quantum effects.

Liquid helium below 2.2 K behaves like a quantum gas because of its low mass and small inter-particle separation. Electrons have very low mass compared to atomic nuclei, so their degeneracy temperature at comparable densities (e.g. in a metal) is very high – their behaviour must be described by quantum statisitics (Fermi-Dirac statistics because they are spin-half particles). Neutrons in neutron stars also form a Bose condensate despite their high temperature because they have been subjected to intense pressure and their inter-particle separation is extremely small.

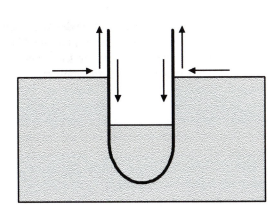

Figure 17.9 Superfluidity in liquid helium. One surprising consequence is the ability of He-II to self-siphon out of a container. This frictionless flow takes place through a thin film of liquid helium that forms on the surface of the container. If the level inside the tube was higher than that of the reservoir it would flow the other way.

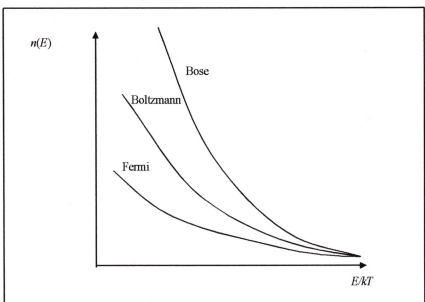

Figure 17.10 Particle energies in a classical gas follow the Boltzmann distribution. If quantum effects have to be taken into account then fermions and bosons have distinct distribution functions. However, at high energies ($E \gg kT$) the probability of two particles occupying the same quantum state is low, so quantum effects are negligible so the quantum distributions converge to the Boltzmann distribution.

17.3.2 Superconductivity

Electrical conductivity in metals depends on the motion of electrons through the metal lattice. Resistance is explained by an interaction between the electrons and the lattice resulting in scattering of the electrons and a transfer of energy to the lattice. Felix Bloch developed a detailed quantum theory of metallic conduction by considering how electron waves are affected as they travel through a periodic potential set up by the lattice. Lattice vibrations (phonons) distort the lattice and scatter the electron waves. The phonons (mechanical analogues of electromagnetic photons) are also quantised and absorb energy in discrete amounts linked to their frequency by the familiar equation $E = hf$). This theory predicts that electrical resistivity will fall with temperature as the lattice vibrations die away (fewer phonons) and resistivity should be zero at absolute zero (which is unattainable).

In 1911 Kamerlingh Onnes was measuring the resistivity of very pure mercury just below the boiling temperature of liquid helium. To his surprise the resistance suddenly began to fall rapidly and soon became too small to measure. This was the first observation of zero resistivity and the fact that it occurred at a finite temperature meant it was an entirely new property – *superconductivity*. Some other metals were also found to exhibit superconductivity, with a distinct transition temperature for each (4.12 K for mercury) but by no means all of them.

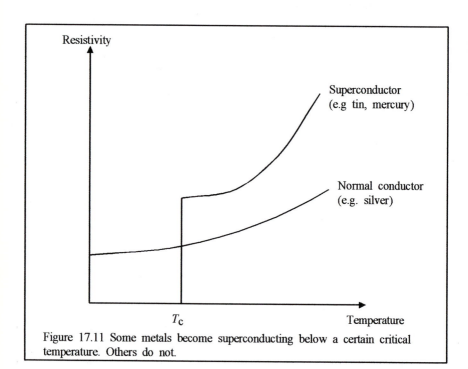

Figure 17.11 Some metals become superconducting below a certain critical temperature. Others do not.

Up to the mid-1980s a few materials had been discovered with higher transitions than mercury, but none above 25 K. Then in 1986 Alex Müller and Georg Bednorz at IBM Zurich discovered evidence of a superconducting transition close to 30 K in a ceramic, La_2CuO_4 doped with a small amount of barium. The following year Chu and Wu discovered superconductivity in $YBa_2Cu_3O_7$ at 123 K, and since then other related compounds have pushed the highest superconducting transition temperature a little higher. These are important discoveries because they allow superconducting properties to emerge at liquid nitrogen temperatures and offer the tantalising hint that room temperature superconductors might be possible. At present, however, there is no consensus on how these materials achieve superconductivity or how closely their super-conducting mechanism is linked to that of the metals (BCS-theory described briefly below). The 1987 Nobel Prize for Physics was awarded jointly to J. Georg Bednorz and K. Alexander Müller *"for their important breakthrough in the discovery of superconductivity in ceramic materials"*. (Felix Bloch is also a physics Nobel Laureate, but the work cited was associated with nuclear magnetic resonance measurements and not the solid state theory of conduction.)

Superconducting samples have zero resistance. This means there is no potential difference across them when current flows through them and a superconducting current set up in a closed circuit will continue to flow indefinitely. Another strange effect associated with superconductivity is the Meisner effect – superconductors exclude magnetic fields. When a superconducting sample is moved into a magnetic field the field lines distort so that they flow around it. On the other hand, if a non-superconducting sample is cooled through its transition temperature while it is in a magnetic field then the sample expels the field. This can be used to create a form of magnetic levitation. If a magnet is lowered toward a superconducting

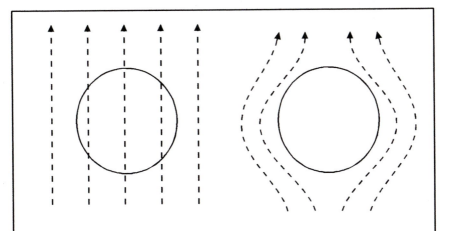

Figure 17.12 The Meissner effect. Above T_C magnetic flux penetrates the material just like a normal conductor. Below T_C the field is completely expelled.

sample there is a transfer of energy from gravitational potential energy of the magnet to the magnetic field as it is expelled. The magnet can only fall further if the change of gravitational potential energy is large enough to pay for the associated change in magnetic energy. Since the latter increases closer to the sample the magnet stops falling before it reaches the sample and floats in the air above it!

One further consequence of the Meisner effect and the quantum nature of superconductivity is the phenomenon of 'flux-quantisation'. If a superconducting ring is placed in a magnetic field then the magnetic flux is again expelled from the superconductor, but some flux can pass through the centre of the ring. But not any amount of flux – the flux through the ring is quantised in units of $h/2e$. This effect has been important in measuring tiny magnetic fields (e.g. from electrical activity in the brain) and in devising new ways to measure fundamental constants and to set voltage standards. It also gives a big hint about the mechanism of superconductivity – if supercurrent is carried by electrons theory suggests that flux ought to be quantised in units of h/e, the extra factor of 2 suggests that the charge carriers have an effective charge $2e$, as if the electrons in some way paired up.

The co-ordinated behaviour of large numbers of electrons in a superconductor and the sudden transition to zero resisitivity are both reminiscent of the transition to superfluidity that takes place in liquid helium below the lambda point. However, the transition in helium is due to Bose-Einstein condensation and electrons are fermions. How can superconductivity be linked to the behaviour of bosons? The present theory is due to John Bardeen, Leon Cooper and Robert Schreiffer in 1957. They came up with the rather strange idea that distant electrons may form bound states using lattice distortions as an intermediary. These bound states are bosons and would undergo a Bose-Einstein condensation at a low enough temperature. This explains a super-current as the flow of 'Cooper-pairs' in a macroscopic quantum state formed by Bose-Einstein condensation. The discovery of the quantum of flux ($h/2e$) gave early experimental support to the model (as mentioned above).

It is interesting to compare the Bose-Einstein condensation of electrons with the behaviour of liquid helium-3 at low temperatures. Helium-3 nuclei are fermions, so they behave as a fermi liquid obeying Fermi-Dirac statistics below their degeneracy temperature. However, at very low temperatures (mK) they to can form Cooper pairs and undergo a Bose-Einstein condensation to become a superfluid, a state observed experimentally for the first time in the 1970s. The mechanism by which the Cooper pairs form in liquid helium-3 is analogous to that in a superconductor, but depends on a distortion of atomic spins in the surrounding liquid rather than atomic positions in a lattice. The helium-3 particles are much more complex internally than electrons and their internal degrees of freedom are constrained when they form Cooper pairs, so having some macroscopic effect on the behaviour of the condensate. This has interesting physical consequences, enabling some weak atomic-scale processes to be amplified so that they manifest themselves in the macroscopic domain. David Lee, Douglas Osheroff and Robert

Richardson shared the 1996 Nobel Prize for Physics *"for the discovery of superfluidity in liquid helium-3"*.

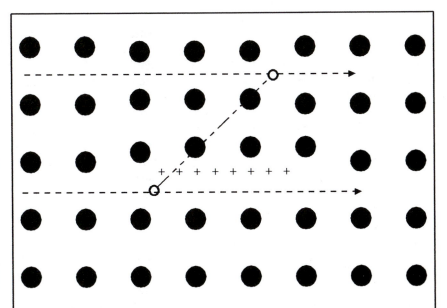

Figure 17.13 Cooper pairs. As the top electron moves through the lattice it distorts the lattice by attracting ions away from their equilibrium positions. This creates temporary regions of positive charge that stay in place for a significant time. These charges can interact with a second electron forming a link between the two electrons so that they behave like a boson.

One of the most important practical applications of superconductivity is in high-energy research at accelerator facilities such as CERN. Large superconducting coils can generate strong magnetic fields (in excess of 20 T) over large volumes (measured in m^3) whilst dissipating negligible amounts of heat.

17.3.3 Laser Cooling

One of the most exciting developments in low temperature physics involves the use of lasers to slow atoms and cool them. In the last few years light has even been used to trap large numbers of atoms in a new state of matter that then exists as a single macroscopic quantum state. These Bose-Einstein condensates form 'blobs' of matter millimetres across in which every atom is in an identical quantum state. These 'BEC blobs' have already been used to probe the implications of quantum theory in macroscopic objects. In future they may be used to create matter lasers, ultra-accurate atomic clocks and for high-resolution lithography.

The idea behind laser cooling is quite simple. If an atom is moving to the right and it absorbs a photon which was moving to the left momentum will be conserved and the atom slows down. To ensure that the atom will absorb photons they must be tuned to a resonant frequency in the atom. The trick is to use a photon frequency in the laboratory that is slightly below the atom's resonant frequency. The Doppler effect as they approach the atoms means that their frequency in the atom's reference frame is shifted to a higher value and they are absorbed. In 1985 William Phillips, at the National Institute of Standards and Technology in Gaithersburg, Maryland used this method of 'Doppler cooling' to cool sodium atoms to 43 μK.

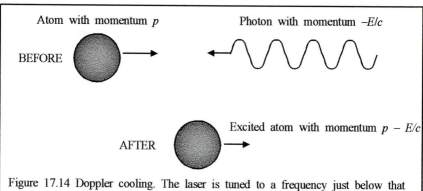

Figure 17.14 Doppler cooling. The laser is tuned to a frequency just below that of a resonance in the atom. The Doppler effect shifts the frequency so that it is absorbed. The atom slows down.

Steven Chu extended the method, using six laser beams from three pairs of lasers arranged along the three cartesian axes. This had the effect of slowing the atoms in all three dimensions. Chu described the effect as an atom passing through 'optical molasses' because the region between the lasers acts like a viscous medium. However, there is a theoretical limit to the lowest temperature that can be reached using Doppler cooling because of the buffeting they get from photons in the laser beams. In 1988 Phillips developed a time-of-flight method to measure the temperature achieved by Dopper cooling and found that it was significantly below the theoretical limit.

The effect was explained by Cohen Tanoudji. He realised that the lasers set up polarised standing waves and the drag resulted from an interaction between the moving atoms and the periodic potential in these standing waves. The wave pattern has an electric field strength that varies with position. This interacts with charges in the atom making the atomic potential energy vary from place to place. As the atom moves through this varying potential it absorbs photons when it is near the peaks and immediately re-radiates them so that it always ends up back in the 'valleys'. It loses more energy as it climbs back up the potential gradient, repeating this process again and again and slowing down. This has been called 'Sysiphus cooling' after the Greek myth in which Sysiphus spends his days pushing a boulder to the top of a hill only to see it roll back down at night so that he must repeat the process again the following day.

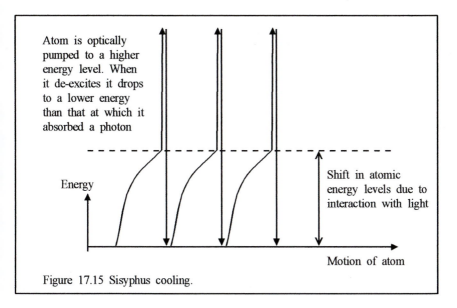

Figure 17.15 Sisyphus cooling.

The laser cooling methods described above are all velocity dependent. This means they cannot be used to trap atoms. If the atoms are to be trapped then the forces acting on them must be position dependent. This is achieved by adding a non-uniform magnetic field to the optical molasses so that the atoms experience a central force and is called a magneto-optical trap (or MOT). Millions of atoms can be trapped so that they form a very low temperature lattice with an atomic separation comparable to the wavelength of light and about 10^3 times greater than the typical separation in a solid or liquid. This is a new form of matter – its density is about a billionth of the density of ordinary matter and, since it is at such a low temperature, the wavefunctions of all the atoms overlap so that it drops into a single macroscopic quantum state. By 1996 visible blobs of this exotic cold matter containing around 10^7 atoms all in an identical quantum state at temperatures measured in tens of nano-kelvin had been created. In 1997 two such 'BEC blobs' were made to collide. The resulting interference fringes suggest that the principle of superposition is still applicable, even for a macroscopic state. This is an important test of fundamental ideas – several theorists have suggested that the superposition principle will break down with large objects. This result seems to suggest it remains valid (so that Schrödinger's cat really is alive and dead at the same time!). Steven Chu, Claude Cohen-Tannoudji and William D. Phillips shared the 1997 Nobel Prize for Physics "*for the development of methods to cool and trap atoms with laser light*".

18

CPT

18.1 INVARIANCE AND SYMMETRY

18.1.1 Conservation Laws

"I asked Dirac why he had not introduced parity in his famous book on quantum mechanics. With characteristic simplicity Dirac answered: "Because I did not believe in it", and he showed me a paper in the Reviews of Modern Physics in which he said so. In that same paper, Dirac expressed doubts about time reversal invariance as well, but, four years ago, I did not pay much attention to that. Unjustifiedly ... ."

(Conversation between Abraham Pais and Paul Dirac which took place in 1959, as reported in *Inward Bound*, Abraham Pais, OUP, 1986)

The arrow of time as we experience it in everyday life seems to be linked to the inexorable evolution of complex systems toward equilibrium states. This has been described in detail and we saw how it led to the introduction of entropy and the microscopic description of the second law of thermodynamics. While this goes some way toward explaining the observed temporal bias in our experiences and distinguishes the past (low entropy) from the future (high entropy) it does not explain how this asymmetry arose in the first place. Neither does it satisfy our fundamental desire to know *what time is*. If we look at the details of thermodynamic time by restricting our attention to the behaviour of just a few particles we find that the arrow of time begins to dissolve. The laws of physics on the smallest scales are time reversible – or are they? Does an electron know the difference between the past and the future? If we accept that an electron is a fundamental particle with no internal structure it is hard to see how we can attach any kind of thermodynamic arrow of time to it (unless, of course, it is intimately connected to a complex environment with many degrees of freedom, but this rather detracts from the idea of the electron as fundamental!). For this reason it was assumed that there is no intrinsic arrow of time operating on the level of individual particles and so all particle interactions should be symmetric with respect to time reversal (an operation we shall simply denote as T). The discovery that T-invariance is violated in certain weak decays came as a shock to most physicists and the fundamental reasons for this behaviour are still not understood. Roger Penrose has commented that it would be surprising if this subtle but highly significant effect is not telling us something very deep about the nature of time.

"Conservation laws have a special significance for classical systems, for they enable us to say something about these systems without a detailed knowledge of their structure. Classical theory without conservation laws in inconceivable. The most obvious example to quote is perhaps classical thermodynamics, where we are able to establish with no great difficulty a set of very general relationships about thermal systems without knowing anything of the supposed molecular or atomic states of matter. We can do this with the help of the conservation of energy. A reflection of this fact is that phenomenological thermodynamics almost alone among nineteenth-century scientific achievements has survived twentieth century criticism. Classical thermodynamics is a set of behavioural laws offering no explanation for what happens. This exemplifies a general characteristic of conservation laws, namely that they enable us to elicit behavioural laws but not explanations."

(D.W. Theobald, The Concept of Energy, E. & F.N. Spon, London, 1966)

The universe is in a state of continual flux, but certain quantities seem to stay the same as everything around them changes. Mass-energy, momentum, angular momentum and charge are all conserved quantities. This means their individual totals are the same now as they were at any stage in the history of the universe. Interactions may result in a transfer between particles or systems, but it never diminishes or increases any of them. Why is this? Conservation laws are linked to underlying symmetry operations and as such they give us an important insight into the fundamental restraints under which the laws of physics operate.

Conserved quantity	Symmetry
Mass-energy	The laws of physics are the same at all times, i.e. invariant with respect to a change in time origin
Linear momentum	The laws of physics are the same in all places, i.e. invariant with repect to displacement of space origin
Angular momentum	The laws of physics are the same in all directions, i.e. invariant with respect to a rotation
Charge	The laws of physics are the same regardless of our choice of the zero of electric potential, i.e. 'gauge invariant'

Maths Box: Conservation Laws in Mechanics

Newton formulated mechanics in terms of *things* – massive particles acted upon by forces. The concept of force was central to this and the action of force through space and time are responsible for transfers of energy and momentum and lead to the conservation laws.

$$F = \frac{dp}{dt} \quad \Delta p = \int F dt$$

$$F = \frac{dE}{ds} \quad \Delta E = \int F ds$$

Later, mechanics was reformulated in a more abstract way in terms of systems and processes particularly by Lagrange and Hamilton (between 1780 and 1840). Their approach, whilst linked to Newton's, did not start from the concept of force (in fact forces need not be introduced) but energy plays a central role. Lagrange introduced a new function called the Lagrangian (L) equal to the difference between kinetic (T) and potential energy (V):

$$L = T - V$$

This can be used to link generalised position (q_i) and momentum (p_i) co-ordinates by a simple equation:

$$p_i = \frac{\partial L}{\partial \dot{q}_i}$$

and showed that this led to an equation of motion for the evolution of a closed system:

$$\frac{d}{dt}\left(\frac{\partial L}{\partial \dot{q}_i}\right) - \frac{\partial L}{\partial q_i} = 0$$

Now if the Lagrangian does not depend on a particular generalised co-ordinate, say $q_j = x$ then this simplifies to:

$$\frac{d}{dt}\left(\frac{\partial L}{\partial \dot{x}}\right) = 0 \quad \text{since} \quad \frac{\partial L}{\partial x_i} = 0$$

and so

$$\frac{\partial L}{\partial \dot{x}} = \text{constant}$$

but
$$\frac{\partial L}{\partial \dot{x}} = p_x$$

so the x-component of momentum is conserved.

This can be generalised. If the Lagrangian is independent of the choice of co-ordinate origin then linear momentum will be conserved in all directions – the linear momentum vector of a closed system is a constant.

In a similar way, if the Lagrangian is independent of time energy is conserved and if it is independent of orientation (choice of origin in polar co-ordinates) angular momentum will be conserved.

Eugene Wigner invented an interesting thought experiment to illustrate why charge is conserved. It also shows what is meant by a 'gauge theory'.

- Assume that charge is not conserved.

- Assume that the laws of electrostatics are independent of the choice of zero potential (i.e. it is a gauge theory).

- Imagine a reversible device that can create or destroy charge, requiring energy W to create a charge Q and returning the same amount of energy when that charge is destroyed.

- Set up one of these machines in a position X at potential V_1.

- Create a charge. This needs energy input W.

- Move the charge and machine to a position Y at potential V_2 ($V_2 < V_1$). The field does work on the charge allowing it to do work $W = Q (V_1 - V_2)$.

- Now destroy the charge. This returns an energy W. This result is independent of potential because we have assumed this gauge invariance.

- Return the machine to X.

- The system has now returned to its original state but there has been a net gain of energy equal to ω. The law of conservation of energy has been violated.

To summarise – if we assume non-conservation of charge and gauge invariance then the law of conservation of energy has to be abandoned. The alternative is to see this as a *reductio ad absurdum*, demonstrating that charge is in fact conserved.

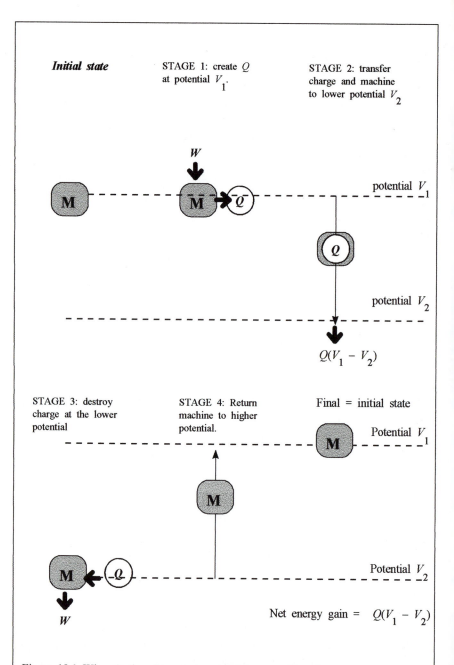

Figure 18.1 Wigner's thought experiment. M is a machine that can create or destroy charge at a cost: work W must be done to create the charge and is returned when it is destroyed. If this is possible it leads to a violation of the law of conservation of energy. The diagram above shows how a cycle of processes could generate unlimited amounts of energy.

18.1.2 CPT Invariance

When Dirac combined relativity with quantum theory in the 1920s he created a relativistic quantum theory of the electron that implied the existence of a new type of particle, a kind of 'mirror-image' of the electron having the same mass but opposite charge. This is, of course, the positron, the first anti-particle. The relation between matter and anti-matter is itself a kind of symmetry operation which involves inverting all the quantum numbers of the original particle. This operation is called charge conjugation (C). We have already seen that Feynman wondered whether anti-matter particles are really matter particles travelling backwards in time, so it will not be surprising to find a link between charge conjugation and time reversal (T). There is a third operation that is linked to these two, parity (P). This transforms all spatial co-ordinates into their mirror image, so it affects the handedness of spins, etc.

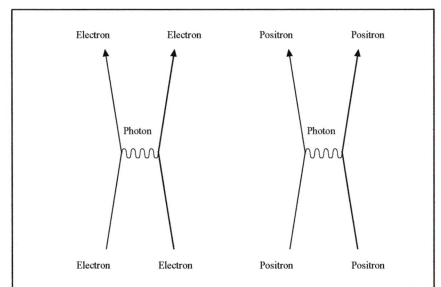

Figure 18.2 Charge conjugation exchanges all particles for their anti-particles. The electromagnetic interaction conserves charge conjugation so both diagrams above represent real physical processes. Photons are their own anti-particle.

It is not difficult to find processes which are invariant with respect to any or all of these symmetry operations. These are processes which, when transformed according to C, P or T operations become new processes that are also allowed by the laws of physics. For example, a time-reversed oscillator is still an oscillator. A time-reversed car crash is strange to look at, but none of the individual processes violate any of the fundamental laws of physics (the strangeness is because of the second law of thermodynamics – we have chosen an example that requires such

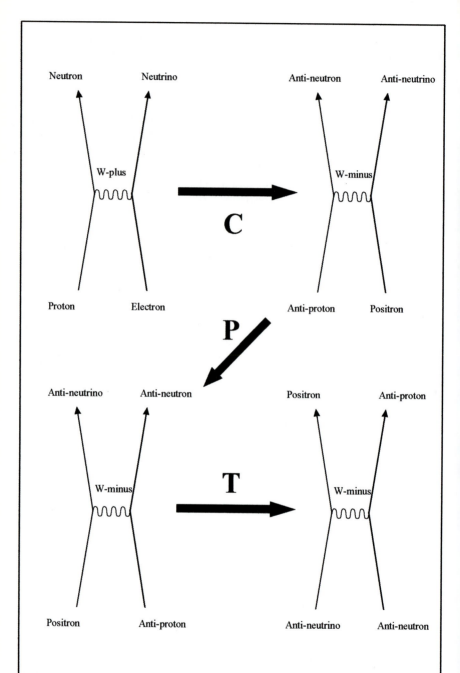

Figure 18.3 CPT invariance implies that the combined operations of charge conjugation parity and time reversal on a real physical process should always generate another possible physical process.

special initial conditions that we are   not likely to ever witness it – but it *is* an allowed process). A mirror image of an alpha decay would still represent a possible alpha decay – the strong interactions responsible for the decay conserve parity (P-invariance). In a similar way, two electrons repel one another by exchanging virtual photons, so do two positrons (C-invariance, photons are their own anti-particle).

In the mid-1950s G. Lüders and W. Pauli showed that the mathematical structure of relativistic quantum field theories lead to the following remarkable conclusion:

If we assume: (1) Locality (i.e. interactions are local in space-time);
                  (2) Lorentz invariance (i.e. the laws of physics look the same
                  from all inertial reference frames);
                  (3) Unitarity (i.e. probabilities for all possibliilities always add
                  up to 1).

Then: All relativistic quantum field theories are invariant with respect
                  to the combined operations of charge conjugation, parity and
                  time-reversal. They are CPT-invariant.

This means that if the operations C, P and T are carried out one after the other on a particular physical process then the resulting process is also allowed by the laws of physics. It also means that if a process violates one of these symmetries it must also violate one of the other two (or violate their combined symmetry) in order to conserve CPT invariance. For example, if a process is found to violate CP conservation then it must also violate T. Such a process would be very interesting, because it implies the existence of an intrinsic arrow of time on the subatomic scale.

18.2 VIOLATING SYMMETRIES

18.2.1 Parity Violation

Experiments have never found a violation of C, P or T symmetries involving either the strong or electromagnetic interactions. Here are some examples:

- *C-invariance in the strong interaction:*

$$^{0}_{-1}\pi^{-} + ^{1}_{1}p \rightarrow ^{0}_{0}\pi + ^{1}_{0}n$$ becomes (under C-operation) $^{0}_{1}\pi^{+} + ^{-1}_{-1}\bar{p} \rightarrow ^{0}_{0}\pi + ^{-1}_{0}\bar{n}$

which is a perfectly valid physical process

- *P-invariance in gamma-emission:*

When an excited nucleus emits a gamma-ray and drops to a lower energy state the gamma-ray photon is emitted in any direction at random. A mirror reflection of this process is another possible gamma-decay. The photon itself, a spin 1 particle can exist as a right-handed or left-handed particle (these states are analogous to the classical polarization states of the same name) so the reflected photon is still a possible photon.

- *T-invariance in the strong interaction:*

$$^1_1\text{p} + ^1_1\text{p} \rightarrow ^0_1\pi^+ + ^2_1\text{D} \qquad \text{becomes (under T-operation)} \qquad ^0_1\pi^+ + ^2_1\text{D} \rightarrow ^1_1\text{p} + ^1_1\text{p}$$

Both collisions have been observed in the laboratory and (when the incident particles in the second reaction have equal but opposite momenta to their counterparts in the first reaction) the cross-sections for both processes are the same.

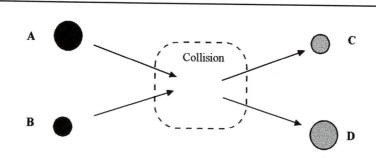

Figure 18.4 Time reversal invariance has been tested for the strong interaction by comparing the rates of nuclear reactions, such as A + B goes to C + D with the rates of the reverse reaction, C + D goes to A + B with incident momenta in the time-reversed reaction adjusted to be the negative of the momenta of the product nuclei in the original collision. The strong force conserves time reversal symmetry.

The weak interaction, however, violates some of these conservation laws. This is not entirely surprising when we consider the nature of the neutrino, a particle involved in many of the weak interactions. It is a fermion, like the electron, but (unlike the electron) its spin axis lies parallel to its direction of motion, so it has an intrinsic handedness. Neutrinos are all left-handed and anti-neutrinos are all right-handed. This means that neutrinos are transformed into impossible particles by an inversion of parity.

Up to 1956 no experiments had been carried out to test whether the weak decays (as involved in beta-decay or the decay of a pion to a muon and a neutrino) conserved parity. Almost everyone assumed that they would, but there was a problem (called the 'theta-tau problem') explaining the way neutral kaons decay, a process governed by the weak interaction. Sometimes they decay to two pions and sometimes to three. The problem with this is that pions have odd parity, so the two decays result in final states of different parity. If the neutral kaon is a unique particle and weak decays conserve parity then the products of its decay should have the same parity as the original particle. One way out was to assume that there are actually two kinds of neutral kaon, distinguished only by their parity (the theta, even, and tau, odd). Another way was to assume that weak interactions do, sometimes, violate parity conservation so that the kaon could decay by either route (the two-pion route would be much more likely anyway, because it involves decay to two rather than three particles).

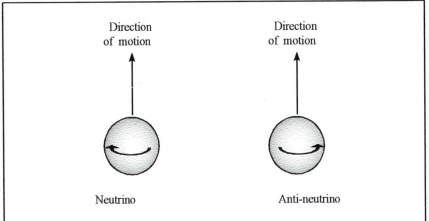

Figure 18.5 Neutrinos are weakly interacting particles. They have an intrinsic 'handedness' because their spin axis is parallel to their momentum vector. This means they will violate parity conservation, since this changes handedness creating non-existent particles. All neutrinos are 'left-handed' and all anti-neutrinos are 'right-handed'.

This idea was taken up by two young Chinese physicists Chen Ning Yang and Tsung Dao Lee who published a paper entitled Question of Parity Conservation in Weak Interactions in the Physical Review in October 1956. In this paper they pointed out that no experiments up to that time had made a direct test of parity conservation in the weak interaction. They also suggested a number of ways in which this could be tested experimentally. The first result came in 1957 from Chien-Shiung Wu, another Chinese-American physicist, at Columbia. She investigated the emission of beta-particles from decaying cobalt-60 nuclei. In beta-decay the electrons are emitted close to the magnetic poles of the spinning nucleus. A mirror reflection of a spinning nucleus changes its handedness and therefore its magnetic polarity. This means that conservation of parity requires an equal rate of beta-emission from north and south poles in a large number of decays. If there were a bias in either direction, say toward the south pole, then the mirror image would show a bias toward the north pole – an impossible result given the original assumption (of a bias to south). In order to test this Wu cooled a large collection of cobalt-60 nuclei close to absolute zero and then applied a strong magnetic field to align their nuclear spin axes. She then counted beta-emissions from each pole. The

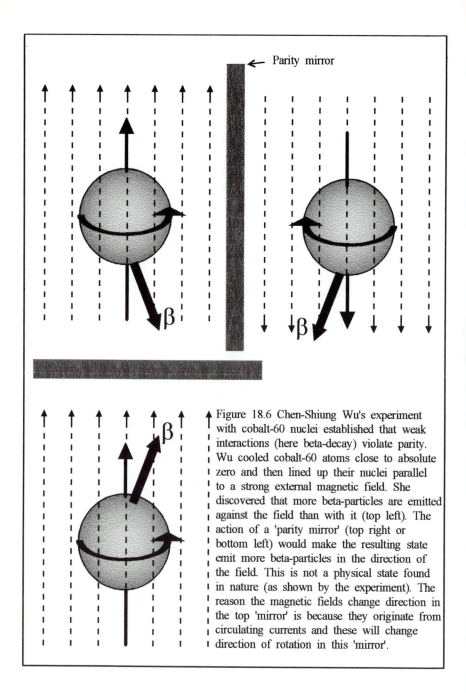

Figure 18.6 Chen-Shiung Wu's experiment with cobalt-60 nuclei established that weak interactions (here beta-decay) violate parity. Wu cooled cobalt-60 atoms close to absolute zero and then lined up their nuclei parallel to a strong external magnetic field. She discovered that more beta-particles are emitted against the field than with it (top left). The action of a 'parity mirror' (top right or bottom left) would make the resulting state emit more beta-particles in the direction of the field. This is not a physical state found in nature (as shown by the experiment). The reason the magnetic fields change direction in the top 'mirror' is because they originate from circulating currents and these will change direction of rotation in this 'mirror'.

results sent a shock-wave through the world of physics – more electrons leave the south pole than the north pole. Parity has been violated. Wu's results reveal an intrinsic handedness in beta-decay.

Other results confirmed these first observations of parity violation. The weak force mediates the decay of mu-mesons to electrons and neutrinos and it was soon shown (also at Columbia) that these decays also have a handedness – the electrons are emitted more frequently in one direction than the other. When neutral pions decay to a pair of photons they do so by the electromagnetic interaction conserving parity. When positive pions decay to an anti-muon and a muon-neutrino parity is violated because the neutrino itself is left-handed (as we have already pointed out) and its reflection does not represent an existing particle.

Chen Ning Yang and Tsung-Dao Lee received the 1957 Nobel Prize for Physics *"for their penetrating investigation of the so-called parity laws which has led to important discoveries regarding the elementary particles."*

Even before the dust had settled physicists began to look for explanations of this startling result. There seemed to be two possibilities. The particles involved in weak decays might have an intrinsic handedness or space itself might be 'twisted' on a very small scale and weak decays may respond locally to the twist. The second of these ideas has an obvious problem – if space itself has a handedness we might expect this to affect gravitational processes even more strongly than those involving the weak decay (because gravity is so much weaker). No such effects have ever been detected (although it is possible that they average out in some way in our macroscopic observations). On the other hand the two-component theory of the neutrino supports the idea that the handedness is built into the particles themselves, perhaps as a kind of hidden space-time structure. The discovery of muon neutrinos and anti-neutrinos and the third generation particles associated with the tau particle show that this handedness is common to all neutrinos and all weak interactions.

Parity violation implies that charge conjugation is also violated by the weak interaction. Think about the decay of a pi-minus:

$$_{-1}^{0}\pi^- \rightarrow \,_{-1}^{0}\mu^- + \,_{0}^{0}\bar{\nu}_\mu$$

The charge-conjugate reaction is:

$$_{1}^{0}\pi^+ \rightarrow \,_{1}^{0}\mu^+ + \,_{0}^{0}\nu_\mu$$

Can this occur? The answer is subtle. Pi-pluses *can* decay to anti-muons and muon-neutrinos, but the process is only allowed if parity is also inverted (see diagram). The reaction is invariant under the combined CP symmetry operation but not under C or P in isolation.

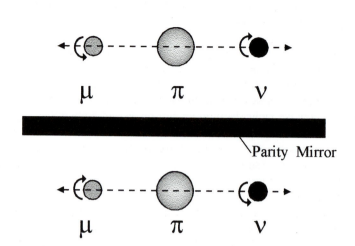

Figure 18.7 Pi-plus and its anti-particle, the pi-minus, can decay to a muon and a neutrino or an anti-muon and anti-neutrino respectively. The diagram above *could* represent either type of decay. It is clear though, looking at the neutrinos, that the top diagram emits a left-handed particle and the bottom one a right-handed particle. The latter cannot be a neutrino since all neutrinos are left-handed. However, anti-neutrinos are all right-handed. This means that both decays violate parity and that the pi-plus must decay by the top mechanism whilst the pi-minus (its antiparticle) must decay by the bottom process. This also shows that the decay violates charge conjugation invariance. We need to apply C *and* P to produce a viable reaction. This sort of observation led physicists to think that CP-symmetry might always be conserved even though C and P were individually violated in weak decays.

The discovery that parity, charge conjugation and time reversal are individually violated in weak interactions came as a shock, but it did not destroy the idea that CPT invariance is universal. In fact it was suggested that the combined CP symmetry operation should be conserved even by the weak interactions. The reason for this suggestion becomes clear if we consider the effect of a 'CP-mirror' on some of the parity violating processes discussed above. In each case the application of CP operations results in a viable reaction.

18.2.2 CP Violation

It is time to say a little more about kaons. They were discovered by Rochester and Butler in 1947 when they noticed a distinctive 'V' shape in cloud chamber photographs. This turned out to be a pair of pions created by kaon decay. Soon four of these new mesons had been discovered – the positive and negative kaons and the neutral kaon and its anti-particle. But the new discoveries brought disturbing problems with them. Kaons are created in strong interactions when hadrons collide. Their production rate is a few per cent of the production rate for pions, so they were expected to interact strongly with matter and have a very short lifetimes as they would decay to the lower mass pions. Strong decays have a characteristic lifetime of about 10^{-23} s, whereas kaons live about 10^{15} times longer! The kaon lifetime is characteristic of the weak decay. This was a rather disturbing result – what is it about kaons that make them behave like this? What prevents them decaying by the strong interaction?

In the early 1950s Murray Gell-Mann, Tadao Nakano and Kazuhiko Nishijima put forward a new scheme to explain the kaons longevity. They suggested that kaons carry a new quantum number, S, associated with a property that is now called 'strangeness'. Furthermore, if strangeness is conserved by the strong interaction this would prevent them decaying rapidly to pions. On the other hand, if the weak interaction does not conserve strangeness then kaons can decay to pions by weak decays. The strangeness scheme was also consistent with the reactions that create kaons. They are always accompanied by other strange particles. E.g.

$$_{-1}^{0}\pi^{-} + {}_{1}^{1}p \rightarrow {}_{-1}^{0}K^{-} + {}_{1}^{1}\Sigma \text{ (strong interaction)}$$

here the negative kaon has strangeness $S = -1$ and the sigma particle has strangeness $S = +1$. This gives a total strangeness zero on the right. Both pions and protons have strangeness zero. The decay of the neutral kaon ($S = +1$) is then:

$$_{0}^{0}K^{0} \rightarrow {}_{-1}^{0}\pi^{-} + {}_{+1}^{0}\pi^{+} \quad \text{(weak interaction)}$$

Here strangeness is not conserved, it changes from +1 to 0. This is acceptable if the weak interaction does not conserve strangeness. The scheme was found to agree with all observed particle interactions and is still accepted today.

But neutral kaons had another trick to play, and this one turns out to be even stranger than strangeness! Apply the charge-conjugation operation to the decay of the neutral kaon above and it becomes:

$$_{0}^{0}\overline{K^{0}} \rightarrow {}_{+1}^{0}\pi^{+} + {}_{-1}^{0}\pi^{-}$$

Notice that the particles on the right-hand side have transformed into themselves whilst the particle on the left-hand side has not. Its strangeness has changed from +1 to −1. This is allowed because weak interactions do not conserve strangeness. It also leads to the possibility that neutral kaons can exist in a mixed state throught the reaction:

$$\,^0_0K^0 \leftrightarrow \,^0_{-1}\pi + \,^0_{+1}\pi \leftrightarrow \,^0_0\overline{K^0} \qquad \text{(weak)}$$

This means that the neutral kaons must be described as a superposition of these two states. In fact there are two distinct ways in which this can be done to produce single particle states:

$$K_1 = \frac{1}{\sqrt{2}}\left(\left|K^0\right\rangle + \left|\overline{K^0}\right\rangle\right)$$

$$K_2 = \frac{1}{\sqrt{2}}\left(\left|K^0\right\rangle - \left|\overline{K^0}\right\rangle\right)$$

What happens when we apply CP operations to these particles? K_1 clearly transforms into itself and K_2 changes sign. This implies that the decay of K_1 to a pi-plus pi-minus pair (which also transform to themselves under CP operations) is allowed by CP symmetry, whereas the decay of K_2 by this mechanism is not. K_2 must decay to a state which changes parity under CP. A possible mode is:

$$\,^0_0K^0 \rightarrow \,^0_0\pi + \,^0_0\pi + \,^0_0\pi$$

(Pi-zeroes transform to themselves under charge conjugation but have parity -1 so three of them give a parity change $\Delta P = (-1)^3 = -1$).

To summarise:

$K_1 \rightarrow 2\pi$ allowed by CP - invariance $K_1 \rightarrow 3\pi$ forbidden by CP - invariance

$K_2 \rightarrow 3\pi$ allowed by CP - invariance $K_2 \rightarrow 2\pi$ forbidden by CP - invariance

A decay to three particles is less likely to occur than a decay to two, so the K_2 decay mechanism is about 100 times slower than the K_1 decay mechanism. Imagine a beam of neutral kaons. The K_1 component will disappear in the first 10^{-10} s producing pairs of pions. Further on, after about 10^{-8} s the K_2 component will decay to triplets of pions. This behaviour has been observed and is in agreement (almost – see below) with the K_1/K_2 scheme.

In 1964 James Christenson, James Cronin, Val Fitch and René Turlay carried out an experiemnt to test the K_1/K_2 scheme. They measured the decays of the K_2 component of neutral kaon beams and looked for CP-violating decays to two pions. These had not turned up in any previous experiment. To everyone's surprise they discovered that about 0.2% of all K_2 decays do in fact result in a pair of pions.

"I was shaken by the news. I knew quite well that a small amount of CP-violation

would not drastically alter the earlier discussions, based on CP-invariance, of the neutral K-complex. Also, the experience of seeing asymmetry fall by the wayside was not new to anyone who had lived through the 1956-7 period. At that time, however, there had at once been the consolation that P- and C- violation could be embraced by a new and pretty concept, CP-invariance. What shook all concerned now was that with CP gone there was nothing elegant to replace it with. The very smallness of the 2π rate, CP-invariance as a near miss, made it even harder to digest."

(Abraham Pais recalling a conversation over coffee with Jim Cronin and Val Fitch in the Berkeley National Laboratories cafeteria in spring 1964 – reported in *Inward Bound*, p.538, Abraham Pais, OUP, 1986)

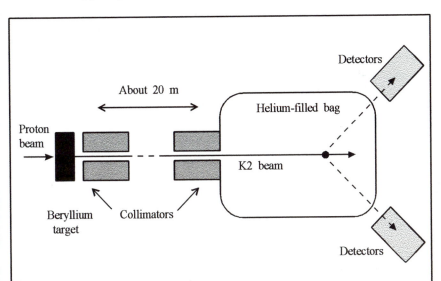

Figure 18.8 CP-violation. Schematic diagram showing the 1964 experiment of Christenson, Cronin, Fitch and Turlay. The detectors were set up to record the decay of the long-lived K2 component into two pions, an event forbidden by CP-invariance.

18.2.3 Time Reversal Asymmetry

Where does this leave CPT-invariance? If it holds then the violation of CP invariance must be accompanied by a violation of time-reversal symmetry. Kaon decays must have an intrinsic arrow of time that reveals itself through these decays. In the 1990s experiments at CERN and at Fermilab have measured a corresponding time reversal asymmetry in the decays of neutral kaons to electrons or positrons and pions.

The CERN test at CPLEAR test led by Panagiotis Pavloupoulos at CERN created kaons and anti-kaons in proton/anti-proton collisions:

$$p + \bar{p} \rightarrow K^+ + \pi^- + K^0$$
$$p + \bar{p} \rightarrow K^- + \pi^+ + \bar{K}^0$$

(strong)

Whether it is a K-nought or K-nought-bar that is actually created can be determined by looking at the sign and hence strangeness of the accompanying charged kaon (since strong interactions conserve strangeness). The CPLEAR group then monitored the so-called 'semi-leptonic decay' of the neutral kaon:

$$K^0 \rightarrow e^+ + \pi^- + \nu_e$$
$$\bar{K}^0 \rightarrow e^- + \pi^+ + \bar{\nu}_e$$

(weak)

Once again the nature of the decaying neutral kaon can be worked out from its decay products – a pi-minus indicates K-nought and a pi-plus indicates K-nought-bar.

The idea of the experiment was to compare the rates of kaons transforming to anti-kaons with the rate of anti-kaons transforming to kaons via intermediate states:

$$K^0 \leftrightarrow \text{intermediates} \leftrightarrow \bar{K}^0$$

If the process is T-invariant these should proceed at the same rate. This implies that the two decays – K-noughts to positrons and K-nought-bars to electrons should proceed at the same rate. They don't. The decays are time asymmetric. Furthermore, the rate at which these processes violate time reversal symmetry is, within the range of experimental errors, equal to the amount of CP violation in the neutral kaon system (as measured by experiments like that of Fitch and Cronin). This makes the CPLEAR experiment a significant test of the CPT-theorem itself.

CPLEAR's results were announced in 1998 and soon the KTeV group at Fermilab confirmed their conclusion. The Fermilab approach was different. In the KTeV experiment the very rare decay (1 in 10^7) of K-noughts to an electron and a pion was compared with the rate of creation of K-noughts from collisions between electrons and pions. Once again T-invariance was violated. These experiments confirm that we can still retain CPT invariance for all quantum field theories including the weak interaction, but C, P and T invariance are all violated by the weak interaction.

As usual, these results raise more questions than they answer. There is still no real understanding of how particles can distinguish the past from the future, but it is quite clear (at least in these kaon decays) that they can. It is particularly surprising that the violations occur in such a small number of experimental cases

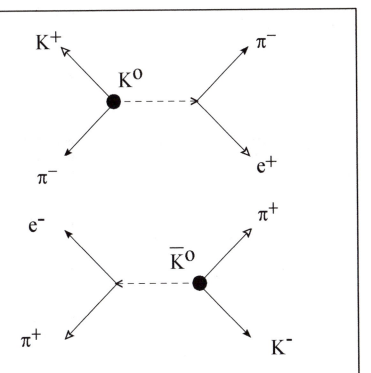

Figure 18.9 Time reversal invariance implies that neutral kaons should transform to their anti-particles and vice versa at the same rate. This implies that the semi-leptonic decays of kaons and anti-kaons should proceed at the same rate. The CPLEAR experiment at CERN compared the rates of the two reactions shown above and showed that they *do not* proceed at the same rate.

– why is particle physics so nearly C, P and T invariant? And what of gravity? How will the incorporation of gravity into a theory of everything affect the underlying symmetry laws? Will even CPT survive quantum gravity? Both CPLEAR and KTeV have shown that the transformation of matter into antimatter is asymmetric in time. This raises the intriguing possibility that there is a link between CP and T violation in these systems and the dominance of matter over antimatter in the universe at large.

The naive view is that the Big Bang should have created equal amounts of matter and anti-matter, so it is surprising to find ourselves in a universe that is so clearly dominated by matter. But it only needed an imbalance of about one part in a billion between the early particles and anti-particles to account for the matter that survived. How could such an asymmetry arise? One possibility is that there was a small asymmetry between the rates of decay to matter and anti-matter in a primordial heavy meson. Unfortunately the small difference between kaon to anti-kaon decays and anti-kaon to kaon decays (about 1 part in 500) is not large enough to account for the imbalance of matter and anti-matter – there has to be something extra and several laboratories around the world are gearing up to look for it including tests on anti-atoms in the AD project at CERN and on B-mesons at Stanford and KEK (Japan).

At the end of the twentieth century the enigma of kaon decay continues to cast light on the most puzzling problems of all – the nature of time and the origin of matter. But it may be that the kaons have already told us all they can and we will have to look at decays of a heavier meson, the B-meson, to learn more about the origin and significance of CP-violation. B-mesons, being much heavier than kaons, have many more alternative ways in which they can decay. These extra channels will provide a variety of ways to check CP-violation rates and look for the characteristic decays that may distinguish matter from anti-matter. The two main centres for these experiments are the B-factories at Stanford and KEK. They are called B-factories because they are designed to create large numbers of B-mesons in collisions of electrons and positrons. This is essential for the success of the project because the decays in which CP-violation is expected to occur are very rare. The most important decay channel for the B-mesons is to two other mesons, a J/Ψ and a K_1 kaon complex. It is expected that Bs and anti-Bs will decay into this channel at different rates. The detector designed to look for this difference is called BaBar and uses a relativistic trick to spread out the decays. The electrons and positrons that collide to create the Bs will not have exactly the same energy, the electrons will be given greater energy. This means the B-mesons and their anti-particles will have a large linear momentum and high velocity away from their creation point. This is important because these mesons have a lifetime of just over 1 ns, so if they are not moving quickly their decays will occur virtually on top of the creation point making it impossible to distinguish between rates of B and anti-B decay. At high speeds relativistic time dilation means their lifetime in the laboratory is extended and they survive long enough to move a measurable distance from the creation point before decaying. If, as expected, the matter and anti-matter decays have different rates the two events will occur at different distances from the creation point.

Conservation Laws in Quantum Field Theories			
Conserved Quantity	**Strong**	**Electromagnetic**	**Weak**
Parity P	Yes	Yes	No
Charge conjugation C	Yes	Yes	No
Time reversal T or CP	Yes	Yes	Usually BUT: 1 in 500 violation in kaon decays
CPT	Yes	Yes	Yes

The advantage of using electron-positron collisions to create B-mesons is that these are 'clean' collisions because electrons and positrons are relatively simple entities. However, the LHC at CERN and other hadron colliders such as the Tevatron at Fermilab can deliver much higher energy and so create even larger numbers of B-mesons together with lots of unwanted debris created in the collision. If it becomes possible to separate the rare but significant B-meson decays from the common but uninteresting showers of events that accompany them it may turn out that the B-factories are eventually overtaken by these larger accelerators.

James Cronin and Val Fitch shared the Nobel Prize for Physics in 1980 *"for the discovery of violations of fundamental symmetry principles in the decay of neutral K-mesons."*

19

APPENDICES

APPENDIX 1: THE BLACK-BODY RADIATION SPECTRUM

The derivation given here is not that of Planck, who focussed on the atomic oscillators responsible for radiating the photons. We focus on the radiation itself in the black-body cavity. The derivation has two main parts:

- Deriving the number density for the radiation modes – this is basically Rayleigh's approach based on the idea of standing waves trapped in an enclosure.
- Deriving the number of photons per mode using Bose-Einstein statistics.

Imagine a cubic cavity of side l whose walls are at temperature T. The problem is to derive an expression for the equilibrium energy spectrum of radiation filling the cavity.

Begin by considering a 1D problem. What modes of vibration would form standing waves on a string of length l? The wavelengths and frequencies of the first four harmonics are listed below:

$$\lambda_1 = 2l = \frac{2}{1}l \qquad f_1 = \frac{c}{\lambda_1} = \frac{1c}{2l}$$

$$\lambda_2 = l = \frac{2}{2}l \qquad f_2 = \frac{c}{\lambda_2} = \frac{2c}{2l}$$

$$\lambda_3 = \frac{2}{3}l = \frac{2}{3}l \qquad f_3 = \frac{c}{\lambda_3} = \frac{3c}{2l}$$

$$\lambda_4 = \frac{1}{2}l = \frac{2}{4}l \qquad f_4 = \frac{c}{\lambda_4} = \frac{4c}{2l}$$

$$\lambda_n = \frac{2}{n}l \qquad f_n = \frac{c}{\lambda_n} = \frac{nc}{2l}$$

Notice how the frequencies are evenly spaced whereas the wavelengths form a more awkward sequence. This suggests that counting modes will be easier if we work in 'frequency space' rather than 'wavelength space'. If we consider all three dimensions of the cavity then there will be a set of similar modes for the x, y and z directions. In addition to these modes there will be composite modes involving more than one dimension, although these will always be able to be broken down into combinations of modes along x, y and z. This leads to a frequency space containing a cubic lattice of allowed modes. Each 'cube' has side $c/2l$ so the 'volume' per mode is $c^3/8l^3$. The number of modes between f and $f + \delta f$ can now be calculated by considering the positive part of a spherical shell of radius f and thickness δf.

$$\text{Volume (in } f\text{-space) of this spherical shell} = 4\pi f^2 \delta f$$

$$\text{Volume in } f\text{-space of positive part} = \frac{\pi f^2 \delta f}{2}$$

$$\text{Volume in } f\text{-space per mode} = \frac{c^3}{8l^3}$$

$$\text{Number of modes between } f \text{ and } f + \delta f = \frac{4\pi f^2 l^3}{c^3}$$

$$\text{Modes per unit volume between } f \text{ and } f + \delta f: \quad n(f)\delta f = \frac{4\pi f^2 \delta f}{c^3}$$

Electromagnetic waves can be polarised, so each mode has two degrees of freedom associated with it. This increases the number of modes per unit volume between f and $f + \delta f$ by a factor of 2 to:

$$n(f)\delta f = \frac{8\pi f^2 \delta f}{c^3} \tag{1}$$

The second part of the problem is to work out the average number of photons per mode per unit volume. This is done using statistical mechanics and Bose-Einstein statistics. Photons of energy $E = hf$ will be emitted when atomic oscillators make a quantum jump with energy difference E. When thermal equilibrium is achieved the probability of emission of a photon by atoms in the wall of the cavity must be equal to the probability of absorption. However, photons are bosons, so the probability of emission depends on the number of photons already present in the mode concerned. Einstein broke this down into three parts:

Spontaneous emission: the probability that an isolated excited atom will emit a photon. Call this probability a.
Stimulated emission: the probability that an excited atom will emit a photon into a cavity containing N similar photons. Call this probability b. The photons are bosons so $b = Na$.

Absorption: the probability of an isolated atom in its ground state absorbing a photon must equal the probability that it emits the same photon from an excited state, a. The probability that an atom in the wall of the cavity absorbs a photon from thermal radiation in the cavity will be $c = Na$, since there are N photons present each of which it could absorb.

The actual absorption and emission rates will depend on the probabilities above and the relative numbers of atoms in their ground states M_g and in excited states M_e. This ratio was derived by Boltzmann and is given by a 'Boltzmann factor':

$$\frac{M_e}{M_g} = e^{-\frac{E}{kT}}$$

If R_e is the rate of emission and R_a the rate of absorption then:

$$R_e = (a+b)M_e$$
$$R_a = cM_g$$

These rates must be equal.

$$(a+b)M_e = cM_g$$

$$\frac{(a+Na)}{Na} = \frac{M_g}{M_e} = e^{\frac{E}{kT}}$$

$$\frac{(N+1)}{N} = e^{\frac{E}{kT}}$$

which can be rearranged to give:

$$N = \frac{1}{e^{\frac{E}{kT}} - 1}$$

This is the expression we need. It gives us the average number of photons of energy $E = hf$ present in the equilibrium spectrum of thermal radiation. The energy carried by these photons is:

$$e(f) = \frac{hf}{e^{\frac{hf}{kT}} - 1} \tag{2}$$

The energy density in the spectrum is then the product of this result and the density of modes from equation (1)

$$u(f)\,\delta f = e(f)n(f)\,\delta f = \frac{8\pi f^2 \delta f}{c^3} = \frac{8\pi h f^3 \delta f}{c^3 \left(e^{\frac{hf}{kT}} - 1 \right)}$$

APPENDIX 2: THE SCHRÖDINGER EQUATION

The general form of a 1D wave equation is:

$$\frac{\partial^2 \Psi}{\partial x^2} = \frac{1}{c^2} \frac{\partial^2 \Psi}{\partial t^2}$$

where solutions have the form: $\quad \Psi = \Psi_0 \exp 2\pi i\left(ft - \frac{x}{\lambda} \right)$

and $c = f\lambda$ (wave speed).

If we are dealing with electron waves f will be replaced by E/h and λ by h/p to give:

$$\Psi = \Psi_0 \exp 2\pi i\left(\frac{Et}{h} - \frac{px}{h} \right)$$

What can be constructed from the space and time derivatives of this equation?

$$\frac{\partial \Psi}{\partial x} = -\frac{2\pi i p}{h} \Psi \tag{1}$$

$$\frac{\partial^2 \Psi}{\partial x^2} = -\frac{4\pi^2 p^2}{h^2} \Psi = -\frac{8\pi^2 m K}{h^2} \Psi \quad \text{where } K \text{ is kinetic energy}$$

If the potential energy is V (a function of position and time) then $K = E - V$. Now the second spatial derivative becomes:

$$\frac{\partial^2 \Psi}{\partial x^2} = -\frac{8\pi^2 m(E - V)}{h^2} \Psi \tag{2}$$

$$\frac{\partial \Psi}{\partial t} = -\frac{2\pi i E}{h} \Psi \tag{3}$$

$$\frac{\partial^2 \Psi}{\partial t^2} = -\frac{4\pi^2 E^2}{h^2} \Psi \tag{4}$$

Equations (2) and (3) both involve the total energy E:

$$E\Psi = -\frac{h^2}{8\pi^2 m} \frac{\partial^2 \Psi}{\partial x^2} + V\Psi \qquad \text{from (2)}$$

$$E\Psi = \frac{ih}{2\pi} \frac{\partial \Psi}{\partial t} \qquad \text{from (3)}$$

These can be combined to obtain a linear differential equation – the time-dependent 1D Schrödinger equation:

$$-\frac{h^2}{8\pi^2 m}\frac{\partial^2 \Psi}{\partial x^2}+V\Psi=\frac{ih}{2\pi}\frac{\partial \Psi}{\partial t}$$

or

$$\left[-\frac{h^2}{8\pi^2 m}\frac{\partial^2}{\partial x^2}+V\right]\Psi=\left[\frac{ih}{2\pi}\frac{\partial}{\partial t}\right]\Psi \tag{5}$$

The bracketed terms in the second version of equation (5) represent mathematical operators that act on the wavefunction Ψ. The 1D equation can be extended to 3D by replacing the second partial derivative with respect to x by the Laplacian operator ∇^2:

$$\nabla^2=\left[\frac{\partial^2}{\partial x^2}+\frac{\partial^2}{\partial y^2}+\frac{\partial^2}{\partial z^2}\right]$$

The 3D time-dependent Schrödinger equation is then:

$$\left[-\frac{h^2}{8\pi^2 m}\nabla^2+V\right]\Psi=\frac{ih}{2\pi}\frac{\partial \Psi}{\partial t} \tag{6}$$

The operator in square brackets is called the Hamiltonian operator (H) and it is obviously linked to total energy in classical mechanics. This guides us in transferring classical problems to the quantum domain. If we can identify the corresponding classical Hamiltonian this can be converted to a quantum mechanical operator and inserted in the Schrödinger equation.

The linearity of the Schrödinger equation is very important. If a differential equation is linear it means that two different solutions, say ψ_1 and ψ_2, can be added together to give another valid solution $\Psi=\psi_1+\psi_2$. This superposition of states accounts for interference effects and leads to some interesting questions of interpretation (as we shall see).

If the electron moves so that its potential energy depends only on its position and not on time, as it does in the static electrostatic potential of the hydrogen nucleus, then equation (2) is completely independent of time and can be solved to give the spatial variation of Ψ. In fact Ψ can now be considered as a product of a spatial wavefunction ψ_s and a temporal wavefunction ψ_t. This is interesting because a time-independent solution for an electron in an atom represents a stationary state analogous to standing waves on a stretched string. In fact equation (2) can be written in a compact form as the time-independent Schrödinger equation:

$$H\psi_s = E\psi_s \quad \text{(where } H \text{ is the 3D Hamiltonian operator from above)}$$

It is worth looking at the time dependent part of Ψ too. This is found by solving equation (3) for ψ_t.

$$\frac{ih}{2\pi}\psi_t = E\psi_t$$

which has solutions of the form: $\psi_t = Ae^{-\frac{2\pi iEt}{h}}$

The general solution represents a superposition of wavefunctions of this form with different values of E. This is linked to the idea that the energy of the electron is indeterminate prior to a measurement. After the measurement the electron wavefunction is forced to collapse into one or other of these superposed states and yield a definite value for E.

APPENDIX 3: THE HYDROGEN ATOM

The first major success of wave mechanics was a solution to the problem of the hydrogen atom – an electron moving in a central potential. The method of solution is outlined below, for details the reader is referred to any standard undergraduate textbook on quantum theory:

$$V(r) = \frac{-e}{r}$$

The time-independent Schrödinger equation is:

$$H\Psi = E\Psi$$

with

$$H = \frac{-\hbar^2}{2m_e} \nabla^2 + V(r)$$

To exploit the spherical symmetry of the potential this is solved in spherical polar co-ordinates (r, θ, ϕ). This simplifies the problem enormously because the dependence on co-ordinates θ and ϕ separates from the radial dependence and we can write:

$$\Psi(r, \theta, \phi) = u(r) v(\theta) w(\phi)$$

The resulting differential equations for $u(r)$ $v(\theta)$ and $w(\phi)$ can (with some effort) be solved to find the allowed wavefunctions and energies for the electron in the hydrogen atom. If the interaction of the electron and proton is taken into account then the reduced mass μ appears in the equation for the energy levels:

$$\mu = \frac{m_e m_p}{m_e + m_p}$$

$$E_n = -\frac{\mu Z^2 e^4}{(4\pi\varepsilon_0)^2 2\hbar^2 n^2}$$

n is the principal quantum number and must be one of the integers:

$$n = l + 1, l + 2, l + 3, \text{etc.}$$

Constraints on solutions restrict allowed values for l to:

$$l = |m|, |m| + 1, |m| + 2, \text{etc.}$$

The absolute values of m are also integers: 0, 1, 2, 3, ...

The three quantum numbers n, l, and m arise from the equations in r, θ, and ϕ respectively. Their physical interpretation is:

n radial quantum number – determines energy;
l total angular momentum quantum number;
m magnetic quantum number – projection of angular momentum.

When $n = 1$ the electron is in the ground state and the energy is 13.6 eV.

For each value of n there are n^2 degenerate (equal energy) levels and each of these can be occupied by a pair of electrons (with opposite spins).

The advantage of Schrödinger's wave mechanical approach is that it provides the wavefunctions for the electron (eigenfunctons) as well as the energy levels (eigenvalues).

APPENDIX 4: THE LORENTZ TRANSFORMATION EQUATIONS

The Lorentz transformation can be derived using just a few assumptions based on the principle of relativity itself. The will look for a simple linear transformation of A's co-ordinates into B's. By 'linear' we mean that x' will depend on a sum of terms involving x and t but not powers of these or more complicated functions of one or both (x will depend in a symmetric way on x' and t'). We can also be guided in the knowledge that, at low velocities ($v \ll c$), the Lorentz transformation must reduce to the Galilean transformation.

The Galilean Transformation

Let an event occur at (x, y, z, t) in frame A. If frame B passes in the positive x direction at velocity v then the co-ordinates of this event in B will be reduced by the distance B travels in the x direction in time t, that is vt. The transformation is therefore:

$$x' = x - vt \qquad\qquad x = x' + vt'$$
$$y' = y \qquad\qquad\qquad y = y'$$
$$z' = z \qquad\qquad\qquad z = z'$$
$$t' = t \qquad\qquad\qquad t = t'$$

For these reasons we shall assume the transformation has the form:

$$x' = ax - bt \tag{1}$$
$$x = ax' + bt' \tag{2}$$

and proceed to determine the unknown coefficients a and b.

If $x = 0$ in equation (2) then the relation between x' and t' must be the co-ordinates of the origin of A as seen in B. This gives us:

$$\frac{\mathrm{d}x'}{\mathrm{d}t'} = -\frac{b}{a}$$

This is clearly a velocity and must be the velocity of A's origin in B's frame as it moves in the negative x' direction. If we put $x' = 0$ in equation (1) this gives us the trajectory of B's origin in A and leads to:

$$\frac{\mathrm{d}x}{\mathrm{d}t} = +\frac{b}{a}$$

Therefore
$$\frac{b}{a} = v \tag{3}$$

Now consider a light pulse that leaves the origin of both reference frames at the instant ($t = t' = 0$) when they coincide. Light is used since we know this will have the same velocity, c, relative to both observers. The x and x' co-ordinates of the light pulse will be:

$$\text{In A:} \qquad x = ct$$

$$\text{In B:} \qquad x' = ct'$$

By substituting for x' in equation (2) and rearranging we obtain:

$$a^2 - \frac{b^2}{c^2} = 1 \qquad\qquad (4)$$

Equations (3) and (4) can be used to derive expressions for a and b:

$$a = \frac{1}{\sqrt{1 - \dfrac{v^2}{c^2}}} \qquad \text{or} \qquad a = \gamma$$

$$b = \gamma v$$

These can now be substituted back into equations (3) and (4) to obtain the Lorentz transformations for x and x' previously derived. From these we can also derive the transformations for t and t' as in the last section.

Comments

1. Starting from these equations we can derive time dilation, the relativity of simultaneity, length contraction, velocity addition, etc.

2. The co-ordinates (x, y, z, t) and (x', y', z', t') are alternative representations of the location of an event in 'space-time'. The co-ordinates are described as 'covariant'. This means that the co-ordinates differ from one reference frame to another but that the law connecting the co-ordinates is the same and is independent of the event considered.

3. The transformation itself can be regarded as equivalent to a rotation of a four dimensional vector in space-time.

APPENDIX 5: THE SPEED OF ELECTROMAGNETIC WAVES

Maxwell's equations for the electromagnetic fields in free space can be written:

$$\nabla \cdot \underline{E} = 0 \qquad \text{(1)} \quad \text{(Gauss's theorem of electrostatics)}$$

$$\nabla \cdot \underline{B} = 0 \qquad \text{(2)}$$

$$\nabla \otimes \underline{E} = -\frac{\partial \underline{E}}{\partial t} \qquad \text{(3)} \quad \text{(Faraday's/Lenz's laws of electromagnetic induction)}$$

$$\nabla \otimes \underline{B} = \mu_0 \varepsilon_0 \frac{\partial \underline{B}}{\partial t} \qquad \text{(4)} \quad \text{(Ampère's circuit law)}$$

The following identity will be useful:

$$\nabla \otimes \nabla \underline{X} = \nabla(\nabla \cdot \underline{X}) + \nabla^2 \underline{X}$$

We shall apply it to equation (4) and use equation (2) to simplify the result:

$$\nabla \otimes \nabla \otimes \underline{B} = \mu_0 \varepsilon_0 \left(\nabla(\nabla \cdot \underline{B}) + \nabla^2 \underline{B} \right) = \mu_0 \varepsilon_0 \nabla^2 \underline{B} = \mu_0 \varepsilon_0 \frac{\partial}{\partial t} (\nabla \otimes \underline{B})$$

$$\nabla^2 \underline{B} = \frac{\partial}{\partial t} (\nabla \otimes \underline{B}) = \mu_0 \varepsilon_0 \frac{\partial}{\partial t} \left(\frac{\partial \underline{B}}{\partial t} \right)$$

$$\nabla^2 \underline{B} = \mu_0 \varepsilon_0 \frac{\partial^2 \underline{B}}{\partial t^2}$$

This final equation has the form of a wave equation. The waves have a phase velocity:

$$c = \frac{1}{\sqrt{\varepsilon_0 \mu_0}}$$

An equation for the same form can be obtained by manipulating equation (3) in an identical way. This shows that Maxwell's equations have solutions representing travelling waves in free space which have a velocity c.

APPENDIX 6: THE NOBEL PRIZE FOR PHYSICS

The Nobel Prize in Physics was set up in 1901 by Alfred Nobel. It was intended for *"the person who shall have made the most important discovery or invention in the domain of physics"*.
The numbers in brackets, such as (Ch3), refer to chapters of the book where that physicist's contributions are discussed.

1901: Wilhelm Conrad Röntgen (1845-1923), Germany, Munich University: for the discovery of X-rays (Ch1).
1902: The prize was awarded jointly to: Hendrik Antoon Lorentz (1853-1928), the Netherlands, Leyden University; and Pieter Zeeman (1865-1943), the Netherlands, Amsterdam University: for their work on the influence of magnetism on radiation (the Zeeman effect).
1903: The prize was divided, one half being awarded to: Antoine Henri Becquerel (1852-1908), France, École Polytechnique, Paris: for the discovery of radioactivity and the other half jointly to: Pierre Curie (1859-1906), France, École municipale de physique et de chimie industrielles (Municipal School of Industrial Physics and Chemistry), Paris; and his wife Marie Curie née Sklodowska (1867-1934), France: for their joint researches on radioactivity (Ch5).
1904: Lord Rayleigh (John William Strutt) (1842-1919), Great Britain, Royal Institution of Great Britain, London: for his investigations of gas densities and the discovery of argon.
1905: Philipp Eduard Anton Lenard (1862-1947), Germany, Kiel University: for his work on cathode rays (Ch5).
1906: Sir Joseph John Thomson (1856-1940), Great Britain, Cambridge University: for the discovery of the electron (Ch5).
1907: Albert Abraham Michelson (1852-1931), U.S.A., Chicago University: for his optical precision instruments and accurate measurements of the speed of light (Ch10).
1908: Gabriel Lippmann (1845-1921), France, Sorbonne University, Paris: for techniques of colour photography.
1909: The prize was awarded jointly to: Guglielmo Marconi (1874-1937), Italy, Marconi Wireless Telegraph Co. Ltd., London, Great Britain; and Carl Ferdinand Braun (1850-1918), Germany, Strasbourg University, Alsace (then Germany): for the development of radio communications systems.
1910: Johannes Diderik Van Der Waals (1837-1923), the Netherlands, Amsterdam University: for the equation of state for gases and liquids.
1911: Wilhelm Wien (1864-1928), Germany, Würzburg University: for laws governing the spectrum of thermal radiation (Ch1).
1912: Nils Gustaf Dalén (1869-1937), Sweden, Swedish Gas-Accumulator Co., Lidingö-Stockholm: for the invention of automatic regulators for use in conjunction with gas accumulators for illuminating lighthouses and buoys.
1913: Heike Kamerlingh-Onnes (1853-1926), the Netherlands, Leyden University:

for investigations on the properties of matter at low temperatures and the production of liquid helium (Ch17).

1914: Max Von Laue (1879-1960), Germany, Frankfurt-on-the Main University: for the discovery and investigation of X-ray crystallography.

1915: The prize was awarded jointly to: Sir William Henry Bragg (1862-1942), Great Britain, London University; and his son Sir William Lawrence Bragg (1890-1971), Great Britain, Victoria University, Manchester: for the investigation of crystalline structures using X-ray diffraction.

1916: No Prize awarded.

1917: Charles Glover Barkla (1877-1944), Great Britain, Edinburgh University: for his discovery of characteristic X-ray spectra of the elements.

1918: Max Karl Ernst Ludwig Planck (1858-1947), Germany, Berlin University: for the discovery of energy quantisation and the explanation of the black-body radiation spectrum (Ch1).

1919: Johannes Stark (1874-1957), Germany, Greifswald University: for the investigation of 'canal rays' and the Stark effect.

1920: Charles Edouard Guillaume (1861-1938), Switzerland, Bureau International des Poids et Mesures (International Bureau of Weights and Measures), Sèvres: for his discovery of anomalies in nickel steel alloys.

1921: Albert Einstein (1879-1955), Germany and Switzerland, Kaiser-Wilhelm Institut (now Max-Planck-Institut) für Physik, Berlin: for his discovery of light quantisation and its use to explain the photoelectric effect (Ch1, 3, 11, 12).

1922: Niels Bohr (1885-1962), Denmark, Copenhagen University: for his atomic model and his explanation of the radiation spectrum of atoms (Ch1, 2, 3, 5).

1923: Robert Andrews Millikan (1868-1953), U.S.A., California Institute of Technology, Pasadena, CA: for his measurement of the electronic charge and investigations of the photoelectric effect.

1924: Karl Manne Georg Siegbahn (1886-1978), Sweden, Uppsala University: for his discoveries and research in the field of X-ray spectroscopy.

1925: James Franck (1882-1964), Germany, Goettingen University and Gustav Hertz (1887-1975) Germany, Halle University: for the discovery of the laws governing collisions between electrons and atoms.

1926: Jean Baptiste Perrin (1870-1942), France, Sorbonne University, Paris: for his work on the discontinuous structure of matter.

1927: The prize was divided equally between: Arthur Holly Compton (1892-1962), U.S.A., University of Chicago: for the discovery of the Compton effect (Ch1), and Charles Thomson Rees Wilson (1869-1959), Great Britain, Cambridge University: for the invention of the cloud chamber to observe particle tracks (Ch7).

1928: Sir Owen Willans Richardson (1879-1959), Great Britain, London University: for the explanation of thermionic emission (Ch5).

1929: Prince Louis-Victor De Broglie (1892-1987), France, Sorbonne University, Institut Henri Poincaré, Paris: for his discovery of the wave nature of electrons (Ch1, 2).

1930: Sir Chandrasekhara Venkata Raman (1888-1970), India, Calcutta

University: for work on the scattering of light and the Raman effect.

1931: No award.

1932: Werner Heisenberg (1901-1976), Germany, Leipzig University: for the creation of quantum mechanics (matrix mechanics) (Ch2, 3, 4).

1933: Erwin Schrödinger (1887-1961), Austria, Berlin University, Germany and Paul Adrien Maurice Dirac (1902-1984), Great Britain, Cambridge University: for their mathematical versions of quantum mechanics (Ch2, 3, 4).

1934: No award.

1935: Sir James Chadwick (1891-1974), Great Britain, Liverpool University: for the discovery of the neutron (Ch5).

1936: The prize was divided equally between: Victor Franz Hess (1883-1964), Austria, Innsbruck University: for the discovery of cosmic rays, and Carl David Anderson (1905-1991), U.S.A., California Institute of Technology, Pasadena, CA: for the discovery of the positron (Ch2).

1937: The prize was awarded jointly to: Clinton Joseph Davisson (1881-1958), U.S.A., Bell Telephone Laboratories, New York, NY; and Sir George Paget Thomson (1892-1975), Great Britain, London University: for their experimental discovery of the diffraction of electrons by crystals (Ch1).

1938: Enrico Fermi (1901-1954), Italy, Rome University: for the creation of artificial radionuclides and nuclear reactions brought about by slow neutrons (Ch5).

1939: Ernest Orlando Lawrence (1901-1958), U.S.A., University of California, Berkeley, CA: for the invention and development of the cyclotron (Ch8).

1940: No award.

1941: No award.

1942: No award.

1943: Otto Stern (1888-1969), U.S.A., Carnegie Institute of Technology, Pittsburg, PA: for his discovery of the magnetic moment of the proton and his development of molecular ray methods.

1944: Isidor Isaac Rabi (1898-1988), U.S.A., Columbia University, New York, NY: for nuclear magnetic resonance.

1945: Wolfgang Pauli (1900-1958), Austria, Princeton University, NJ, U.S.A.: for the discovery of the Pauli Exclusion Principle (Ch2).

1946: Percy Williams Bridgman (1882-1961), U.S.A., Harvard University, Cambridge, MA: for experimental investigations and discoveries in very high-pressure physics.

1947: Sir Edward Victor Appleton (1892-1965), Great Britain, Department of Scientific and Industrial Research, London: for investigations of the physics of the upper atmosphere and the discovery of the Appleton layer.

1948: Lord Patrick Maynard Stuart Blackett (1897-1974), Great Britain, Victoria University, Manchester: for the development of the Wilson cloud chamber method, leading to discoveries in nuclear physics and cosmic radiation (Ch7).

1949: Hideki Yukawa (1907-1981), Japan, Kyoto Imperial University and Columbia University, New York, NY, U.S.A.: for his prediction of the existence of mesons to explain the strong nuclear forces (Ch5).

1950: Cecil Frank Powell (1903-1969), Great Britain, Bristol University: for the photographic method to record particle interactions and discoveries made using these techniques (Ch7).

1951: The prize was awarded jointly to: Sir John Douglas Cockcroft (1897-1967), Great Britain, Atomic Energy Research Establishment, Harwell, Didcot, Berks., and Ernest Thomas Sinton Walton (1903-1995), Ireland, Dublin University: for their work on the transmutation of atomic nuclei by artificially accelerated atomic particles using the 'Cockcroft-Walton accelerator' (Ch8).

1952: The prize was awarded jointly to: Felix Bloch (1905-1983), U.S.A., Stanford University, Stanford, and Edward Mills Purcell (1912-1997), U.S.A., Harvard University, Cambridge, MA: for new methods for nuclear magnetic precision measurements and discoveries made with them.

1953: Frits (Frederik) Zernike (1888-1966), the Netherlands, Groningen University: for the invention of the phase contrast microscope.

1954: The prize was divided equally between: Max Born (1882-1970), Great Britain, Edinburgh University: for the statistical interpretation of the wavefunction (Ch2,3), and Walther Bothe (1891-1957), Germany, Heidelberg University, Max-Planck Institut (former Kaiser-Wilhelm-Institut) für medizinische Forschung, Heidelberg: for the invention of the coincidence method and discoveries made using it (Ch7).

1955: The prize was divided equally between: Willis Eugene Lamb (1913-), U.S.A., Stanford University, Stanford, CA: for the fine structure of the hydrogen spectrum, in particular the Lamb shift; and Polykarp Kusch (1911-1993), U.S.A., Columbia University, New York, NY: for precision measurement of the magnetic moment of the electron (Ch4).

1956: The prize was awarded jointly, one third each, to: William Shockley (1910-1989), U.S.A., Semiconductor Laboratory of Beckman Instruments, Inc., Mountain View, CA; John Bardeen (1908-1991), U.S.A., University of Illinois, Urbana, IL, b; and Walter Houser Brattain, (1902-1987), U.S.A., Bell Telephone Laboratories, Murray Hill, NJ: for semiconductor research and the discovery of the transistor effect.

1957: The prize was awarded jointly to: Chen Ning Yang (1922-), China, Institute for Advanced Study, Princeton, NJ, U.S.A.; and Tsung-Dao Lee (1926-), China, Columbia University, New York, NY, U.S.A.: for theoretical work on symmetry principles leading to the discovery of the violation of parity (Ch18).

1958: The prize was awarded jointly to: Pavel Alekseyevich Cerenkov (1904-1990), USSR, Physics Institute of USSR Academy of Sciences, Moscow; Il'ja Mikhailovich Frank (1908-1990), USSR, University of Moscow and Physics Institute of USSR Academy of Sciences, Moscow; and Igor Yevgenyevich Tamm (1885-1971), USSR, University of Moscow and Physics Institute of USSR Academy of Sciences, Moscow: for the discovery and the interpretation of the Cerenkov effect (Ch7).

1959: The prize was awarded jointly to: Emilio Gino Segrè (1905-1989), U.S.A.,

University of California, Berkeley, CA; and Owen Chamberlain (1920-), U.S.A., University of California, Berkeley, CA: for their discovery of the antiproton.

1960: Donald A. Glaser (1926-), U.S.A., University of California, Berkeley, CA: for the invention of the bubble chamber (Ch7).

1961: The prize was divided equally between: Robert Hofstadter (1915-1990), U.S.A., Stanford University, Stanford, CA: for his studies of electron scattering in nuclei and work on the structure of nucleons; and Ludwig Rudolf Mössbauer (1929-), Germany, Technische Hochschule, Munich, and California Institute of Technology, Pasadena, CA, U.S.A.: for the Mössbauer effect (Ch12).

1962: Lev Davidovich Landau (1908-1968), USSR, Academy of Sciences, Moscow: for theoretical work on condensed matter especially liquid helium (Ch17).

1963: The prize was divided, one half being awarded to: Eugene P Wigner, (1902-1995), U.S.A., Princeton University, Princeton, NJ: for his work in nuclear and particle physics based on fundamental symmetry principles; and the other half jointly to: Maria Goeppert-Mayer (1906-1972), U.S.A., University of California, La Jolla, CA; and J. Hans D Jensen (1907-1973), Germany, University of Heidelberg: for discoveries concerning the nuclear shell structure (Ch5).

1964: The prize was divided, half being awarded to Charles H Townes (1915-), U.S.A., Massachusetts Institute of Technology (MIT) , Cambridge, MA; and the other half jointly to: Nicolay Gennadiyevich Basov (1922-), USSR, Lebedev Institute for Physics, Akademija Nauk, Moscow, and Aleksandr Mikhailovich Prokhorov (1916-), USSR, Lebedev Institute for Physics, Akademija Nauk, Moscow: for their work in quantum electronics.

1965: The prize was awarded jointly to: Sin-Itiro Tomonaga (1906-1979), Japan, Tokyo, University of Education, Tokyo; Julian Schwinger (1918-), U.S.A., Harvard University, Cambridge, MA; and Richard P. Feynman (1918-1988), U.S.A., California Institute of Technology, Pasadena, CA: for the discovery of quantum electrodynamics (QED) (Ch4).

1966: Alfred Kastler (1902-1984), France, École Normale Supérieure, Université de Paris: for the discovery and development of optical methods for studying hertzian resonances in atoms.

1967: Hans Albrecht Bethe (1906-), U.S.A., Cornell University, Ithaca, NY: for the explanation of energy sources in stars (Ch14).

1968: Luis W Alvarez (1911-1988), U.S.A., University of California, Berkeley, CA: for contributions to particle physics, in particular the discovery of a large number of resonance states.

1969: Murray Gell-Mann (1929-), U.S.A., California Institute of Technology, Pasadena, CA: for the classification of elementary particles and forces (Ch6).

1970: The prize was divided equally between: Hannes Alfvén (1908-1995), Sweden, Royal Institute of Technology, Stockholm: for his work in magneto-hydrodynamics and plasma physics; and Louis Néel (1904-), France, University of Grenoble, Grenoble: for his work in solid state physics on antiferromagnetism and ferrimagnetism.

1971: Dennis Gabor (1900-1979), Great Britain, Imperial College of Science and Technology, London: for the invention of holography.

1972: The prize was awarded jointly to: John Bardeen (1908-1991), U.S.A., University of Illinois, Urbana, IL; Leon N Cooper (1930-), U.S.A., Brown University, Providence, RI; and J. Robert Schrieffer (1931-), U.S.A., University of Pennsylvania, Philadelphia, PA: for the BCS-theory of superconductivity.

1973: The prize was divided, one half being equally shared between: Leo Esaki (1925-), Japan, IBM Thomas J. Watson Research Center, Yorktown Heights, NY, U.S.A.; and Ivar Giaever (1929-), U.S.A., General Electric Company, Schenectady, NY: for experimental discoveries regarding quantum tunnelling phenomena in semiconductors and superconductors, respectively, and the other half to: Brian D. Josephson (1940-), Great Britain, Cambridge University, Cambridge: for the theoretical investigation of supercurrents through tunnel barriers and the invention of the Josephson junction.

1974: The prize was awarded jointly to: Sir Martin Ryle (1918-1984), Great Britain, Cambridge University, Cambridge; and Antony Hewish (1924-), Great Britain, Cambridge University, Cambridge: for pioneering research in radio astrophysics: Ryle for his observations and inventions, in particular of the aperture synthesis technique, and Hewish for the discovery of pulsars (Ch13).

1975: The prize was awarded jointly to: Aage Bohr (1922-), Denmark, Niels Bohr Institute, Copenhagen; Ben Mottelson (1926-), Denmark, Nordita, Copenhagen; and James Rainwater (1917-1986), U.S.A., Columbia University, New York: for work on the structure of the nucleus (Ch5).

1976: The prize was divided equally between: Burton Richter (1931-), U.S.A., Stanford Linear Accelerator Center, Stanford, CA; Samuel C. C. Ting (1936-), U.S.A., Massachusetts Institute of Technology (MIT), Cambridge, MA, (European Center for Nuclear Research, Geneva, Switzerland): for the discovery of the J/Ψ particle (Ch6).

1977: The prize was divided equally between: Philip W. Anderson (1923-), U.S.A., Bell Laboratories, Murray Hill, NJ; Sir Nevill F. Mott (1905-1996), Great Britain, Cambridge University, Cambridge; and John H. Van Vleck (1899-1980), U.S.A., Harvard University, Cambridge, MA: for fundamental theoretical investigations of the electronic structure of magnetic and disordered systems.

1978: The prize was divided, one half being awarded to: Pyotr Leonidovich Kapitsa (1894-1984), USSR, Academy of Sciences, Moscow: for his basic inventions and discoveries in the area of low-temperature physics (Ch17); and the other half divided equally between: Arno A. Penzias (1933-), U.S.A., Bell Laboratories, Holmdel, NJ, and Robert W. Wilson (1936-), U.S.A., Bell Laboratories, Holmdel, NJ: for the discovery of the cosmic microwave background radiation (Ch15).

1979: The prize was divided equally between: Sheldon L. Glashow (1932-), U.S.A., Lyman Laboratory, Harvard University, Cambridge, MA; Abdus Salam (1926-1996), Pakistan, International Centre for Theoretical Physics, Trieste, and Imperial College of Science and Technology, London, Great Britain, and Steven

Weinberg (1933-), U.S.A., Harvard University, Cambridge, MA: for electro-weak unification (Ch9).

1980: The prize was divided equally between: James W. Cronin (1931-), U.S.A., University of Chicago, Chicago, IL; and Val L. Fitch (1923-), U.S.A., Princeton University, Princeton, NJ: for the discovery of violations of fundamental symmetry principles in the decay of neutral K-mesons (Ch18).

1981: The prize was awarded jointly to: Nicolaas Bloembergen (1920-), U.S.A., Harvard University, Cambridge, MA; and Arthur L. Schawlow (1921-), U.S.A., Stanford University, Stanford, CA: for contributions to laser spectroscopy; and the other half to: Kai M. Siegbahn (1918-), Sweden, Uppsala University, Uppsala: for high-resolution electron spectroscopy.

1982: Kenneth G. Wilson (1936-), U.S.A., Cornell University, Ithaca, NY: for the theory of critical phenomena in connection with phase transitions.

1983: The prize was awarded by one half to: Subramanyan Chandrasekhar (1910-1995), U.S.A., University of Chicago, Chicago, IL: for his theoretical work on the structure and evolution of the stars (Ch14); and by the other half to: William A. Fowler (1911-1995), U.S.A., California Institute of Technology, Pasadena, CA: for theoretical and experimental studies of nucelosynthesis in the universe (Ch14).

1984: The prize was awarded jointly to: Carlo Rubbia (1934-), Italy, CERN, Geneva, Switzerland; and Simon Van Der Meer (1925-), the Netherlands, CERN, Geneva, Switzerland: for the discovery of the W and Z particles (Ch6,9).

1985: Klaus Von Klitzing (1943-), Federal Republic of Germany, Max-Planck-Institute for Solid State Research, Stuttgart: for the discovery of the quantized Hall effect.

1986: The prize was awarded by one half to: Ernst Ruska (1906-1988), Federal Republic of Germany, Fritz- Haber-Institut der Max-Planck-Gesellschaft, Berlin: for fundamental work in electron optics, and for the design of the first electron microscope; and the other half jointly to: Gerd Binnig (1947-), Federal Republic of Germany, IBM Zurich Research Laboratory, Rüschlikon, Switzerland; and Heinrich Rohrer (1933-), Switzerland, IBM Zurich Research Laboratory, Rüschlikon, Switzerland: for the design of the scanning tunneling microscope.

1987: The prize was awarded jointly to: J. Georg Bednorz (1950-), Federal Republic of Germany, IBM Research Laboratory, Rüschlikon, Switzerland; and K. Alexander Müller (1927-), Switzerland, IBM Research Laboratory, Rüschlikon, Switzerland: for the discovery of superconductivity in ceramic materials (Ch17).

1988: The prize was awarded jointly to: Leon M. Lederman (1922-), U.S.A., Fermi National Accelerator Laboratory, Batavia, IL; Melvin Schwartz (1932-), U.S.A., Digital Pathways, Inc., and Jack Steinberger (1921-), U.S.A., CERN, Geneva, Switzerland: for the neutrino beam method and the demonstration of the doublet structure of the leptons through the discovery of the muon neutrino (Ch6).

1989: The prize was awarded by one half to: Norman F. Ramsey (1915-), U.S.A., Harvard University, Cambridge, MA: for the invention of the separated oscillatory fields method and its use in the hydrogen maser and other atomic clocks; and the other half jointly to: Hans G. Dehmelt (1922-), U.S.A., University of Washington,

Seattle, WA; and Wolfgang Paul (1913-1993), Federal Republic of Germany, University of Bonn, Bonn: for the development of the ion trap technique.

1990: The prize was awarded jointly to: Jerome I. Friedman (1930-), U.S.A., Massachusetts Institute of Technology, Cambridge, MA; Henry W. Kendall (1926-), U.S.A., Massachusetts Institute of Technology, Cambridge, MA; and Richard E. Taylor (1929-), Canada, Stanford University, Stanford, CA, U.S.A.: for the investigation of deep inelastic scattering and its relevance to the quark model (Ch6).

1991: Pierre-Gilles de Gennes (1932-), France, Collège de France, Paris: for discovering that methods developed for studying order phenomena in simple systems can be generalized to more complex forms of matter, in particular to liquid crystals and polymers.

1992: Georges Charpak (1924-), France, École Supérieure de Physique et Chimie, Paris and CERN, Geneva, Switzerland: for the invention and development of particle detectors, in particular the multiwire proportional chamber (Ch7).

1993: The prize was awarded jointly to: Russell A. Hulse (1950-), U.S.A., Princeton University, Princeton, NJ, and Joseph H. Taylor Jr. (1941-), U.S.A., Princeton University, Princeton, NJ: for the discovery of a new type of pulsar and investigations of the energy loss due to gravitational radiation (Ch12).

1994: The prize was awarded by one half to: Bertram N Brockhouse (1918-), Canada, McMaster University, Hamilton, Ontario: for the development of neutron spectroscopy, and by the other half to: Clifford G. Shull (1915-), U.S.A., Massachusetts Institute of Technology, Cambridge, MA: for the development of the neutron diffraction technique.

1995: The prize was awarded by one half to: Martin L. Perl (1927-), U.S.A., Stanford University, Stanford, CA, U.S.A.: for the discovery of the tau lepton; and with one half to: Frederick Reines (1918-1998), U.S.A., University of California at Irvine, Irvine, CA, U.S.A.: for the detection of the neutrino (Ch6).

1996: The prize was awarded jointly to: David M. Lee (1931-), U.S.A., Cornell University, Ithaca, NY, U.S.A.; Douglas D. Osheroff (1945-), U.S.A., Stanford University, Stanford, CA, U.S.A.; and Robert C. Richardson, U.S.A., Cornell University, Ithaca, NY, U.S.A: for the discovery of superfluidity in helium-3 (Ch17).

1997: The prize was awarded jointly to: Steven Chu (1948-), U.S.A., Stanford University, Stanford, California, U.S.A.; Claude Cohen-Tannoudji (1933-), France, Collège de France and École Normale Supérieure, Paris, France; and William D. Phillips (1948-), U.S.A., National Institute of Standards and Technology, Gaithersburg, Maryland, U.S.A: for the development of methods to cool and trap atoms with laser light (Ch17).

1998: The prize was awarded jointly to: Robert B. Laughlin (1950-), U.S.A., Stanford University, Stanford, CA, U.S.A.; Horst L. Störmer (1949-), U.S.A., Columbia University, New York, NY and Bell Labs, NJ, USA; and Daniel C. Tsui (1939-), U.S.A., Princeton University, Princeton, NJ, U.S.A: for the discovery of a new form of quantum fluid with fractionally charged excitations.

APPENDIX 7: GLOSSARY OF IMPORTANT IDEAS

Absolute zero – classically this is the temperature (–273.15 °C or 0K) at which all molecular motion ceases. In quantum theory particles retain a zero point energy that becomes significant at very low temperatures (e.g. preventing helium-4 from solidifying under normal pressures). Close to absolute zero the effects of thermal agitation become negligible and new macroscopic quantum phenomena are observed. In particular, electrons in some metals form Cooper pairs, act like bosons and condense into a single macroscopic quantum state (Bose condensate) in which they can move through the lattice without resistance. This is called superconductivity. A related effect results in zero viscosity and superfluidity.

Anti-matter – Dirac's relativistic quantum equation for the electron (the Dirac equation) allowed for positive and negative energy solutions. If the negative energy states were vacant they would be filled by electrons making quantum jumps down to lower energy levels. Dirac assumed they must all be full. He realised that one of these negative energy electrons could be promoted to a positive energy state by absorbing energy from a gamma-ray. This results in the annihilation of the gamma-ray and the creation of an electron. It also leaves a 'hole' in the negative energy electron sea. Against this background the hole appears like a positively charged particle with the same mass as an electron – an anti-electron (or positron). An extension of the argument led to the prediction of anti-particles for all particles (although in some cases particles and anti-particles are indistinguishable). The collision of a particle with its anti-matter partner results in mutual annihilation and the emission of energy in the form of gamma-rays.

Arrow of Time – the identification of an asymmetry in physics that distinguishes the future from the past. There are several arrows, the most important being the thermodynamic arrow in which the future has high entropy and the past low entropy. This connects the arrow of time with the second law of thermodynamics. On a microscopic scale the arrow arises because the universe began in an unlikely macroscopic state and random processes then made it inevitable that it evolves toward more probable macroscopic states (i.e. those that can be achieved in a larger number of micro-states). Other arrows include a cosmological arrow linked to the expansion of the universe and an arrow linked to time reversal violation in some weak interactions (e.g. kaon decays). So far there is no known connection between these different arrows.

Baryons – hadrons made of three quarks (like protons or neutrons). Baryons are fermions. Baryon number is conserved in all known interactions.

Big Bang – if the galaxies are all moving away from us now then they must have been closer together in the past. Hawking proved that they must have come from a singularity at some time in the past. This was the Big Bang in which space-time and matter exploded into being from a point of infinite density and zero scale. Recent theories suggest that the hot Big Bang may have followed an even more violent period of exponential expansion called 'inflation'.

Black Hole – there is a limiting mass above which degeneracy pressure in a neutron star is insufficient to prevent further gravitational collapse. This is about two solar masses. For more massive neutron cores the collapse continues and a black hole is formed. The defining characteristic of a black hole is the formation of an event horizon at which the escape velocity equals the speed of light. This is formed at the Schwarzschild radius and matter and radiation that falls inside this radius cannot escape. Work by Hawking in the 1970s suggests that black holes can in fact radiate as a result of vacuum fluctuations close to the event horizon. He showed that the radiation of particles emitted from a black hole has a black-body spectrum and associated a temperature with the black hole. This work linked general relativity, quantum theory and thermodynamics for the first time.

Black-body Radiation Spectrum – spectrum of radiation from an ideal emitter. Classical physics provided Wien's Law (for peak wavelength) and Stefan's law (for intensity of radiation) but could not explain the shape of the spectrum, especially at high frequencies. This problem was solved in 1900 by Max Planck when he introduced the idea that oscillators themselves can only have discrete amounts of energy. This was the beginning of quantum theory.

Bose-Einstein Condensate – new state of matter in which a large number of bosons occupy a single macroscopic quantum state.

Bosons – particles with integer spin that obey Bose-Einstein statistics. All force-carrying particles (e.g. photons) are bosons.

Colour – an intrinsic property that distinguishes quarks and is the source of the force that binds them together. There are three quarks colours, red, green and blue (and their anti-colours) and quarks always bind to form 'colourless' (or 'white') particles. Anti-quarks carry anti-colour and gluons (the carriers of the colour force) carry colour differences so that the exchange of a gluon changes the colours of the two quarks involved in the interaction.

Complementarity – Niels Bohr argued that, however weird the quantum domain may be, we can only discuss it in the context of macroscopic classical models and measuring apparatus. In this large-scale world the nature of the phenomena we observe is determined by an interaction between the quantum system and a classical measuring device that amplifies the effects. In such a context we must choose (by our experimental arrangement) what aspects of reality to reveal. The results of mutually exclusive experiments are themselves mutually exclusive. For example, Young's slits experiment reveals the wave-like interference of light, but if we set up any device that can track photons so we know which of the two slits they pass through, then this measurement of particle-like properties makes the interference effects disappear.

Copenhagen Interpretation – a philosophy and interpretation for quantum theory built around: Heisenberg's Uncertainty Principle, Bohr's Correspondence Principle and Complementarity, and Born's statistical interpretation of Schrödinger's wave mechanics.

Correspondence Principle – Niels Bohr realised that the quantum description of the atom must merge into the classical description in the limit of large quantum numbers – when the energy involved is large compared to individual quantum jumps. He used this as a guiding principle in constructing quantum models of classical systems.

Cosmic Background Radiation – electromagnetic radiation created soon after the Big Bang when most of the matter and anti-matter created in the Big Bang recombined and annihilated. Expansion of the universe since that time has red-shifted the radiation to form the microwave background with its characteristic black-body spectrum corresponding to thermal radiation at about 2.7 K. This radiation is almost perfectly uniform from all parts of the sky (after subtracting Doppler effects due to our own local motion) and provides very strong support for the Big Bang theory. It was discovered by Penzias and Wilson in 1963-4. In 1992 results from the Cosmic Background Explorer satellite (COBE) showed that the background radiation does have tiny fluctuations in intensity. These were expected and correspond to the differences in density in the early universe that resulted in the present distribution of galaxies and gaps.

CPT Invariance – Charge conjugation (C) replaces all particles with their anti-particles, parity inversion (P) inverts the co-ordinates of all particles, time reversal (T) runs particle reactions backwards. It is a fundamental theorem of quantum field theory that the combined symmetry operation of CPT applied to any particle interaction should produce another interaction that is also a possible process. Various experiments have shown that C, P, CP and T operations are all individually violated in certain particle decays, but no experiment has ever shown a violation of CPT invariance.

Critical density – the average matter density in the universe that results in flat space-time and an ever-slowing expansion that just stops after an infinite time. Inflation theory predicts that the universe should have this critical density. Present measurements of the density of observable matter give a result around 20% of the critical value, prompting a search for the 'missing mass'.

Degeneracy Pressure – if particles are confined in a small volume of space the uncertainty in their position is small and the uncertainty in their momentum is large. This results in an increasing degeneracy pressure as they are compressed. Electron degeneracy pressure supports white dwarf stars against further gravitational collapse and neutron degeneracy pressure supports neutron stars.

Degeneracy temperature – the temperature at which the quantum effects due to de Broglie waves should become important in a fluid. This is when the de Broglie wavelength is comparable to the inter-particle separation. For atoms this occurs close to absolute zero, for electrons (which have a much smaller mass and so a longer wavelength at any given temperature) the degeneracy temperature is much higher so that quantum distributions (the Fermi-Dirac distribution) are essential to understand their behaviour in materials (e.g. metallic conduction).

Einstein Locality – in relativity events that have a space-like separation cannot be causally connected. Einstein was disturbed by quantum theory because the collapse of the wave-function implies that separate parts of a quantum system remain connected even when they are spatially separated (i.e. the distance between two events exceeds the distance light could travel during the time that separates them).

Electroweak Unification – unified quantum field theory of electromagnetism and the weak nuclear force (responsible for beta-decays). The theory predicted three new massive force-carrying particles to accompany the photon: the W-plus, W-minus and Z-nought particles, all of which were discovered at CERN in 1983.

Entanglement – two-particles are entangled when their combined state cannot be written as a linear superposition of single particle states. This implies that the individual particles do not have definite states prior to an observation. This is a purely quantum mechanical situation. Consider two photons emitted from a common source and correlated so that their polarisation directions must be perpendicular. An entangled state would be a linear superposition of (1) photon A polarised vertically and photon B polarised horizontally, with (2) photon A polarised horizontally and photon B polarised vertically. In such an entangled state neither photon exists in a definite state prior to measurement. However, once measurement has been made the state of the other photon is fixed – so if we measure photon A and find it to be vertically polarised then we know that B must be horizontally polarised. Entanglement implies non-locality and was described by Einstein as 'spooky action-at-a-distance'.

Entropy – an important concept invented to make sense of the second law of thermodynamics. Macroscopic entropy is calculated from heat supplied divided by temeprature at which it is supplied. However, entropy is linked (loosely) to the degree of disorder in a system and increases when the number of micro-states available to the system increases (this may be because the system is heated or expands). The entropy of the universe never decreases and all macroscopic irreversible processes result in an increase of entropy. Boltzmann derived a crucial equation linking macroscopic entropy S to the number of microscopic configurations W corresponding to the macroscopic state, it is: $S = k \ln W$ where k is the Boltzmann constant. As a system is heated there are more quanta of energy to distribute among available states so W and hence S increase.

Equivalence – the idea that the laws of physics inside a uniformly accelerated reference frame in otherwise empty space are the same as those in a uniform gravitational field.

Expanding Universe – the idea that space-time itself is increasing in scale. This implies that the Big Bang was not an explosion into pre-existing space, but was the explosion of space-time and matter from absolutely nothing. In this context the recession of distant galaxies is not because they are moving away through space but because the amount of space between us and them is increasing.

Feynman diagram – this is a picture that represents particle interactions in such a way that it facilitates calculations in quantum electrodynamics. It is not a direct space-time representation of what happens on the subatomic scale so should always be interpreted with care.

Gauge Bosons – in quantum field theory the forces of interaction arise as a result of an exchange of force-carrying particles, all of which are bosons.

General Relativity – Einstein's theory of gravitation. Mass, energy and momentum distort 4D space-time geometry and free bodies follow geodesics as they move through this non-Euclidean geometry: "matter tells space how to curve and space tells matter how to move". General relativity explained the advance of perihelion of Mercury (a problem for classical astronomy) and predicted the deflection of light by gravity. It also predicted gravitational red-shifts and time dilation.

Grand Unified Theory (GUT) – a single theory that unifies all interactions except gravity.

Gravitational Red-Shift – light carries energy, so it has mass and is affected by a gravitational field. Light leaving a massive body loses energy and so (since $E = hf$) has reduced frequency and longer wavelength.

Gravitational Time Dilation – clocks at higher gravitational potential run faster than clocks at lower potential. This means that time passes more slowly at sea level than at the top of a mountain.

Gravitational Waves – periodic disturbances in space-time geometry that carry energy away from certain fluctuating mass distributions, e.g. binary stars. Passing gravitational waves induce tidal effects in objects. These changes can be exploited in gravitational wave detectors. There are two main types of detector, Weber bars and interferometers, but the effects expected are tiny and so far no direct observations have been made. However, there is good independent evidence to support the idea – Taylor and Hulse have made detailed measurements of a binary system containing a pulsar and have shown that the orbital period is changing at a rate consistent with energy losses due to gravitational waves.

Graviton – hypothetical force-carrier in quantum gravity.

Group Theory – general mathematical method dealing with symmetry transformations that has become increasingly important to theorists searching for unified theories.

Hadrons – particles that interact by the strong nuclear force. Hadrons are made of quarks.

Heisenberg Uncertainty (Indeterminacy) Principle – certain properties (conjugate variables) such as position and momentum or energy and time are inextricably linked so that any attempt to make a precise measurement of one will increase the uncertainty in the other. This is not just a matter of our knowledge of these properties, the properties themselves are indeterminate; an electron, for example, does not have a well-defined position and momentum in an atom. The idea of a classical trajectory or orbit must be abandoned. This interpretation of the uncertainty principle refutes naive realism.

Herzsprung-Russell Diagram – a graphical classification scheme in which stellar magnitudes are plotted against spectral classes (related to surface temperatures). Most stars fit on a diagonal band called the main sequence running from hot bright stars down to cool dark stars. However, there are several other distinct regions including the red giants (bright and cool) and the white dwarfs (dim but hot). The bands on the diagram are not evolutionary paths for individual stars.

Hidden Variables – Einstein and others thought that a more fundamental theory would explain the probabilistic nature of quantum theory in terms of a deeper layer of reality. This description would bear a similar relation to quantum theory as statistical mechanics does to classical thermodynamics. David Bohm succeeded in creating a hidden variables theory, but he had to endow these variables with properties (such as non-locality) that are every bit as counter-intuitive as the original features of quantum theory.

Higgs particle – predicted as a quantum of the Higgs field. The Higgs Field is thought to endow other particles with mass.

Hubble constant H_0 – one of the most controversial constants in physics and still uncertain by perhaps ± 50%. Its value is linked to the rate of expansion of the universe, and its reciprocal gives an estimate of the age of the universe (the Hubble time). Present estimates suggest that the universe is about 15 billion years old.

Hubble's Law – red-shift or recession velocity of galaxies is proportional to their distance. $z = H_0 d$ where H_0 is the Hubble constant. This discovery, by Hubble and Slipher in the 1920s, led to the idea of the expanding universe and the Big Bang.

Inertial reference frame – in which free objects move in straight lines obeying Newton's first law. Freely falling reference frames are inertial frames in a region of gravitational field (but only over a vanishingly small volume).

Inflation – an idea proposed by Alan Guth and others in the 1980s to explain how the present universe came to be so homogeneous. Big Bang theory cannot explain how distant parts of the universe came into equilibrium despite moving apart too rapidly for light signals to pass between them. In the inflationary scenario the Big Bang was preceded by a period of rapid exponential inflation during which time all parts were causally connected. Inflation theory also explains why the large-scale geometry of the universe is so close to being flat and predicts that the universe should have a critical density. This last prediction is in conflict with the observed density of matter and is one of the reasons astronomers continue to search for the 'missing mass'.

Kaluza-Klein Theory – Kaluza and Klein extended 4D general relativity to 5D in order to incorporate electromagnetism. The idea that unification might be achieved in higher dimensional spaces has become important with the advent of string theory.

Length Contraction – moving objects contract (relative to a 'stationary' observer) parallel to their direction of motion.

Leptons – particles with half-integer spins that obey Fermi-Dirac statistics and the Pauli Exclusion Principle. The particles that make up matter (electrons and quarks) are all fermions.

Lorentz transformation – set of equations that convert the co-ordinates of events seen by one inertial observer into co-ordinates of the same events as they would be measured by another inertial observer in motion with respect to the first.

Luminiferous Ether – hypothetical all-pervading medium proposed to support electromagnetic waves and as a 'seat' for electromagnetic energy. Attempts to reveal the Earth's motion relative to this medium have always generated null results (the most famous being the Michelson-Morley experiment). Einstein's theory of relativity removed the need for the ether.

Many Worlds Interpretation – Hugh Everitt III published his 'Relative State Interpretation of Quantum Theory' in 1957. He suggested that the act of measurement projects a quantum system into *all* possible states, but that each state is realised in a unique world. Schrödinger's Cat, for example, is alive in one world and dead in another. The act of opening the box to observe the cat results in a bifurcation so that one observer finds the dead cat and a 'copy' of that observer, in a 'parallel universe', observes a live cat. This interpretation gets around some of the logical problems of the Copenhagen Interpretation, but at the expense of an infinite number of additional parallel worlds.

Mass-Energy Equivalence – soon after publishing the special theory of relativity Einstein realised that the new physics implied that all energy has mass and linked the two by the equation $E = mc^2$. This is a universal relation which applies as much to chemical combustion as it does to nuclear reactions. The converse is also true, implying that there is an enormous amount of energy tied up in the rest mass of matter. It is often convenient to distinguish between a particle's rest-mass m_0 and its total mass m which includes a contribution from its kinetic energy.

Maxwell's Demon – hypothetical creature who controls a valve in a partition between two halves of a container of gas. By observing molecules the demon opens the valve selectively, filtering high-energy molecules to one end and low-energy molecules to the other, so that a temperature gradient is established and can be used to run a heat engine. (Leo Szilard proposed a simplified version in which the gas contains just a single molecule.) Such a demon could violate the second law of thermodynamics by creating conditions in which a machine can convert heat to work with 100% efficiency. This thought experiment was analysed many times from both classical and quantum viewpoints and many important ideas came out of the discussion. In particular it emphasised the link between entropy and information. The demon can only operate the valve if he can observe the molecules and store information about them in some kind of memory. He must then erase this memory in order to store information about the next molecule. The process of erasure is essential to return the entire system to its starting condition but results in an increase in entropy of the universe that at least balances the reduction brought about by the actions of the demon.

Maxwell's Equations – four equations defining the behaviour of electric and magnetic fields in space. The Maxwell equations lead to an equation for electromagnetic waves and the speed of light $c = 1/\sqrt{\varepsilon_0\mu_0}$.

Mesons – hadrons made of quark-antiquark pairs. They got their name because the first mesons to be discovered had masses intermediate between nucleons and electrons. However, mesons much heavier than nucleons also exist. All mesons are bosons.

Naïve Realism – the assumption that physical objects possess well-defined objective properties regardless of whether we choose to observe or measure them. This was at the root of Einstein's objection to quantum theory and was taken as an *a priori* assumption in the EPR paper.

Neutron Star – if the core left behind after a supernova explosion is greater than about 1.4 solar masses then electron degeneracy pressure cannot prevent further collapse and the electrons are 'squeezed' into protons to create a neutron star of extremely high density. This collapse conserves angular momentum so neutron stars spin rapidly. Their intense magnetic field creates strong electric fields that accelerate charged particles so that a beam of radiation is emitted from their magnetic poles. This is similar to a searchlight beam and if the Earth is in the line of this beam periodic pulses of radiation are received. These sources are called pulsars and their regularity is comparable to that of atomic clocks.

Noether's Theorem – links symmetry operations with conservation laws. For example, the symmetry that leaves the laws of physics unchanged by translation from one point to another is linked to conservation of linear momentum along the translation axis.

Non-Euclidean Geometries – self-consistent geometries based on a variation of Euclid's axiom about parallel lines. Euclid asserted that there is a unique parallel through a point at some distance from a straight line. Two alternatives, that result in negative and positive curvature, respectively are that (i) there are an infinite number of parallels and (ii) there are none. These ideas were explored by mathematicians such as Reimann, Bolyai Lobatchevsky and Gauss and incorporated into general relativity by Einstein.

Nuclear Fission – some heavy nuclei (e.g. U-235, Pu-239) will spontaneously break in two with the release of a large amount of energy (spontaneous nuclear fission). Fission can also be induced by the absorption of neutrons. All commercial nuclear reactors use nuclear fission. The 'atom bombs' were fission weapons (e.g. the bombs dropped on Hiroshima and Nagasaki).

Nuclear Fusion – the joining together of light nuclei to form heavier nuclei and release large amounts of energy. Since all nuclei are positively charged and the strong forces are extremely short range fusion has a large activation energy and only occurs under extreme conditions at very high temperatures. Fusion reactions are responsible for nucleosynthesis in stars and provide the energy source which allows them to shine. H-bombs rely on fusion. It is hoped that we will one day be able to build a commercial fusion reactor to generate electricity, but this prospect still seems several decades away.

Nucleosynthesis – the Big Bang created all the hydrogen and most of the helium in the universe, but little else. Elements up to iron-56 were made by a sequence of exothermic fusion reactions in the core of stars and the heavier nuclei were created in endothermic fusion reactions in supernovae.

Pauli Exclusion Principle – no two fermions (spin-half particles) can have an identical set of quantum numbers. This explains the existence of the Periodic Table of elements. Atoms with larger atomic numbers stack up their electrons in ever-higher orbits, they are prevented from occupying lower orbitals because these are already occupied.

Planck length – a fundamental length defined by the three constants important in quantum gravity: the speed of light (c), the Planck constant ($\hbar$) and the gravitational constant (G): $(\hbar G/c^3)^{\frac{1}{2}} \approx 10^{-35}$ m.

Quantum Electrodynamics (QED) – the quantum theory of the interaction of light and matter worked out by Feynman, Schwinger and Tomonoga in 1947.

Quantum Gravity – electromagnetism, the weak nuclear force and the colour force are all described by quantum field theories and the interactions are all mediated by force-carrying bosons. So far no one has discovered a way to quantise the gravitational field, but it is assumed that this will be possible. One of the problems is the weakness of gravity. This means that the effects of quantum gravity will only really become important on a very small scale, the Planck length of about 10^{-35} m.

Quantum Statistics – quantum particles are identical, so the exchange of two particles in a pair has no effect on the physical state of the system and does not count as an alternative micro-state. However, there are two ways in which the wavefunction for such a paired state can be written down, as a symmetric function (no change of sign on exchange) or as an asymmetric function (sign changes). In both cases there will be no observable consequence because observable resutls are linked to the square of the absolute value of the wavefunction and $(-1)^2 = 1^2$. Fermions (half-integer spin particles) have asymmetric wavefunctions, bosons (integer spins) have symmetric wavefunctions. One fundamental consequence is that fermions obey the Exclusion Principle whereas the probability for bosons to occupy the same quantum state is enhanced.

Quarks – fundamental particles that make up all hadrons. There are six flavours of quark in three generations: up/down, strange/charmed, bottom/top. They possess a colour charge (red, green, blue) and only combine in 'colourless' combinations (e.g. rgb or r and anti-r). The colour force between quarks strengthens as they are pulled apart and in violent collisions the energy of the field results in jets of new hadrons. Free quarks are never observed. At short range the colour force becomes weak (asymoptotic freedom). The three generations of quarks mirror the three generations of leptons.

Qubits – bits of information encoded in a quantum system. Many two-state systems can be used to carry the information, the most common being the correlated polarisation states of a pair of photons, and the quantum system can be prepared in a variety of ways. It is possible to use quantum systems in a classical way so that vertical polarisation might represent a 1 and horiziontal polarisation might represent 0. This allows four bits of information to be encoded on a pair of photons. However, the most interesting processes involve entangled states encoded in such a way that neither photon is in a definite state (although the state of one is always fixed once an observation has been made on its partner). The four 'Bell states' are a much-used example and are made from linear superpositions of pairs of classical qubits. For example, one of the Bell states is a superposition of particle 1 vertical and particle 2 horizontal with the state that is formed by swapping polarisations.

Red-shift – the increase in wavelength of radiation received from distant galaxies. This could be interpreted as a Doppler shift caused by their motion away from us, but is thought to be due to the expansion of space-time as the scale of the universe increases. Hubble's law shows that the red-shifts are proportional to distance and gives a good way to work out how far way very distant objects like quasars are.

Renormalisation – calculations in quantum electrodynamics result in unwanted infinites associated with self-interactions and zero-point energies. These are cancelled out by subtracting other infinite quantities to leave a finite residue that represents the observable quantities. The detailed procedure by which this is carried out is called renormalisation.

Simultaneity – separate events that occur at the same time in a particular inertial reference frame will occur at different times when viewed from another inertial reference frame moving with respect to the first.

Space-time – Minkowski realised that Einstein's theory of relativity described kinematics in a 4D world in which time plays a similar role to the three spatial dimensions. This crucial insight helped Einstein toward the general theory of relativity in which gravity enters as a geometric effect.

Special Relativity – Einstein's 1905 theory based on the Principle of Relativity: the laws of physics are the same for all inertial observers, regardless of their motion. A second postulate (or consequence of the first, depending on how you think about it) is that the speed of light is the same for all inertial observers. Relativity denied the existence of absolute space or time (or indeed a luminiferous ether) and led to the (initially uncomfortable) conclusion that the different inertial observers will measure different times and distances between the same events.

Spin – the Schrödinger equation led to three quantum numbers associated with electrons in atoms. However, atomic spectra could only be explained by adding a new quantum number. In 1925 Goudsmit and Ulhlenbeck introduced a half-integer quantum number called spin so that each energy level in an atom can be occupied by a pair of electrons with spin $s = \pm\frac{1}{2}\hbar$. The name 'spin' is misleading – it is a purely quantum mechanical property and should not be visualised by analogy with the rotation of the Earth. Spin arises naturally out of the Dirac relativistic equation for the electron, showing its intimate connection with space-time physics. It is the intrinsic angular momentum of a particle.

Standard Model – a combination of quantum field theories including QED, QCD and electroweak unification that describes all the particles and their interactions. It is not a final theory as it contains around 20 constants that have to be put in from experimental measurements and as yet there is no satisfactory quantum theory of gravity, which described separately by Einstein's general relativistic field equations.

Statistical Interpretation – as the wavefunction itself is unobservable, Max Born proposed that its magnitude-squared is proportional to the probability of finding a particle per unit volume at a particular point in space. If an unobserved system is represented by a wavefunction which is itself a superposition of states then, when an observation is carried out, the wavefunction collapses at random into one of the possible states and the probability of each state is given by $\left|\Psi_{state}\right|^2$

Statistical Mechanics – the explanation of macroscopic phenomena in terms of random energy shuffling between particles on a microscopic scale. One of the underlying assumptions is that all micro-states are equally accessible, equally probable and occur as a result of random molecular chaos.

String Theory – the idea that particles are not point-like structures in space-time but extended objects (strings) in a higher dimensional space. The mathematics of super-string theory incorporates super-symmetry that requires all known particles to have super-symmetric partners (none of which have yet been discovered). One advantage of these theories is that general relativity emerges as a low-energy approximation and so they offer the possibility of a genuine theory of everything in which gravity is included with a grand unified theory of all interactions.

Superposition – this is a property of solutions to linear differential equations. If X and Y are separate solutions then $(X + Y)$ is also a solution. This has profound implications in Schrödinger's wave mechanics where the state of a system is described by wavefunctions that are solutions to the Schrödinger equation. Prior to an observation a system might exist as a superposition of several possible states, only one of which will be observed when a measurement is carried out. The transition from a superpostion of states into a single state under measurement or observation is called the collapse of the wavefunction. This discontinuous process can occur at random and contrasts with the continuous deterministic time evolution of the wavefunctions themselves.

Super-symmetry – a hypothetical higher-order symmetry that relates fermions to bosons (and explains the quark-lepton symmetry). One consequence of this theory is that all existing particles should have super-symmetric partners (none of which have yet been discovered).

Symmetry – a system possesses symmetry if it is unchanged by some operation. For example, the laws of physics are unchanged when viewed from a moving reference frame; this symmetry is described by the Lorentz transformation equations. There is a deep link between symmetry principles and conservation laws.

Theory of Everything (TOE) – a single theory that combines a GUT with quantum gravity. Some theorists think super-string theory will provide a TOE.

Time dilation – the relative change in clock rate due to motion. If two inertial observers pass one another they will each see the rate of the other's clock slow down. The amount by which the rate changes is given by the 'gamma-factor'

$$\gamma = \sqrt{1 - v^2/c^2} \ .$$

Vacuum – A classical vacuum is simply empty. Quantum theory and relativity give a very different picture of the vacuum in modern physics. The Uncertainty Principle allows for fluctuations in energy resulting in the continual creation and annihilation of virtual particles. These virtual particles can interact with real particles to produce shifts in spectral lines (e.g. Lamb shift) and even to exert forces on macroscopic objects (Casimir effect). Special relativity implies that the vacuum must look the same to all observers and this puts constraints on the pressure and energy density of this seething sea of virtual particles so that the always cancel out. General relativity links gravitational effects to both mass-energy density and pressure and leads to the possibility of repulsive gravitational effects. Guth's inflation theory assumes that the very early universe contained an unstable vacuum that collapsed and generated powerful repulsive gravitational forces that caused an exponential expansion by a factor of about 10^{50} in about 10^{-32} s. Such gravitational repulsion may be responsible for recent claims that the rate at which the universe is expanding is actually accelerating.

Wave-Particle Duality – radiation and matter exhibit both wave-like properties (e.g. diffraction and interference) and particle-like properties (e.g. indivisibility, quantisation). Neither of these classical models is adequate to fully describe quantum objects (e.g. electrons, photons, atoms).

Yang-Mills theory – general theory of quantum fields in which forces arise from an exchange of force-carrying bosons.

APPENDIX 8: TIMELINE OF MAJOR IDEAS

Pre-C19th: Newtonian mechanics and gravitation

C19th: Laws of thermodynamics / spectroscopy / atomic theory /
observational astronomy / statistical mechanics /
Periodic Table / electromagnetism / ether hypothesis

1895	Radioactivity
1896	X-rays
1897	The electron

C20th:

1900	Quanitisation of energy (Planck)
	Gamma-rays (Villard)
1902	Radioactivity as transformation (Rutherford)
1903	'Plum-pudding' atom (Thomson and Kelvin)
1905	Quantisation of radiation (Einstein)
	Special theory of relativity (Einstein)
	Mass-energy equivalence (Einstein)
1908	Space-time geometry (Minkowski)
	Liquid helium (Onnes)
1909	Alpha scattering (Geiger and Marsden, Rutherford)
1911	Nuclear atom (Rutherford)
	Superconductivity (Onnes)
1912	Cosmic rays (Hess)
1913	Quantum atom (Bohr)
	Periodic Table explained in terms of atomic structure
1914	Hertzsprung-Russell diagram
1915	General relativity (Einstein)
1917	Cosmological constant (Einstein)
1918	Noether's theorem,
	3rd law of thermodynamics (Nernst)
1919	5th dimension (Kaluza and Klein)
	First hint of strong nuclear force (alpha scattering)
1923	Matter waves (de Broglie)
	Slipher finds galactic spectra have red-shifts
	Photons scattering from electrons (Compton)
1924	Bose-Einstein statistics
1925	Exclusion Principle (Pauli)
	Spin (Goudsmit and Uhlenbeck)
	Matrix mechanics (Heisenberg, Jordan and Born)
1926	Wave mechanics (Schrödinger)
	Fermi-Dirac statistics
	Statistical interpretation (Born)

1927	Uncertainty Principle (Heisenberg)
	Complementarity (Bohr)
1928	Dirac equation (Dirac)
1929	Prediction of 'holes' or antimatter (Dirac)
	Neutrino hypothesis (Pauli)
	Hubble's law established
1930	Chandrasekhar limit
1931	Positron discovered (Anderson)
	Radio signals from Milky Way (Jansky)
	Van der Graaf generator
	Cyclotron
1932	Neutron discovered (Chadwick)
	Cockcroft and Walton accelerator
	Time reversal in quantum theory (Wigner)
1934	Vacuum polarisation (Dirac)
	Meson theory of strong interaction (Yukawa)
1936	Muon discovered (Neddermeyer and Anderson)
	Solar fusion reactions – proton cycle (Bethe)
1939	Nuclear fission (Hahn and Meitner)
	Black holes (Oppenheimer and Snyder)
1939-45	Development of radar and nuclear weapons
1947	Lamb shift
	Kaons, first strange particles (Rochester and Butler)
1947-9	QED (Feynman, Schwinger, Tomonoga)
1948	Renormalisation
1953	Bubble chamber images (Glaser)
1954	Yang-Mills field equations
	CERN
1956	Neutrino discovered (Cowan and Reines)
1957	P and C violations in weak decays (Lee, Yang, Wu)
	BCS theory of superconductivity
	CPT theorem
1964	Cosmic background radiation (Penzias and Wilson)
	Omega-minus discovered (Barnes)
	CP violation (Cronin and Fitch)
	Higgs mechanism – spontaneous symmetry breaking
	Quark model of hadrons (Gell-Mann and Zweig)
1966	SLAC
1968	First pulsar/neutron star (Bell and Hewish)
1969	Deep inelastic scattering
	Singularity theorems (Hawking and Penrose)
	Manned Moon landing
1972	Fermilab
	Black hole thermodynamics (Bekenstein and Hawking)

1973	QCD and early ideas about GUTs
	Neutral currents
1974	J/Ψ charmed particle (Richter/SLAC and Ting/BNL)
1975	Tau particle (Perl at SLAC)
	Jets (SLAC)
1976	Supergravity
1979	Inflation (Guth)
	Gluons (DESY)
1983	Wand Z particles discovered at CERN
1986	High-temperature superconductivity
1990	Hubble Space Telescope
1992	COBE reveals fluctuations in background radiation
1994	Top quark (Fermilab)
1996	Bose-Einstein condensates
1998	Neutrino mass (Super-Kamiokande)
	Accelerating expansion (Perlmutter)

APPENDIX 9: FURTHER READING

Quantum Revolutions

A. Pais, *Niels Bohr's Times*, OUP, 1991.

A. Rae, *Quantum Physics, Illusion or Reality?* CUP, 1986.

D. Bohm and B.J. Hiley, *The Undivided Universe*, Routledge 1993.

D. Bohm, *Causality and Chance in Modern Physics*, Routledge 1957.

D. ter Haar, *The Old Quantum Theory*, Pergamon, 1967.

Ed. A. Mann and M. Revzen, *The Dilemma of EPR – 60 Years Later*, IOP/Israel Physical Society, 1996.

Ed. B.J. Hiley and D. Peat, *Quantum Implications*, Routledge, 1988.

Ed. J.A. Wheeler and W.H. Zurek, *Quantum Theory and Measurement*, Princeton, 1983.

Ed. P. Davies, *The New Physics*, CUP, 1989.

J.C. Polkinghorne, *The Quantum World*, Longman, 1984

J. Gleck, *Genius: Richard Feynman and Modern Physics*, Abacus, 1992.

J. Mehra, *The Beat of a Different Drum, The Life and the Science of Richard Feynman*, OUP, 1994.

L. de Broglie, *Physics and Microphysics*, Harper, 1955.

M. Born, *Atomic Physics*, Blackie, 1935.

M. Born, *Physics in My Generation*, Longman, 1970.

M.S. Longair, *Theoretical Concepts in Physics*, CUP, 1984.

N. Bohr, *Atomic Physics and Human Knowledge*, Wiley, 1958.

N. Herbert, *Quantum Reality*, Rider, 1985.

P.A.M. Dirac, *The Principles of Quantum Mechanics*, OUP 1930.

P.T. Matthews, *Quantum Mechanics*, McGraw-Hill, 1968.

Physics World, Vol. 11, No. 3, March 1998, Special issue: Quantum Information.

R. Eisberg and R. Resnick, *Quantum Physics*, Wiley, 1985.

R.P. Crease and C.C. Mann, *The Second Creation*, Macmillan, 1986.

R.P. Feynman, *QED*, Princeton, 1989.

S.F. Adams, Quantum Bombing Reality, Physics Education, November 1998, Vol 33, No. 6, IOP.

S. Toulmin, *Physical Reality*, Harper and Row, 1970.

T. Hey and P. Walters, *The Quantum Universe*, CUP, 1987.

W. Heisenberg, M. Born, E. Schrödinger and P. Auger, *On Modern Physics*, Orion Press, 1961.

W. Heisenberg, *Physics and Philosophy*, Harper and Row, 1962.

W. Heisenberg, *The Physical Principles of Quantum Theory*, Dover 1949.

Explaining Matter

A. Pais, *Inward Bound*, OUP, 1986.

E.A. Davis and I.J. Falconer, *J.J. Thomson and the Discovery of the Electron*, Taylor and Francis, 1997.

E.J. Burge, *Atomic Nuclei and their Particles*, OUP, 1977.

Ed. A. Romer, *The Discovery of Radioactivity and Transmutation*, Dover, 1964.

Ed. C. Sutton, *Building the Universe*, Blackwell/New Scientist, 1985.

F. Close, M. Marten and C. Sutton, *The Particle Explosion*, OUP, 1987.

F. Close, *The Cosmic Onion*, Heinemann, 1983.

J. Polkinghorne, *The Rochester Roundabout*, Longman, 1989.

R.P. Feynman and S. Weinberg, *Elementary Particles and the Laws of Physics*, CUP, 1987.

S.F. Adams, *Particle Physics*, Heinemann, 1997.

S. Weinberg, *Dreams of a Final Theory*, Radius, 1992.

Space and Time

A. Einstein and L. Infield, *The Evolution of Physics*, CUP, 1978.

A. Einstein, *Relativity, The Special and the General Theories*, Methuen, 1979.

A. Einstein, *The Meaning of Relativity*, Methuen, 1946.

A. Einstein, H.A. Lorentz, H. Weyl and H. Minkowski, *The Principle of Relativity* (collection of original papers), Dover, 1952.

A.P. French, *Special Relativity*, Nelson, 1981.

A. Grünbaum, *Philosophical Problems of Space and Time*, Alfred Knopf, 1963.

A. Pais, *Subtle is the Lord, The Science and the Life of Albert Einstein*, OUP, 1982.

B.F. Schutz, *A First Course in General Relativity*, CUP, 1985.

C.M. Will, *Was Einstein Right?* OUP, 1986.

E.F. Taylor and J.A. Wheeler, *Spacetime Physics*, Freeman, 1963.

H.A. Lorentz, *Problems of Modern Physics*, Dover, 1967.

J.A. Wheeler, *A Journey Through Gravity and Spacetime*, Scientific American Library, 1990.

K.S. Thorne, *Black Holes and Time Warps*, Picador, 1994.

L.C. Epstein, *Relativity Revisualised*, Insight Press, 1987.

M. Born, *Einstein's Theory of Relativity*, Dover, 1962.

M. Kaku, *Hyperspace*, OUP 1995.

O. Lodge, *The Ether of Space*, Harper, 1909.

O. Lodge, *Ether and Reality*, Hodder and Stoughton, 1925.

P.A.M. Dirac, *General Theory of Relativity*, John Wiley, 1975.

P.C.W. Davies, *The Search for Gravity Waves*, CUP, 1980.

P.G. Bergmann, *The Riddle of Gravitation*, Scribners, 1968

P. Rowlands, A Simple Approach to the Experimental Consequences of General Relativity, Physics Education, January 1997, Vol. 32, No. 1, IOP.

R.A. Laing, Faster than Light: Superluminal Motion and Light Echoes, Physics Education, Vol. 32, No. 1 January 1997.

S.F. Adams, *Relativity: An Introduction to Space-Time Physics*, Taylor and Francis, 1997.

W. Pauli, *Theory of Relativity*, Pergamon, 1956.

Astrophysics and Cosmology

A.S. Eddington, *Stars and Atoms*, OUP, 1942.
D.W. Sciama, *The Unity of the Universe*, Doubleday, 1961.
E.R. Harrison, *Cosmology*, CUP, 1989.
Ed. N. Henbest, *Observing the Universe*, Blackwell and New Scientist, 1984
G.J. Whitrow, *The Structure and Evolution of the Universe*, Hutchinson, 1959.
H. Bondi, *Cosmology*, CUP, 1961.
H. Friedman, *Sun and Earth*, Scientific American Library, 1986.
J.B. Kaler, *Stars*, Scientific American Library, 1992.
J.D. Barrow and F.J. Tipler, *The Anthropic Cosmological Principle*, OUP, 1988.
J.V. Narlikar, *The Primeval Universe*, OUP, 1987.
J.V. Narlikar, *Violent Phenomena in the Universe*, OUP, 1982.
P. Morrison and P. Morrison, *Powers of Ten*, Scientific American Library, 1982.

Time, Temperature and Chance

A.H.W. Beck, *Statistical Mechanics, Fluctuations and Noise*, Arnold, 1976.
C. Cercignani, *Ludwig Boltzmann, The Man Who Trusted Atoms*, OUP, 1998.
Ed. H.S. Leff and A.F. Rex, *Maxwell's Demon*, Adam Hilger, 1990.
Ed. N. Hall, *Exploring Chaos*, New Scientist, 1993.
Ed. Predrag Cvitanovic, *Universality in Chaos*, Adam Hilger, 1984.
F.E. Simon *et al*, *Low Temperature Physics*, Pergamon, 1952.
G.J. Whitrow, *The Natural Philosophy of Time*, Nelson, 1961.
H.A. Bent, *The Second Law*, OUP, 1963.
H. Reichenbach, *The Philosophy of Space and Time*, Dover, 1958.
I. Prigogine and I. Stengers, *Order Out of Chaos*, Heinemann, 1984.
I. Prigogine, *From Being to Becoming*, W.H. Freeman, 1980.
J. Gleick, *Chaos*, Cardinal, 1987.
L.C. Jackson, *Low Temperature Physics*, Methuen, 1950.
P. Coveney and R. Highfield, *The Arrow of Time*, 1990.
P.K. Kabir, *The CP Puzzle*, Academic Press, 1968.
P.T. Landsberg, *The Enigma of Time*, Adam Hilger, 1982.
P.W. Atkins, *The Second Law*, Scientific American Library, 1984.

General

Biographical Encyclopaedia of Scientists, IOP, 1994.
Twentieth Century Physics, IOP, 1995.
M.S. Longair, *Theoretical Concepts in Physics*, CUP, 1984.
Ed. P. Davies, *The New Physics*, CUP, 1989.
R.P. Feynman, *Lectures in Physics*, Addison Wesley, 1961.
R.P. Feynman, *The Character of Physical Law*, MIT, 1967.
Ed. S.R. Weart and M. Phillips, *History of Physics*, AIP, 1985.

INDEX

A

AAT...320
Abraham-Lorentz theory...................282
Abraham, Max.................................110
absolute Zero 49, 424
absorber theory 111, 129
absorption lines...............................347
abundance of the elements........345, 378
accelerating expansion 360
Adams, John 353
adaptive optics................................323
age of the Earth...............................341
age of the solar system 342
age of the Sun 340
age of the Universe 378
Airy's rings.....................................327
ALEPH..................................204, 219
alpha particles.................................148
Alvarez 196, 210
amplitudes, quantum.........................116
Anderson, Carl..................................64
Andromeda galaxy 363
antimatter 126, 217, 386
anti-particles....................................64
anti-Periodic Table............................64
antiproton 198
aperture synthesis............................331
apparent magnitude.......................338-9
Arecibo radio telescope 330
Aspect, Alain 85
Aspect Experiment.......................86-90
astronomical parallax 337
Astronomical Unit (AU)....................335
atmospheric scattering......................318
atom models....................................137
atomists 133
atoms 133, 416
axiom about the parallels 374

B

BaBar experiment 460
background radiation, cosmic 334
Balmer series....................................24
Balmer, Johann 22
barrier penetration........................54-5

baryon number................................ 386
baryonic matter............................... 377
baryons and mesons.......................... 175
BEBC, bubble chamber 196-7
BEC blobs....................................... 439
Becquerel, Henri 142, 147
Bednorz, Johannes............................ 437
Bekenstein, Jacob............................. 408
Bell inequality.............................. 85-90
Bell states....................................... 113
Bell-Burnell, Jocelyn 352
Bennett, Charles............................... 113
Bessel, Friedrich.............................. 336
beta rays 148, 152
Betelgeuse....................................... 349
Bethe, Hans.....................56, 165, 343
bevatron 198
B-factories....................................... 460
Big Bang theory................. 333, 375-86
Big European Bubble Chamber..... 196-7
binary star systems 281
binding energy................................. 154
black-body radiation 414
black-body spectrum 6-10, 31, 376-7
Blackett, Patrick.............................. 194
black hole..................310-14, 361, 406
black hole laws 408
black hole thermodynamics.............. 406
Bloch, Felix..................................... 436
B-mesons.. 460
Bohm, David 66, 88, 95
Bohr, Aage 160
Bohr, Niels......... 16, 22, 45, 65, 74, 161
Bohr radius...................................... 53
Bohr's quantum condition..................24
Boltzmann constant 402
Boltzmann, Ludwig..........11, 29-31, 399
                                        402-5
Boltzmann's H-theorem................... 403
Boltzmann's suicide 134
Bolyai, Janos 374
Born's statistical interpretation.......... 52
Bose condensate 434
Bose condensation 432

Bose, Satyendra 31
Bose-Einstein condensation.............438
Bose-Einstein statistics 31, 432
bosons............................ 145, 234, 431
Bothe, Walther...............................191
Bradley, James...............................251
Brahe, Tycho...................................336
Brillouin, L. 414
broken symmetry............................380
Brownian motion............... 11, 134, 417
bubble Chamber............................196-7
Bunsen burner.................................135
Bunsen, Robert 135

C

Cannon, Annie 347
carbon cycle 343
Carnot cycle....................................395
Carnot engine..................................395
Carnot, Sadi....................................394
Casimir effect 409
Cassegrain reflector.........................320
Cassini..336
cepheid variables 340, 362
Cerenkov Detectors..........................198
Cerenkov radiation...........................259
Cerenkov, Pavel...............................198
CERN 204, 218, 238
Chadwick, James 146, 154
Chandrasekhar limit.....................355-7
Chandrasekhar mass.........................351
Chandrasekhar, Subrahmanyan........349
chaos..419
charge conjugation 126, 447, 455
Charpak, Georges............................201
Christenson, James...........................456
Chu, Paul Ching-Wu.......................437
circular accelerators 211
    classical radius of electron 144
Clausius, Rudolf 395, 397
cloud Chamber.................................194
COBE....................................333, 376-9
Cockcroft, John 207, 209
Cockcroft–Walton device 154, 209
collapse of the wavefunction.........70, 75
collective model of nucleus 158
colour force..............................182, 188
complementarity....................44, 82, 95
completeness of quantum theory........79

Compton, Arthur...............................12
compton wavelength 16
conductivity, of metals.....................433
conservation laws 382, 442
    in mechanics 444
    in quantum field theories.............460
Cooper pairs.....................................438
Copenhagen interpretation 43, 65-6, 129
correspondence principle 44
cosmic background radiation............376
cosmic rays.............................. 191, 194
cosmological constant 360, 367
cosmological density parameter........368
cosmology..363
Coulomb barrier 56
Coulomb's law 229
Cowan, Clyde...................................172
CP symmetry....................................127
CP violation 454
CPLEAR 458
CPT invariance.......................... 442, 447
CPT symmetry..................................127
creation and annihilation operators 58
critical density 369, 383
Cronin, James...................................127
Crooke's tube 138
cryostats 429
Curie, Madame and Pierre...............149
curvature of space............................374
cyclotron..211
cyclotron resonance..........................213
Cygnus A..327

D

Darwin, Charles 340
Davisson, Clinton 14
de Broglie, Louis............13, 17, 95, 355
de Broglie wavelength 55, 434
de Broglie waves 38
de Broglie's hypothesis......................14
de Sitter universe..............................367
de Sitter, Willem 281, 367
Debye, Peter......................................25
deflection of starlight.............. 297, 305
degeneracy..355
degeneracy temperature 434
DELPHI 219
Democritus..134
density parameter 360

deuterium.................................. 154, 167
Deutsch, David 106, 113
diffraction.....................................14, 39
diffraction limits 327
dilution refrigerator..........................429
Dirac equation................................ 60-2
Dirac, Paul... 57-64, 109-13, 120-1, 442
Doppler cooling...............................440
Doppler effect..................................300
Dyson, Freeman 124, 236

E

Eddington, Arthur 165, 232, 297, 342,
353, 373, 391
Eddington's paradox 353
Edison, Thomas 326
Edlefsen, Niels................................211
eigenstates......................................42
eightfold way 177-80, 237-40
Einstein 11, 21-2, 31, 65, 73, 81, 101
149, 161, 249, 268, 270, 292-4
303, 373, 417, 425
Einstein locality 93
Einstein-Podolsky-Rosen paper 76-86
Ekert, Artur.....................................106
electromagnetic field.........................111
electron..144
electron degeneracy pressure............354
electron positron colliders 219
electron shell.....................................28
electron sub-shells..............................28
electronic configuration.......................27
electroweak interaction 187
electroweak unification 185, 237
energeticists....................................416
energy and momentum 289
energy of the universe 388
energy-momentum four-vector..........289
energy-time uncertainty relation 156
entanglement...................................101
entropy...............29, 397, 402, 414, 424
Eötvos, Baron Roland von 296
equipartition of energy 4, 425
equivalence principle 294-6
Euclid......................................303, 373
Euclidean geometry..........................303
event horizon 409
events in space-time.........................288
Everett III, Hugh 96

exclusion principle ..27-9, 182, 351, 431
expanding universe......................367-8

F

false vacuum....................................381
Faraday, Michael 136, 228
Faraday's laws.................................229
faster than light 257
Fermi, Enrico........ 22, 31, 150, 155, 161
Fermi–Dirac statistics................ 31, 434
Fermilab.................................. 221, 458
fermions 27, 31, 145, 234, 431
Feynman diagram 122, 126
Feynman, Richard.. 60, 109-11, 180, 226
Feynman-Wheeler absorber theory... 129
fifth dimension 231
fine structure constant......................123
fission 161
Fitch, Val 127
Fizeau, Armand 252
flavour..181
flux-quantisation..............................438
four-vector......................................288
fourth dimension..............................285
Fowler, William 354
Fraunhofer, Josef.............................347
free-fall 293
free-will 419
Friedmann, Aleksandr......................368
Friedmann's solutions......................411
Frisch, Otto 161
fusion 161, 340
fusion reactions 167
fusion reactions in stars 344
fusion weapons................................293

G

Gabor, Dennis 414
Galilean Relativity...........................263
Galileo, Galilei.................250, 293, 319
Galois, Evariste 233
gamma rays 148
Gamow, George.......... 56, 207, 343, 376
gauge bosons 186
gauge invariance..............................445
gauge symmetry...............................237
Gauss, Karl 291
Geiger and Marsden experiment 142
Gell-Mann, Murray......177-80, 236, 455

Gemini project 324
general relativity 188, 243, 292-314
generation 169
geodesics 373
geometry of the universe 373
geosynchronous orbit 332
Germer, Lester 16
Glaser, Donald 196
Glashow, Sheldon 184, 227
global positioning system 303
gluon 188
God 65
Goeppert-Mayer, Maria 158
Goudsmit, Samuel 27
gravitational collapse 344
gravitational interaction.... 188, 310, 336
gravitational length contraction 304
gravitational microlenses 298
gravitational potential 299
gravitational red shift 299-300, 373
gravitational waves 310
gravity and geometry 303
gravity and light 296
gravity and time 299
Greenberg, Oscar 182
group theory 232, 267
Guth, Alan 381

H

hadrons 169, 175-80
Hafele and Keating 278
Hahn, Otto 161
Halley, Edmund 336
Hamilton, William 115, 444
Hamiltonian 45, 57, 114
Hamiltonian method 110
Hawking radiation 409
Hawking, Stephen 363, 367
heat death of the universe 404
heat engines 393
Heisenberg Uncertainty Principle . 34, 58
                                            155, 198, 385
Heisenberg's microscope 37
helium abundance 378
helium-4 430
helium-burning 345
Helmholtz, Hermann von 139
Hertz Gustav 228, 326
Hertzsprung, Ejnar 348

Hertzsprung-Russell diagram... 348, 362
Hess, Victor 191
Hewish, Antony 352
hidden variables 81, 88-90
Higgs boson 219, 223
Higgs field 219, 380-6
Higgs, Peter 219, 381
high-z (large red-shift) supernovae... 359
Hipparchus 338
Hiroshima 162
horizon 379
Hubble constant 378
Hubble, Edwin 363
Hubble law 366-72
Hubble period 379
Hubble Space Telescope ...298, 314, 324
Hulse, Russell 312
Huygens, Christiaan 11
hydrogen atom 22-4, 50
hypercharge 236

I

indeterminacy 32-40, 48, 74
indistinguishability 431
inertial frames 293
inflation 380-83
information 411
Infrared Astronomical Satellite 332
Infrared Space Observatory (ISO) 333
interferometer 310, 330
invariance 442
ionising radiation 190
irreversibility 403-4
isotopes 146

J

Jansky, Karl 326
jets 184
Jodrell Bank 328
Joliot-Curie, Pierre 155
Joly, John 341
Jordan, Ernst Pascual 34
Joule, James 392

K

Kaluza, Theodor 231
Kaluza-Klein theory 232
Kamiokande 150, 385
Kamiokande II detector 173

Kant, Immanuel 391
kaons 454
Kapitsa, Pyotr 431
Kaufmann, Walter...........................282
Keck telescope 317, 323, 359
Keck..359
Kelvin, Lord............................340, 395
Kepler's laws of planetary motion.....335
kinetic theory.................................398
Kirchoff, Gustav..............................135
Klein, Oskar 231
Kolhörster, Werner 191
Kramers, Hendrick............................16

L

Lagrange, Comte.....................115, 444
Lagrangian.............................115, 444
Lamb Shift...................................120-4
Lamb shift......................................409
Lamb 121
Landau-Oppenheimer-Volkov limit ..406
Laplace..74
Laplace's Demon 419
Large Electron Positron Collider 218
Large Hadron Collider (LHC)...........219
laser cooling....................................439
Lawrence, Ernest 211
Leavitt, Henrietta.............................362
Lee, Tsung-Dao...............................451
Lemaître, Abbé Georges...................368
Lenard, Philipp 12
length contraction............................279
LEP collider.............................203, 218
lepton number.................................171
leptons...169
Leucippus 134
Leverrier, Urbain 307
LHC..219, 220
light clock......................................291
light nuclei.....................................377
linear accelerators 210
liquid drop model.............................161
liquid helium 429-30
LISA..311
Lobachevski, Nikolai........................374
locality...449
Lodge, Oliver..................................298
London, Fritz 429
long baseline interferometer.............330

Lorentz invariance 449
Lorentz transformation.................282-3
Lorentz, Hendrik 25, 110, 266-9
Lorentz-Fitzgerald contraction.. 256, 280
Los Alamos 165
Loschmidt paradox 403
luminiferous ether 269
luminosity 338-346

M

Mach, Ernst............................ 134, 416
Mach-Zehnder interferometer 110
macroscopic entropy 397
macroscopic quantum states.............431
magic numbers 158
magnetic moment of the electron...... 124
main sequence.................................349
many-times hypothesis..................... 110
many-worlds hypothesis................96-9
mass increase with velocity.............283
mass-energy equivalence... 149, 293, 342
matrix.. 34
matrix mechanics...................... 32, 65
matter era 372
matter waves 13, 21
Maxwell, James Clerk......228-30, 252-4
Maxwell's demon................... 411-13
Maxwell's equations... 22, 135, 231, 263
Maxwell-Boltzmann distn........ 399, 412
megaparsec.....................................337
Meisner effect 437
Merlin..331
Michelson, Albert................. 250, 349
Michelson interferometer................. 17
Michelson-Morley experiment.254, 266,
                                                       277, 280
microelectronic refrigerator..............429
microscopic entropy........................402
micro-states 400
Mie scattering.................................318
Millikan, Robert..............................191
mini-black holes 409
Minkowski, Hermann 232, 270, 285
Minkowski's geometrical vision 290
missing mass 383
models of the nucleus 157
Mössbauer effect 302
Mössbauer, Rudolph.........................300
Mount Wilson observatory....... 253, 363

Müller, Walther 191, 437
multiwire detectors 201
muon...170
muon clock.....................................293

N

Nagasaki..162
Nakano, Tadao...............................455
Ne'eman, Yuval...............................177
Nedermeyer, Seth...........................170
negative curvature..........................374
negative energy states......................63
negative pressure 387
Nernst, Hermann............................425
neutrino 171, 450
neutrino oscillations.......................175
neutron 145, 434
neutron degeneracy pressure............351
neutron star............................351, 434
Newton, Isaac................................336
Newton's laws 74
Nishijima, Kazuhiku 455
Nobel Prize
    1901, Röntgen 147
    1902, Lorentz, Zeeman 25
    1903, Curies and Becquerel 149
    1906, Thomson 138
    1907, Michelson 250
    1911, Wien................................ 8
    1913, Onnes..............................430
    1919, Planck.............................. 8-10
    1922, Einstein, Bohr 10-13, 22-6
    1927, Compton, Wilson 16, 194
    1929, de Broglie 13-16
    1933, Heisenberg...................... 32-9
        Schrödinger....................40-56
        Dirac 57-64
    1935, Chadwick........................145-7
    1936, Anderson, Hess 64, 191
    1937, Davisson, Thomson 13-16
    1938, Fermi 161
    1945, Pauli 27-9
    1948, Blackett...........................194
    1949, Yukawa...........................155-7
    1950, Powell............................196
    1951, Cockcroft and Walton.........209
    1954, Born............................... 43-5
    1955, Lamb.............................120-4
    1957, Yang, Lee.........................453

    1958, Cerenkov, Frank, Tamm..... 199
    1959, Sègre, Chamberlain 199
    1960, Glaser.............................. 198
    1961, Mössbauer........................ 300
    1962, Landau ....
    1965, Feynman, Tomonoga,
        Schwinger 109-30
    1967, Bethe............................... 344
    1969, Gell-Mann........................ 178
    1974, Ryle, Hewish..................... 329
    1975, A. Bohr, Moffelson,
        Rainwater........................ 160
    1976, Richter, Ting 185
    1978, Penzias and Wilson........... 376
    1979, Glashow, Salam,
        Weinberg 237-9
    1980, Cronin, Fitch..................... 461
    1983, Chandrasekhar, Fowler 352
    1984, Rubbia, van der Meer 187
    1987, Bednorz, Müller................ 437
    1988, Lederman, Schwartz,
        Steinberger..................... 171
    1990, Friedman, Kendall,
        Taylor............................181-2
    1992, Charpak............................202
    1993, Taylor, Hulse.................... 312
    1995, Perl, Reines 153, 171
    1996, Lee, Osheroff,
        Richardson..................... 439
    1997, Chu, Tannoudji,
        Phillips 441
Noether, Emmy............................. 235
no-hair theorem 407
non-baryonic matter...................... 384
non-Euclidean geometries........ 303, 374
non-locality................................ 73
nonseparability 94
Norris, Henry............................... 348
nuclear atom............................... 142
nuclear forces 153
nuclear resonance 345
nuclear transformation................... 149
nucleons 144
nucleosynthesis...................... 344, 377

O

observer...................................... 73
Occhialini, Giuseppe 194
Onnes, Heike Kammerlingh..... 430, 436

ontological interpretation....................94
OPAL 219
operators..62
optical astronomy............................319
optical commutators..........................90
Ostwald, Friedrich 417

P

parametric down-conversion.............102
parity..447
parity violation.................................449
particle detectors.............................190
partons..180
path Integrals 114
Pauli Exclusion Principle
                        63, 182, 351, 431
Pauli, Wolfgang 22, 150
Penrose, Roger............. 333, 410, 443
Penzias, Arno...................................333
Periodic Table............................27, 28
Perl, Martin 170
Perlmutter, Saul...............................358
perpetual motion machines.............392-4
photographic emulsions....................195
photomultiplier 88
photon generators..............................91
photon model 12, 37
pion 156
Planck length 231, 243
Planck, Max..5, 10
Planck's law 8
plum pudding model 139, 149
plutonium 165
Poincaré, Henri 267, 374, 420
polar co-ordinates..............................50
polarisation..88
    polonium and radium 149
positive curvature.............................374
Prigogine, Ilya..................................421
primeval atom...................................375
primordial black holes......................409
principle of relativity....................263-91
probability 73
probability amplitude 117-22
proper motion....................................336
proton antiproton collider 222
proton cycle......................................343
proton-neutron model.......................149
pseudo-Euclidean geometry..............286

Pythagoras's theorem................. 288-91

Q

quanta 10
quantisation............................. 8, 12, 24
quantum chromodynamics........ 188, 217
quantum computing 106
quantum correlations 94
quantum cryptography................... 102-3
quantum electrodynamics....58, 125, 237
quantum fields................................. 113
quantum fluctuations 386-8
quantum genesis 386
quantum harmonic oscillator............. 45
quantum harmonic oscillator............. 49
quantum information 99
quantum jump.................................... 64
quantum jumps.................................. 40
quantum number................................ 25
quantum statistics..................... 31, 431
quantum teleportation...................... 113
quantum tunnel effect 55
quark model 210
quark-lepton symmetry 169, 184
quarks 180-2
quasars 258, 362
qubits 99-101

R

radiation era 372
radio astronomy.............................. 326
Rayleigh criterion 327
Rayleigh, Lord.................................... 6
realism 79, 89
reality.................................. 70, 75, 82
red giant 351
red-shift................................. 258, 365
red-shifts of galactic spectra............. 364
refractive index of space-time 306
refrigerator 426
Reines, Frederick............................172
relative state formulation 97
renormalisation...............60, 125-6, 227
resistivity... 436
resonances....................................... 198
Riemann Georg................................ 374
Römer, Ole...................................... 251
Röntgen, Wilhelm 147
ROSAT 334

Rossi, Bruno 191
Rutherford scattering experiment 142
Rutherford, Ernest.....200, 207, 215, 341
Rydberg-Ritz constant 22
Ryle, Martin....................................331

S

Sakharov, Andrei 165
Salam, Abdus...................................184
satellite telescopes 318
scale factor......................................371
Schmidt telescopes..........................320
Schödinger, Erwin...... 21, 32, 39, 40, 51
Schrödinger equation 40, 42, 46, 47
Schrödinger's Cat 75
Schrödinger's equation......................60
Schwarzschild radius........ 313, 361, 373
Schwarzschild solution.............312, 373
Schwarzschild, Karl 312, 406
Schwinger, Julian.... 60, 109, 121-2, 226
scintillators.....................................200
second law 391, 397, 412
Sègre, Emilio 198
semi-empirical mass formula............159
Shannon, Claude..............................415
Shapiro time delay 307
Shapley's period-luminosity law.......363
shell model of nucleus.....................158
Shelter Island conference.................120
Simon, Francis 422
simultaneity 270, 274
SLAC............................... 181, 203, 210
SLC..215
Slipher, Vesto 364
Smoot, George 334
solar neutrinos 173
Sommerfield, Arnold..........................25
space-time approach.....................285-6
space-time geometry........................291
space-time of the universe...............373
sparticles..................................227, 241
SPEAR 170
special relativity..........................263-91
spectra 135
spectral classes of stars 347
spectroscopic parallax 361
speed of light 249
spin................................... 27, 62, 144
spinthariscope 200

Standard Model...................... 168, 243
Stanford linear accelerator 210-15
Stanford linear collider............ 211, 215
static universe.................................375
statistical interpretation 20, 43
statistical mechanics........................400
Stefan's law....................................346
Stefan-Boltzmann law........................8
Steinberger, Jack 171
stellar magnitudes...........................338
stellar spectra 346
Stokes, George 134
strangeness............................. 176, 455
Strassman, Fritz...............................161
SU(3)..240
SU(5)..240
superconducting quantum device......429
superconductivity............................436
superfluidity 431
superfluidity in liquid helium-3 439
supergiant layered star.....................357
Super Kamiokande 175
super-luminal Jets............................259
super-luminal velocities...................257
supernova 345-62
supernova 1987a..................... 259, 357
supernova cosmology project...........358
superposition 70, 85, 100
superstring theory.............188, 227, 243
supersymmetry 227, 241
symmetry.....50, 126, 232, 235, 380, 442
symmetry groups.............................233
synchro-cyclotron............................213
synchronisation...............................274
synchrotron radiation 211, 216
synthesis of heavy nuclei.................345
Sysiphus cooling.............................440
Szilard engine..............................413-6
Szilard, Leo 413

T

Tamm, Igor.....................................165
Tanoudji, Cohen 440
Taylor, Joseph 19, 312
temperature 391, 422, 423
Tevatron.................................. 221, 222
thermodynamic arrow of time .. 398, 403
theta-tau problem............................450
third law of thermodynamics...........425

Thomson, James (J.J.). 138
Thomson, George (G.P.)....................14
time dilation.............................276-7
time projection chambers.................203
time reversal asymmetry..........447, 457
tokamak................................165, 293
Tomonoga, Sin-itiro 110, 124, 226
transactional interpretation..............127
transuranic elements 160
triple-alpha reaction 345
tritium.......................................167
tuning fork diagram.........................364
tunnel effect.................................54
Turlay, René 127, 456
twin paradox 277

U

ultra-violet astronomy 332
ultra-violet catastrophe...................... 7
uncertainty principle 82, 122, 155
unified field theory....................230-2
unitarity 449
Urey, Harold 25, 345
Ussher, Bishop...............................340

V

vacuum 385
vacuum fluctuations 409
vacuum polarization 109
Van de Graaf................................207
variable stars................................340
velocity addition............................280-1
very large telescope (VLT)..............324
Very Long Baseline Interferometry
    (VLBI)   329
Villard, Paul 148
virial theorem................................383
virtual electrons 59
virtual particles.......................385, 409
virtual photons 59, 123, 186
virtual states 59
vortex rings...................................136

W

wavefunction.................... 31, 40-56, 99
weak interaction.............................450
weak nuclear force 147
Weber bars...................................310
Weinberg, Steven...........................184

Wheeler, John................................129
white dwarf 353-4
Wideröe, Rolf................................210
Wien's displacement law............. 8, 346
Wigner, Eugene..............................445
Wilson, Charles 194
work function 13
Wu, Chieng-Shiung 451
Wulf, Theodor 191

X

X-ray Astronomy 332
X-ray background 332
X-rays 147, 148, 335

Y

Yang, Chen-Ning.....................239, 451
Yang-Mills field theories......... 239, 245
Young's experiment 119
Yukawa, Hideki..............................155

Z

Zeeman effect.................................25
Zeilinger, Anton 113
zero point energy 48, 430
zeroth law of thermodynamics.......... 423
Zweig, George................................180
Zwicky, Fritz 351, 356, 383